T0134949

Studies in Computational Intelligence

Volume 671

Series editor

Janusz Kacprzyk, Polish Academy of Sciences, Warsaw, Poland
e-mail: kacprzyk@ibspan.waw.pl

About this Series

The series "Studies in Computational Intelligence" (SCI) publishes new developments and advances in the various areas of computational intelligence—quickly and with a high quality. The intent is to cover the theory, applications, and design methods of computational intelligence, as embedded in the fields of engineering, computer science, physics and life sciences, as well as the methodologies behind them. The series contains monographs, lecture notes and edited volumes in computational intelligence spanning the areas of neural networks, connectionist systems, genetic algorithms, evolutionary computation, artificial intelligence, cellular automata, self-organizing systems, soft computing, fuzzy systems, and hybrid intelligent systems. Of particular value to both the contributors and the readership are the short publication timeframe and the worldwide distribution, which enable both wide and rapid dissemination of research output.

More information about this series at http://www.springer.com/series/7092

Vicenç Torra · Anders Dahlbom
Yasuo Narukawa
Editors

Fuzzy Sets, Rough Sets, Multisets and Clustering

Springer

Editors
Vicenç Torra
School of Informatics
University of Skövde
Skövde
Sweden

Yasuo Narukawa
Toho Gakuen
Tokyo
Japan

Anders Dahlbom
School of Informatics
University of Skövde
Skövde
Sweden

ISSN 1860-949X ISSN 1860-9503 (electronic)
Studies in Computational Intelligence
ISBN 978-3-319-83767-3 ISBN 978-3-319-47557-8 (eBook)
DOI 10.1007/978-3-319-47557-8

Printed on acid-free paper

This Springer imprint is published by Springer Nature
The registered company is Springer International Publishing AG
The registered company address is: Gewerbestrasse 11, 6330 Cham, Switzerland

To Prof. Sadaaki Miyamoto

Professor Sadaaki Miyamoto in MDAI 2010, in Perpinyà

This book is a token of appreciation to Prof. Sadaaki Miyamoto for his scientific work that has influenced us much in our own research, for his unconditional support to MDAI conference series, and last but not least for his friendship.
Long and interesting scientific discussions as well as long and interesting non-scientific discussions have helped us to understand the fuzzy world better.

Preface

In April 2016, Prof. Sadaaki Miyamoto retired from the University of Tsukuba. He was born in Osaka in 1950. He obtained his Ph.D. from Kyoto University in March 1978. He joined the University of Tsukuba in 1978 (until 1990) and then again from 1994. During 1990–1994, he was Professor in the University of Tokushima. During these years, he had a fruitful career producing some inspiring works.

We have prepared this book as a token of appreciation to him. The book collects chapters written by colleagues and friends of Miyamoto. Authors of these chapters have met Miyamoto in different occasions and all wanted to express their appreciation.

The book includes a first chapter that presents an introduction on when we met Miyamoto, a personal outline of some of his works, and a description of the contents of the book. Then, the book contains its chapters divided into four parts that roughly correspond to the terms in the title of the book.

Skövde, Sweden — Vicenç Torra
Skövde, Sweden — Anders Dahlbom
Tokyo, Japan — Yasuo Narukawa
June 2016

Contents

On This Book: Clustering, Multisets, Rough Sets and Fuzzy Sets

Vicenç Torra, Yasuo Narukawa and Anders Dahlbom

Abstract This chapter gives an overview of the content of this book, and links them with the work of Prof. Sadaaki Miyamoto, to whom this book is dedicated.

1 Introduction

In October 2000, the first editor of this volume visited Prof. Miyamoto's lab in the University of Tsukuba (Japan) for the first time, he stayed in the lab for two weeks. This was the beginning of a fruitful collaboration that is at present 16 years old.

Later, in 2004, we started with the second editor in Barcelona the conference series Modeling Decisions for Artificial Intelligence (MDAI 2004). Prof. Miyamoto collaborated with the conference and gave one of the plenary talks. He has participated and supported the conference since then. He also organized MDAI 2005. The second author visited Miyamoto's lab with the first author in 2004. Prof. Miyamoto gave him some of his papers on multisets, that later lead to some papers published by IEEE.

Still later, the third editor enter the set of regular participants in the MDAI conference. It was also in Japan, where he participated in MDAI 2009, held in Awaji Island.

V. Torra (✉) · A. Dahlbom
School of Informatics, University of Skövde, Skövde, Sweden
e-mail: vtorra@his.se

A. Dahlbom
e-mail: anders.dahlbom@his.se

Y. Narukawa
Toho Gakuen, 3-1-10, Naka, Kunitachi, Tokyo 186-0004, Japan
e-mail: narukawa@d4.dion.ne.jp

Y. Narukawa
Department of Computational Intelligence and Systems Science,
Tokyo Institute of Technology, 4259 Nagatuta, Midori-ku, Yokohama 226-8502, Japan
e-mail: narukawa@d4.dion.ne.jp

V. Torra et al. (eds.), *Fuzzy Sets, Rough Sets, Multisets and Clustering*,
Studies in Computational Intelligence 671, DOI 10.1007/978-3-319-47557-8_1

After the last MDAI 2015 (Fig. 1) conference in Skövde (Sweden) we decided to prepare this volume and dedicate it to Prof. Miyamoto. We have invited collaborators and friends of Prof. Miyamoto. We also invited Miyamoto himself (without explaining that it was a dedicated book) to contribute to the book with a chapter. The plan and content of the book is detailed below.

Prof. Miyamoto has largely contributed in his scientific career on the topics of clustering, fuzzy sets, multisets and rough sets. That is why the title of this book includes all these keywords. Prof. Miyamoto is well known and recognized in all these areas, and we have used and applied some of his results in our own research.

It is difficult and arbitrary to select works from the long list of Prof. Miyamoto's publications, but based on what we have read and used from his work we try to do so here.

On clustering, he wrote the monographs [1, 2, 8] and edited the book where these references [3, 11] are found. The first author bought a copy of [2], a nice, concise and to-the-point book, in one of his first trips to Japan. This copy has travelled back and forth (Europe–Japan) several times since then. We have also extensively used papers on entropy-based fuzzy c-means [9] and variable-size fuzzy c-means [10] in our own work.

On multisets, fuzzy sets, and fuzzy multisets, we underline [3, 5–7, 12]. We used their definitions and results in our work. [7], a paper published in the journal Fuzzy Sets and Systems on occasion of the 40th Anniversary of Fuzzy Sets, includes some interesting remarks on fuzzy sets and fuzzy multisets.

Any of us have worked so much on rough sets. So on this area our arbitrary selection is still more arbitrary. But in connection with multisets, we have used [4].

2 Book Content

At the end the book is divided in four parts, containing a total of 19 chapters. The first part is on clustering and classification. The second part is on multisets or bags, fuzzy bags and other fuzzy extensions. Then, follows a part with chapters on rough sets (although we have included one on clustering for rough sets in the first part on clustering). The last part is about fuzzy sets and decision making.

2.1 Clustering and Classification

This part starts with a chapter by Miyamoto. The chapter tries to answer which are the main contributions of fuzzy clustering to the theory of clustering, from a theoretical viewpoint.

The second chapter by Katsuhiro Honda reviews some fuzzy clustering models induced by probabilistic mixture. The author discusses the effects of introducing

Fig. 1 Prof. Sadaaki Miyamoto in MDAI 2007 in Osaka with Masahiro Inuiguchi and Vicenç Torra (*top left*), in MDAI 2010 in Perpinyà with Yasunori Endo (*top right*), in MDAI 2012 with Yasuo Narukawa (*bottom left*), in MDAI 2015 with Yasuo Narukawa, Masahiro Inuiguchi, Yuji Yoshida, Aoi Honda, Katsushige Fujimoto and Vicenç Torra (*bottom right*)

adjustable fuzziness penalties into the statistical models. Among the models studied, the entropy regularization-based FCM introduced by Miyamoto is considered.

The third chapter by Yuchi Kanzawa proposes a new approach for semi-supervised clustering with soft pairwise constraints. FCM is used for clustering in this approach.

The fourth chapter by Endo and Kinoshita focuses on rough clustering. It reviews a few rough clustering algorithms that are defined through an objective function to optimize.

The next chapter by Hamasuna and Endo is about clustering algorithms based on tolerance. The underlying idea is that uncertainty in the data can be expressed in terms of a certain tolerance around the data points. That is, data with tolerance as the authors write. Clustering is then formalized for this type of data and algorithms are defined.

Szilágy in the following chapter focuses on some robust clustering algorithms. The chapter presents several clustering algorithms defined with the goal of being robust and based on the fuzzy-possibilistic product partition.

This chapter is followed by another one by García-Lapresta and Pérez-Román about an agglomerative hierarchical clustering method. The authors introduce a consensus measure that is later used to define a similarity function used in the clustering process.

Owsiński, Kacprzyk, Opara, Stańczak and Zadrożny propose in their paper a problem related to clustering. Given the clustering results, the goal is to find which is the clustering algorithm and the parameters that led to those results. It is thus, a kind of reversal engineering problem. The authors propose the use evolutionary programming for this purpose.

The following chapter, by Aliahmadipour, Torra and Eslami, is a review and a discussion on clustering algorithms dealing with hesitant fuzzy sets.

This part on clustering and classification finishes with a chapter on decision trees, and about their use for knowledge discovery. It is a chapter written by Armengol, García–Cerdaña, and Dellunde.

2.2 *Bags, Fuzzy Bags, and Some Other Fuzzy Extensions*

This part includes two chapters. The first one by Kouchakinejad, Mashinchi, and Mesiar is about L-fuzzy bags. The authors study this type of bags when L is a complete lattice.

The second chapter of this part is a discussion by Szmidt and Kacprzyk on two extensions of fuzzy sets: Atanassov's Intuitionistic Fuzzy Sets (A-IFS, for short) and Interval-valued Fuzzy Sets (IVFSs, for short). The authors discuss the differences between these two types of fuzzy sets.

2.3 *Rough Sets*

This part includes two chapters. The first one by Inuiguchi is about attribute reduction. The author propose several ways to do the attribute reduction. The author also study attribute importance and interaction indices. Game theory and the Shapley value is used for this purpose.

The last chapter of this part by Kudo and Murai is a review on rough set-based interrelationship mining. The authors focus on theoretical aspects.

2.4 *Fuzzy Sets and Decision Making*

The last part of the book is about methods for decision making, mainly based on fuzzy sets and other fuzzy-related approaches.

The first chapter by Yager is about aggregation functions, and more particularly, on the use of the OWA operator. It focuses on the case of finite probability distributions, and how to deal with their linear ordering.

The second chapter by Yoshida is about a dynamic average in the context of fuzzy random variables. The motivation of this work is in economics and in the context of a value-at-risk portfolio model.

Cabrerizo, Pérez, Chiclana and Herrera–Viedma present in the next chapter consensus approaches based on soft consensus measures. They are for group decision making. The paper is an overview of this topic.

The fourth chapter by Sanz et al. is about construction of capacities, also known as fuzzy measures and non-additive measures. The authors show how overlap functions and overlap indices can be used to build this type of measures.

The fifth chapter by Franco, Hougaard and Nielsen is about the application of multi-criteria modeling and its application to clustering and preference modeling.

References

1. Miyamoto, S. (1990) Fuzzy Sets in Information Retrieval and Cluster Analysis, Theory and Decision Library, Series D, Kluwer Academic Publishers.
2. Miyamoto, S. (1999) Introduction to fuzzy clustering (in Japanese), Ed. Morikita, Japan.
3. Miyamoto, S. (2000) Multisets and fuzzy multisets, in Z.-Q. Liu, S. Miyamoto (Eds.), Soft Computing and Human-Centered Machines, Springer-Tokyo.
4. Miyamoto, S. (2004) Generalizations of multisets and rough approximations, Int. J. of Intel. Syst. 19 639–652.
5. Miyamoto, S. (2004) Multisets and fuzzy multisets as a framework of information systems, MDAI 2004, Lecture Notes in Artificial Intelligence 3131 1–28.
6. Miyamoto, S. (2009) Generalized bags, bag relations, and applications to data nalysis and decision making, MDAI 2009, Lecture Notes in Artificial Intelligence 2009 37–54.
7. Miyamoto, S. (2005) Remarks on basics of fuzzy sets and fuzzy multisets, Fuzzy Sets and Systems 156 427–431.
8. Miyamoto, S., Ichihashi, H., Honda, K. (2008) Algorithms for fuzzy clustering, Springer.
9. Miyamoto, S., Mukaidono, M. (1997) Fuzzy c-means as a regularization and maximum entropy approach, Proc. of the 7th IFSA Conference, Vol.II 86–92.
10. Miyamoto, S., Umayahara, K. (2000) Fuzzy c-means with variables for cluster sizes (in Japanese), Proc. of the 16th Fuzzy System Symposium, 537–538.
11. Miyamoto, S., Umayahara, K. (2000) Methods in Hard and Fuzzy Clustering, in Z.-Q. Liu, S. Miyamoto (eds.) Soft Computing and Human-Centered Machines, Springer-Tokyo, 85–129.
12. Mizutani, K., Inokuchi, R., Miyamoto, S. (2008) Algorithms of nonlinear document clustering based on fuzzy multiset model, Int. J. of Intel. Syst. 23:2 176–198.

Part I
Clustering and Classification

Contributions of Fuzzy Concepts to Data Clustering

Sadaaki Miyamoto

Abstract This chapter tries to answer the fundamental question of what main contributions of fuzzy clustering to the theory of cluster analysis from theoretical viewpoints. While fuzzy clustering is thought to be clearly useful by users of this technique, others think that the concept of fuzziness is not needed in clustering. Thus the usefulness of fuzzy clustering is not trivial. The discussion here is divided into two: one is on fuzzy c-means which is best-known fuzzy method of clustering. However, there is another techniques, discussed by Zadeh, in hierarchical clustering which is equivalent to the old technique of the single linkage. This chapter overviews the both techniques, beginning from basic discussion of fuzzy c-means, and introducing the fundamental concept of fuzzy classifiers and its usefulness. A concept of inductive clustering is introduced which means that a result of clustering can be extended to a partition of the whole space. Moreover hierarchical fuzzy clustering is briefly discussed where the transitive closure gives a simple algebraic form of clusters.

Keywords Fuzzy clustering · Fuzzy c-means · Fuzzy classifier · Hierarchical clustering · Inductive clustering

1 Introduction

Data clustering alias cluster analysis, which generates groups of objects from a set of data using mutual similarity (or dissimilarity) between a pair of data, has been known for a long time [8, 12, 19, 20]. According as the subjects of data mining becomes more and more popular, many different techniques of clustering have been developed.

Fuzzy c-means clustering [2–4, 9, 10, 16, 30] is now considered to be a standard technique in cluster analysis by many researchers. There is, however, another method

S. Miyamoto (✉)
University of Tsukuba, Tsukuba, Ibaraki 305-8573, Japan
e-mail: miyamoto@risk.tsukuba.ac.jp

V. Torra et al. (eds.), *Fuzzy Sets, Rough Sets, Multisets and Clustering*,
Studies in Computational Intelligence 671, DOI 10.1007/978-3-319-47557-8_2

of fuzzy clustering in hierarchical cluster analysis which uses the transitive closure of a symmetric fuzzy relation [42].

The usefulness of fuzzy c-means clustering is considered as a matter of fact by many researchers, while other researchers still think that non-fuzzy methods including statistical models [25, 34, 35] are enough for clustering data.

A notable feature of fuzzy c-means is said to be its robustness, i.e., results are stable clusters for different initial clusters. Such discussion is empirical and theoretical background for robustness is unclear.

In this paper we try to consider how the robustness of fuzzy c-means is explained from a theoretical viewpoint, for which a natural fuzzy classifier defined on the whole space is introduced. When such a partition of the whole space is naturally induced from a result of clustering, the method is called here *inductive clustering*. Kernel-based clustering is considered here, where both a non-inductive algorithm and an inductive algorithm are studied.

Moreover another method of hierarchical fuzzy clustering is discussed, which uses the transitive closure of symmetric fuzzy relations [42]. This methods has been shown to be equivalent to the well-known method of the single linkage of agglomerative hierarchical clustering [26]. Here the significance of fuzzy relation is the algebraic form of clusters instead of those generated by an algorithm.

We first consider fuzzy c-means and its variations, then hierarchical fuzzy clustering is briefly discussed.

2 Fuzzy *c*-Means

We begin with notations and then introduce the method of fuzzy c-means by Dunn [9, 10] and Bezdek [2, 3].

Let the set of objects for clustering be denoted by $X = \{x_1, \dots, x_N\}$ where each object is a point of p-dimensional Euclidean space $\boldsymbol{R}^p$. Thus $x_k = (x_k^1, \dots, x_k^p)^\top \in \boldsymbol{R}^p$, $k = 1, \dots, N$. Clusters are denoted either by G_i or simply by i. A similarity or dissimilarity measure between two objects is assumed. For fuzzy c-means, a standard dissimilarity measure is the squared Euclidean distance:

$$D(x, y) = \|x - y\|^2 = \sum_{j=1}^{p} (x^j - y^j)^2. \tag{1}$$

In fuzzy c-means and related methods, the number of clusters denoted by c is assumed to be given beforehand. The membership of object x_k to cluster i is assumed to be given by u_{ki}. Moreover the collection of all memberships is denoted by matrix $U = (u_{ki})$. It is natural to assume that $u_{ki} \in [0, 1]$ for all $1 \leq i \leq c$ and $1 \leq k \leq N$, and moreover $\sum_{j=1}^{c} u_{kj} = 1$, for all $1 \leq k \leq N$.

The method of fuzzy c-means as well as crisp c-means uses a center for a cluster, which is denoted by $v_i = (v_i^1, \dots, v_i^p)^\top \in \boldsymbol{R}^p$ for cluster i. For simplicity, all cluster centers are summarized into matrix $V = (v_1, \dots, v_c)$.

Crisp c-Means Algorithm

Many studies of clustering handle K-means [24], also called crisp c-means, of which the basic algorithm is as follows [3]:
CCM: Crisp c-means algorithm.

CCM0: Generate randomly c cluster centers.
CCM1: Allocate each object $x_k (k = 1, \dots, N)$ to the cluster of the nearest center.
CCM2: Calculate new cluster centers v_i as the centroid (alias the center of gravity). If all cluster centers are convergent, stop. Otherwise go to **CCM1**.

End CCM.
The center of a cluster G_i is given by

$$v_i = \frac{1}{|G_i|} \sum_{x_k \in G_i} x_k, \tag{2}$$

where $|G_i|$ is the number of objects in G_i.

2.1 *Fuzzy c-Means Algorithm*

The fundamental idea of fuzzy c-means is an alternative optimization of an objective function, which is proposed by Dunn [9, 10] and Bezdek [2, 3]:

$$J(U, V) = \sum_{i=1}^{c} \sum_{k=1}^{N} (u_{ki})^m D(x_k, v_i) \quad (m \geq 1), \tag{3}$$

where $D(x_k, v_i)$ is the squared Euclidean distance (1).

Using this objective function, the following alternative optimization [3] is carried out.
FCM (fuzzy c -means) algorithm.

FCM0: Generate randomly initial fuzzy clusters. Let the solutions be $(\bar{U}, \bar{V})$
FCM1: Minimize $J(U, \bar{V})$ with respect to U. Let the optimal solution be a new $\bar{U}$.
FCM2: Minimize $J(\bar{U}, V)$ with respect to V. Let the optimal solution be a new $\bar{V}$.
FCM3: If the solution $(\bar{U}, \bar{V})$ is convergent, stop. Else go to **FCM1**.

End FCM.

A criterion for convergence is omitted here, see, e.g., [3]. Optimization with respect to U is with the constraint:

$$u_{ki} \in [0, 1], \quad 1 \le i \le c, 1 \le k \le N, \quad \sum_{j=1}^{c} u_{kj} = 1, \quad 1 \le k \le N, \tag{4}$$

while optimization with respect to V is without any constraint.

It is well-known that, when $m = 1$, the solution U is reduced to the allocation to the cluster of the nearest center:

$$u_{ki} = 1 \iff i = \arg \min_{1 \le j \le c} D(x_k, v_i),$$

and the center is given by (2). Thus the algorithm is equivalent to **CCM** when $m = 1$.

Hence we assume $m > 1$ hereafter, in order to have fuzzy solutions, where the optimal solutions are as follows:

$$\bar{u}_{ki} = \left\{ \sum_{j=1}^{c} \left(\frac{D(x_k, \bar{v}_i)}{D(x_k, \bar{v}_j)} \right)^{\frac{1}{m-1}} \right\}^{-1}, \tag{5}$$

$$\bar{v}_i = \frac{\sum_{k=1}^{N} (\bar{u}_{ki})^m x_k}{\sum_{k=1}^{N} (\bar{u}_{ki})^m}. \tag{6}$$

The derivations are omitted; the readers should refer to [3] or other textbooks.

Equation (5) does not seem to work when $x_k = v_i$, In such a case (5) should be interpreted as

$$\bar{u}_{ki} = \left\{ 1 + \sum_{j \ne i} \left(\frac{D(x_k, \bar{v}_i)}{D(x_k, \bar{v}_j)} \right)^{\frac{1}{m-1}} \right\}^{-1} \tag{7}$$

by eliminating two terms which appear to have a singular point.

Sometimes we omit the bars like

$$u_{ki} = \left\{ \sum_{j=1}^{c} \left(\frac{D(x_k, v_i)}{D(x_k, v_j)} \right)^{\frac{1}{m-1}} \right\}^{-1}, \tag{8}$$

$$v_i = \frac{\sum_{k=1}^{N} (u_{ki})^m x_k}{\sum_{k=1}^{N} (u_{ki})^m}, \tag{9}$$

for simplicity and without confusions.

2.2 A Natural Classifier

Let us consider the next function defined on $\boldsymbol{R}^p$ with a given set of cluster centers V:

$$U_i(x; V) = \left\{ \sum_{j=1}^{c} \left(\frac{D(x, v_i)}{D(x, v_j)} \right)^{\frac{1}{m-1}} \right\}^{-1}, \tag{10}$$

or

$$U_i(x; V) = \left\{ 1 + \sum_{j \neq i} \left(\frac{D(x, v_i)}{D(x, v_j)} \right)^{\frac{1}{m-1}} \right\}^{-1}. \tag{11}$$

It is clear that $U_i(x; V)$ has been derived from u_{ki} simply by replacing object symbol x_k by variable x.

This replacement appears trivial and it also appears that $U_i(x; V)$ has no further information than u_{ki}. On the contrary, this function of fuzzy classifier is important if we wish to observe theoretical properties of fuzzy c-means.

We have the following proposition that shows how the solutions of fuzzy c-means classify the given space.

Proposition 1 *The function $U_i(x_k; V)$ with a given V has the following properties.*

(i) $U_i(x_k; V) = u_{ki}$, *i.e., the fuzzy classifier interpolates the membership value* u_{ki}.
(ii) *When $|x|$ tends to infinity, $U_i(x; V), i = 1, \ldots, c$, approaches the same value of $1/c$:*

$$\lim_{\|x\| \to \infty} U_i(x; V) = \frac{1}{c}.$$

(iii) *The maximum value of $U_i(x; V), i = 1, \ldots, c$, is at $x = v_i$:*

$$\max_{x \in \boldsymbol{R}^p} U_i(x; V) = U_i(v_i, V) = 1.$$

Proof Property *(i)* is trivial. Property *(ii)* is easily obtained by observing

$$\lim_{\|x\| \to \infty} \frac{D(x, v_i)^{\frac{1}{m-1}}}{D(x, v_j)^{\frac{1}{m-1}}} = 1$$

and $U_i(x; V)$ is given by (11). Finally, the third property is almost trivial since the denominator of (11) is reduced to 1 when $x = v_i$. □

Note the significance of the function $U_i(x; V)$ in these propositions. An object x_k is a fixed point, while x is a variable that can be moved toward infinity or can be a cluster center. Without such a classifier, we cannot observe theoretical properties of fuzzy c-means.

Classifier of K-Means

The crisp classifier $U_i^{\rm ccm}(x; V)$ of K-means (CCM) is obviously the nearest center allocation:

$$U_i^{\rm ccm}(x; V) = 1 \iff i = \arg\min_{1 \le j \le c} D(x; v_j). \tag{12}$$

We define

$$\mathcal{V}_i = \{x \in \boldsymbol{R}^p \ : \ U_i^{\rm ccm}(x; V) = 1 \text{ and } U_j^{\rm ccm}(x; V) = 0, \ \forall j \neq i\ \}, \tag{13}$$

We note that $\mathcal{V}_i$ is a Voronoi region [21] with center v_i and other cluster centers.

We moreover note the next proposition.

Proposition 2 *If we define*

$$\mathcal{V}_i' = \{x \in \boldsymbol{R}^p \ : \ U_i(x; V) > U_j(x; V), \ \forall j \neq i\} \tag{14}$$

for classifiers of fuzzy c-means, we have

$$\mathcal{V}_i' = \mathcal{V}_i, \quad 1 \le i \le c.$$

The proof is easy by direct calculation and omitted. This proposition means that when the result of clustering by fuzzy c-means is made crisp using the maximum membership reallocation by (14), it leads to Voronoi regions.

Note 1 The above Voronoi regions are open sets and the boundary of two or more regions are left unclassified. The problem of a point on the boundary is not essential here and it can belong to all neighboring regions or be left unclassified. Note also the next relation:

$$\bigcup_{i=1}^{c} \overline{\mathcal{V}}_i = \boldsymbol{R}^p, \qquad \mathcal{V}_i \cap \mathcal{V}_j = \emptyset \quad (i \neq j), \tag{15}$$

where $\overline{\mathcal{V}}_i$ is the closure of $\mathcal{V}_i$.

2.3 A Method of Entropy

Other objective functions of fuzzy c-means have also been proposed among which we discuss the use of entropy [23, 27]:

$$J_{\rm ent}(U, V) = \sum_{i=1}^{c} \sum_{k=1}^{N} \{u_{ki} D(x_k, v_i) + \lambda^{-1} u_{ki} \log u_{ki}\}, \quad (\lambda > 0). \tag{16}$$

We easily have the solutions for alternative minimization of $J_{\text{ent}}(U, V)$:

$$u_{ki} = \frac{\exp(-\lambda D(x_k, v_i))}{\sum_{j=1}^{c} \exp(-\lambda D(x_k, v_j))}, \tag{17}$$

$$v_i = \frac{\sum_{k=1}^{N} u_{ki} x_k}{\sum_{k=1}^{N} u_{ki}}, \tag{18}$$

from which the classifier is given as follows:

$$U_i^{\text{ent}}(x; V) = \frac{\exp(-\lambda D(x, v_i))}{\sum_{j=1}^{c} \exp(-\lambda D(x, v_j))}. \tag{19}$$

These solutions are sometimes called the *entropy method* in contrast to the fuzzy c-means using (3).

We consider properties of $U_i^{\text{ent}}(x; V)$, which are more complicated than those of $U_i(x; V)$.

Proposition 3 *Define*

$$\mathcal{V}_i'' = \{x \in \boldsymbol{R}^p \ : \ U_i^{\text{ent}}(x; V) > U_j^{\text{ent}}(x; V), \ \forall j \neq i \ \} \tag{20}$$

for classifiers of fuzzy c-means using entropy term. We then have

$$\mathcal{V}_i'' = \mathcal{V}_i, \quad 1 \leq i \leq c.$$

Proposition 4 *Assume that matrix* $V = (v_1, \ldots, v_c)$ *has full rank* (rank $V = \min\{c, p\}$). *If* $\mathcal{V}_i''$ *is unbounded, then*

$$\lim_{\|x\| \to \infty; x \in \mathcal{V}_i''} U_i^{\text{ent}}(x; V) = 1,$$

whereas if $\mathcal{V}_i''$ *is bounded, then*

$$\lim_{\|x\| \to \infty} U_i^{\text{ent}}(x; V) = 0.$$

On the other hand, we have

$$0 < U_i^{\text{ent}}(x; V) < 1.$$

The proof is given in [28, 30] and omitted here.

Robustness of Fuzzy c-Means

Let us compare the solutions of the three methods of fuzzy c-means with (3), the K-means of CCM algorithm, and the entropy method using (16). For this purpose we compare the functions $U_i(x; V)$, $U_i^{\text{ccm}}(x; V)$, and $U_i^{\text{ent}}(x; V)$.

Suppose that x is very far away from centers $v_1, \ldots, v_c$, then the membership of x by fuzzy c-means is $U_i(x; V) \approx \frac{1}{c}$ for all $1 \le i \le c$, whereas $U_i^{\text{ccm}}(x; V) = 1$ and $U_j^{\text{ccm}}(x; V) = 0$ $(j \neq i)$ for $x \in \mathcal{V}_i'$ by CCM. The result by the entropy method is similar to CCM; $U_i^{\text{ent}}(x; V) \approx 1$ and $U_j^{\text{ccm}}(x; V) \approx 0$ $(j \neq i)$ for $x \in \mathcal{V}_i''$. This means that the results by the K-means (CCM) and the entropy method are strongly influenced by outliers, i.e., objects far from cluster centers.

Moreover, the function $U_i(x; V)$ has the maximum value of unity when $x = v_i$, while the entropy method does not have this property.

Thus the fuzzy c-means has the desirable properties than the K-means and the entropy method.

3 Generalization of Fuzzy c-Means

Many variations of fuzzy c-means have been studied, e.g., fuzzy c-varieties [3], fuzzy c-regressions [15], noise clustering [6], and possibilistic clustering [22]. We, however, limit ourselves to the discussion of the method of Gustafson and Kessel [14] and its extension [30] to take clusterwise covariance and another variable for cluster size into account.

In this section we introduce

$$D(x, v; S) = (x - v)^\top S^{-1}(x - v)$$

which is the squared Mahalanobis distance.

3.1 *The Method of Gustafson and Kessel and Its Generalization*

The method of Gustafson and Kessel incorporate clusterwise covariance variables denoted by $S_1, \ldots, S_c$. The objective function is

$$J(U, V, \boldsymbol{S}) = \sum_{i=1}^{c} \sum_{k=1}^{N} (u_{ki})^m D(x_k, v_i; S_i) \quad (m > 1), \tag{21}$$

where a simplified symbol $\boldsymbol{S} = (S_1, \ldots, S_c)$ and the clusterwise squared Mahalanobis distance $D(x_k, v_i; S_i)$ is used.

Miyamoto et al. introduced an objective function

$$J(U, V, S, A) = \sum_{i=1}^{c} \sum_{k=1}^{N} (\alpha_i)^{1-m} (u_{ki})^m D(x_k, v_i; S_i) \quad (m > 1), \tag{22}$$

with an additional variable $A = (\alpha_1, \ldots, \alpha_c)$ with the constraint

$$\sum_{i=1}^{c} \alpha_i = 1, \qquad \alpha_j \geq 0, \quad 1 \leq j \leq c. \tag{23}$$

Note also that S_i is with the constraint

$$|S_i| = \rho_i \quad (\rho_i > 0) \tag{24}$$

where ρ_i is a fixed parameter and $|S_i|$ is the determinant of S_i. We assume, for simplicity, $\rho_i = 1$ [16].

The solutions are as follows:

$$u_{ki} = \left\{ \sum_{j=1}^{c} \left(\frac{D(x_k, v_i; S_i)}{D(x_k, v_j; S_j)} \right)^{\frac{1}{m-1}} \right\}^{-1} \tag{25}$$

$$v_i = \frac{\sum_{k=1}^{N} (u_{ki})^m x_k}{\sum_{k=1}^{N} (u_{ki})^m} \tag{26}$$

$$S_i = \frac{1}{|\hat{S}_i|^{\frac{1}{p}}} \sum_{k=1}^{N} (u_{ki})^m (x_k - v_i)(x_k - v_i)^\top. \tag{27}$$

$$\alpha_i = \left[\sum_{j=1}^{c} \left\{ \frac{\sum_{k=1}^{N} (u_{kj})^m D(x_k, v_j; S_j)}{\sum_{k=1}^{N} (u_{ki})^m D(x_k, v_i; S_i)} \right\}^m \right]^{-1} \tag{28}$$

where

$$\hat{S}_i = \sum_{k=1}^{N} (u_{ki})^m (x_k - v_i)(x_k - v_i)^\top.$$

Since four types of variables are used for the augmented method of Gustafson and Kessel, the alternative optimization iteratively calculates (25), (26), (27), and (28) until convergence.

3.2 K–L Information Method

The K–L (Kullback–Leibler) information method by Ichihashi et al. [17, 18, 30] is another generalized version of fuzzy c-means which uses the entropy method. The objective function is as follows.

$$J_{\mathrm{KL}}(U, V, S, A) = \sum_{i=1}^{c} \sum_{k=1}^{N} u_{ki} D(x_k, v_i; S_i) + \sum_{i=1}^{c} \sum_{k=1}^{N} \{\nu u_{ki} \log \frac{u_{ki}}{\alpha_i} + \log |S_i|\}. \tag{29}$$

The solutions are given by the following:

$$u_{ki} = \frac{\dfrac{\alpha_i}{|S_i|} \exp\left(-\dfrac{D(x_k, v_i; S_i)}{\nu}\right)}{\displaystyle\sum_{j=1}^{c} \dfrac{\alpha_j}{|S_j|} \exp\left(-\dfrac{D(x_k, v_j; S_i)}{\nu}\right)}, \tag{30}$$

$$v_i = \frac{\displaystyle\sum_{k=1}^{N} u_{ki} x_k}{\displaystyle\sum_{k=1}^{N} u_{ki}} \tag{31}$$

$$S_i = \frac{1}{\displaystyle\sum_{k=1}^{N} u_{ki}} \sum_{k=1}^{N} u_{ki} (x_k - v_i)(x_k - v_i)^\top \tag{32}$$

$$\alpha_i = \frac{1}{N} \sum_{k=1}^{N} u_{ki} \tag{33}$$

The method of K–L information is very similar to the solution of EM algorithm of the Gaussian mixture [25, 34] and moreover generalizes the latter statistical model. The G-K method, when compared with the K–L method, seems to have the robustness property discussed in the previous section.

4 Kernel Based Fuzzy c-Means

The support vector machines [38, 39] with positive definite kernel functions are now one of the most popular methods of supervised classification. Apart from support vector machines, kernel functions themselves are considered to be useful by many researchers (e.g., [13, 36]). Such positive definite kernels can be used for fuzzy c-means, as we see in this section.

The reason why we use kernels for clustering is that essentially the K-means and fuzzy c-means have linear boundaries between clusters of Voronoi regions, as we have seen above.

The introduction of the covariance variables in the last section enables the cluster boundaries to be quadratic, but more flexible nonlinear boundaries cannot be obtained.

In order to have clusters with nonlinear boundaries, we can use positive definite kernels. Kernels are introduced by using a high-dimensional mapping $\Phi : \boldsymbol{R}^p \to H$, where H is generally a Hilbert space with the inner product $\langle \cdot, \cdot \rangle_H$ and the norm $\| \cdot \|_H$.

Given objects $x_1, \ldots x_N$, we consider its images by the mapping $\Phi : \Phi(x_1), \ldots, \Phi(x_N)$. Note that the method of kernels does not assume that an explicit form of $\Phi(x_1), \ldots, \Phi(x_N)$ is known, but their inner product $\langle \Phi(x_i), \Phi(x_j) \rangle_H$ is assumed to be given using a known kernel function $K(x, y)$:

$$K(x, y) = \langle \Phi(x_i), \Phi(x_j) \rangle_H.$$

A well-known example is the Gaussian kernel:

$$K(x, y) = \exp(-C\|x - y\|^2).$$

In this case,

$$\langle \Phi(x), \Phi(y) \rangle_H = \exp(-C\|x - y\|^2).$$

We consider kernel-based fuzzy c-means [29]. The objective function uses $\Phi(x_1), \ldots, \Phi(x_N)$ and cluster centers $w_1, \ldots, w_c$ of H:

$$J(U, V) = \sum_{i=1}^{c} \sum_{k=1}^{N} (u_{ki})^m \|\Phi(x_k) - w_i\|_H^2 \quad (m > 1), \tag{34}$$

where $W = (w_1, \ldots, w_c)$. We have

$$u_{ki} = \left\{ \sum_{j=1}^{c} \left(\frac{\|\Phi(x_k) - w_i\|_H^2}{\|\Phi(x_k) - w_j\|_H^2} \right)^{\frac{1}{m-1}} \right\}^{-1} \tag{35}$$

$$w_i = \frac{\sum_{k=1}^{N} (u_{ki})^m \Phi(x_k)}{\sum_{k=1}^{N} (u_{ki})^m} \tag{36}$$

Note, however, that the explicit form of $\Phi(x_k)$ and hence w_i is not available.

We have two ways to handle this situation. First way is to make $\Phi(x_k)$ explicit, whereas the second way is to eliminate $\Phi(x_k)$ and express them in terms of $K(x, y)$.

Use of Gram Matrix

First way to handle $\Phi(x_k)$ is the use of the Gram matrix. Let the Gram matrix be

$$\mathcal{K} = (K(x_k, x_l)), \quad 1 \le k, l \le N. \tag{37}$$

Since $\mathcal{K}$ is positive semi-definite and expressed as

$$\mathcal{K} = T^\top \Lambda T,$$

where $\Lambda = diag(\lambda_1, \dots, \lambda_N)$ is the diagonal matrix of nonnegative eigenvalues and T is the orthogonal matrix, we can define

$$\mathcal{K}^{\frac{1}{2}} = T^\top \Lambda^{\frac{1}{2}} T,$$

where $\Lambda^{\frac{1}{2}} = diag(\sqrt{\lambda_1}, \dots, \sqrt{\lambda_N})$. Let e_k be kth elementary vector: $e_1 = (1, 0, \dots, 0)^\top$, $e_2 = (0, 1, 0, \dots, 0)^\top$, and so on. Put

$$\Phi(x_k) = \mathcal{K}^{\frac{1}{2}} e_k, \quad k = 1, 2, \dots, N. \tag{38}$$

In other words, $\Phi(x_k)$ is kth column (or row) vector of $\mathcal{K}^{\frac{1}{2}}$. Then solutions (35) and (36) are used.

Note that

$$\begin{aligned}\langle \Phi(x_k), \Phi(x_l) \rangle &= (\mathcal{K}^{\frac{1}{2}} e_k)^\top (\mathcal{K}^{\frac{1}{2}} e_l) = e_k^\top \mathcal{K}^{\frac{1}{2}} \mathcal{K}^{\frac{1}{2}} e_l \\ &= e_k^\top \mathcal{K} e_l = K(x_k, x_l),\end{aligned}$$

hence (38) is appropriate.

Note also that Φ is defined on the finite set X ($\Phi : X \to \boldsymbol{R}^p$).

Updating Dissimilarity

Second way is to eliminate w_i from the iterative calculation: updating formula of w_i by Eq. (36) is replaced by the update of dissimilarity

$$D_H(x_k, w_i) = \|\Phi(x_k) - w_i\|^2.$$

We have

$$\begin{aligned}D_H(x_k, w_i) = K(x_k, x_k) &- \frac{2}{\sum_{k=1}^N (u_{ki})^m} \sum_{j=1}^N (u_{ji})^m K(x_j, x_k) \\ &+ \frac{1}{(\sum_{k=1}^N (u_{ki})^m)^2} \sum_{j=1}^N \sum_{\ell=1}^N (u_{ji} u_{\ell i})^m K(x_j, x_\ell).\end{aligned} \tag{39}$$

Using (39), we calculate

$$u_{ki} = \left\{ \sum_{j=1}^{c} \left(\frac{D_H(x_k, w_i)}{D_H(x_k, w_j)} \right)^{\frac{1}{m-1}} \right\}^{-1} \tag{40}$$

Thus the alternative optimization of u_{ki} by (35) and w_i by (36) is replaced by the iteration of (39) and (40) until convergence.

Fuzzy classifiers of kernel-based fuzzy c-means can also be derived by substituting variable x into x_k [30]: we have

$$\begin{aligned} D(x, w_i) = K(x, x) &- \frac{2}{\sum_{k=1}^{N}(u_{ki})^m} \sum_{j=1}^{N} (u_{ji})^m K(x, x_j) \\ &+ \frac{1}{(\sum_{k=1}^{N}(u_{ki})^m)^2} \sum_{j=1}^{N} \sum_{\ell=1}^{N} (u_{ji} u_{\ell i})^m K(x_i, x_\ell), \end{aligned} \tag{41}$$

$$U_i(x, W) = \left\{ \sum_{j=1}^{c} \left(\frac{D_H(x, w_i)}{D_H(x, w_j)} \right)^{\frac{1}{m-1}} \right\}^{-1} \tag{42}$$

Note that $\Phi(x)$ in the second method is defined for an arbitrary point $x \in \boldsymbol{R}^p$ $(\Phi: \boldsymbol{R}^p \to H)$ that is different from the function in the first method.

5 Inductive Clustering Versus Non-inductive Clustering

Supervised classification method such as the standard Bayesian classification and the support vector machines provide classification rules defined on the whole space. If the space is $\boldsymbol{R}^p$ and suppose that the Bayesian rule is $P(G_i|x)$ and the SVM rule is SVM, then they are functions of $P(G_i|\cdot): \boldsymbol{R}^p \to [0, 1]$ and $SVM: \boldsymbol{R}^p \to \{-1, +1\}$. Thus the Bayesian rule is probabilistic, while SVM rule is crisp.

Recently semi-supervised learning has been studied and accordingly the concept of transductive learning [5] has been proposed which means that classification of a finite set of new objects is derived but a classification rule of the whole space is not required. In contrast to transductive learning, a former conventional method of a classification rule of the whole space is called inductive learning.

Turning to the original topic of clustering, the author suppose that many researchers think that clustering is 'transductive' in the above sense, i.e., classification of a given set of objects is enough and nothing more is needed. The above discussed fuzzy classifiers are defined on the whole space, contrary to this general understanding. In short, we discussed inductive properties of fuzzy c-means clustering and related methods.

We now try to make the concept of inductive clustering clearer. If a method of clustering has an *intrinsic classification rule* defined on the whole space, we call the method *inductive clustering*, while the method does not have such a classification rule on the whole space and it gives a classification result on a given set of objects alone, the we call the method *non-inductive clustering* (We avoid the name of *transductive clustering*, as non-inductive property implies nothing in particular).

In this sense, the K-means, fuzzy c-means, and the statistical model of mixture distributions [25] are all inductive clustering, and we can study theoretical properties of them, as we have seen above. Note also that kernel-based fuzzy c-means has the both versions of non-inductive clustering and inductive clustering, since the first method gives $\Phi: X \to \boldsymbol{R}^p$ which is non-inductive clustering, since we cannot use $\Phi(x)$ for $x \notin X$. In contrast, the second way is to use $\Phi: \boldsymbol{R}^p \to H$, which leads us to an inductive version, since we have $\Phi(x)$ and hence $U_i(x, W)$ for any $x \in \boldsymbol{R}^p$.

We emphasize that an advantage of inductive clustering is that we can study its theoretical properties more easily than non-inductive clustering. Indeed, methods of inductive clustering seem to have better or simpler behaviors when generating clusters, like the nearest prototype property of the K-means. On the other hand, if we give up inductiveness in clustering, we have more choice of clustering algorithms, and this attitude has been taken by researchers of clustering, since a proposal of a clustering algorithm does not lead to inductiveness in general.

In spite of this general understanding, we emphasize again the importance of inductive clustering in order to have greater progress in studies of cluster analysis.

Next subject is hierarchical clustering, where we observe again inductiveness of a method, although hierarchical clustering is generally non-inductive.

6 Hierarchical Fuzzy Clustering

Zadeh [42] discussed a fuzzy similarity relation which is reflexive, symmetric, and transitive. In other words, an arbitrary α-cut of a fuzzy similarity relation is reflexive, symmetric, and transitive as a crisp relation.

6.1 Transitive Closure of Fuzzy Relation

We assume that object set $X = \{x_1, \ldots, x_N\}$ are not necessarily in an Euclidean space. Rather, a relation $S(x, y)$ of X satisfying reflexivity and symmetry

$$S(x, x) = 1, \quad \forall x \in X, \tag{43}$$

$$S(x, y) = S(y, x), \quad \forall x, y \in X \tag{44}$$

is assumed, where a larger value of $S(x, y)$ means that x and y are more similar and a smaller value of it implies they are less similar.

Fuzzy transitivity means that

$$S(x, z) \geq \min\{S(x, y), S(y, z)\}, \quad \forall y \in X, \tag{45}$$

but we do not assume this property of transitivity for a given S, since transitivity is a too strong condition in real applications.

Moreover we do not use the term of 'similarity relation' as Zadeh used, but similarity is a general term to show two objects are similar, in order to keep compatibility of terms between the two fields of fuzzy systems and statistical data analysis. We use the term of *fuzzy equivalence relation* instead of similarity relation when a fuzzy relation is reflexive, symmetric, and transitive.

An α-cut $[S]_\alpha$ of S is a crisp relation:

$$[S]_\alpha(x, y) = \begin{cases} 1 & (S(x, y) \geq \alpha), \\ 0 & (S(x, y) < \alpha). \end{cases}$$

Note the next proposition.

Proposition 5 *If a fuzzy relation S is reflexive, symmetric, and transitive, then every α-cut of it is a crisp equivalence relation:*

$$[S]_\alpha(x, x) = 1, \quad \forall x \in X, \tag{46}$$

$$[S]_\alpha(x, y) = [S]_\alpha(y, x), \quad \forall x, y \in X, \tag{47}$$

$$[S]_\alpha(x, y) = 1,\ [S]_\alpha(y, z) = 1 \Rightarrow [S]_\alpha(x, z) = 1. \tag{48}$$

Proof The first two equations are trivial. The third relation is also easily proved by observing that if $S(x, y) \geq \alpha$ and $S(y, z) \geq \alpha$, then (48) follows from (45). □

Note that we do not assume the transitivity. In order to have a transitive relation from a reflexive and symmetric fuzzy relation, we calculate the transitive closure. For this purpose we introduce the max-min composition of fuzzy relations:

$$(S \circ T)(x, z) = \max_{y \in X} \min\{S(x, y), T(y, z)\},$$

where S and T are fuzzy relations of X. Using the max-min composition, we can define the transitive closure S^* of S:

$$S^*(x, y) = \max\{S(x, y), S^2(x, y), S^3(x, y), \ldots\},$$

where $S^2 = S \circ S$ and $S^k = S \circ S^{k-1}$. It also is not difficult to see $S^* = S^{N-1}$ when S is reflexive and symmetric.

When S is reflexive and symmetric, the transitive closure S^* is also reflexive and symmetric, and moreover transitive. The proof that S^* is transitive is omitted here. Readers can refer to, e.g., [26].

Note that Proposition 5 holds for S^*. Then each α-cut of S^* induces an equivalence class of X, and moreover if α decreases, the equivalence class becomes coarser, and when it increases, the equivalence class becomes finer. Thus S^* defines hierarchical clusters.

6.2 Single Linkage and Transitive Closure

We describe general algorithm of agglomerative hierarchical clustering as follows:
AHC (Algorithm of Agglomerative Hierarchical Clustering).

AHC1: Let initial clusters be individual objects: $G_i = \{x_i\}, i = 1, \ldots, N$. $S(G_i, G_j) = S(x_i, x_j)$, $1 \le i, j \le N$, and put $K = N$.

AHC2: Find pair of clusters of maximum similarity:

$$(G_p, G_q) = \arg\max_{i,j} S(G_i, G_j). \tag{49}$$

Merge $G_r = G_p \cup G_q$. $K = K - 1$ and if $K = 1$, stop.

AHC3: Update $S(G_r, G')$ for all other clusters G'. Go to **AHC1**.

End AHC.

The updating step of **AHC3** admits different choices of similarity between clusters, among which the single linkage, the complete linkage, and the average linkage use the followings:

Single Linkage:

$$S(G_r, G') = \max_{x \in G_r, y \in G'} S(x, y) \tag{50}$$

$$= \max\{S(G_p, G'), S(G_q, G')\} \tag{51}$$

Complete Linkage:

$$S(G_r, G') = \min_{x \in G_r, y \in G'} S(x, y) \tag{52}$$

$$= \min\{S(G_p, G'), S(G_q, G')\} \tag{53}$$

Average Linkage:

$$S(G_r, G') = \frac{\sum_{x \in G_r, y \in G'} S(x, y)}{|G_r||G'|} \tag{54}$$

$$= \frac{|G_p|}{|G_r|} S(G_p, G') + \frac{|G_q|}{|G_r|} S(G_q, G') \tag{55}$$

Discussion in this section is mostly focused upon the single linkage.

We have the proposition of equivalence between the transitive closure and the single linkage [26].

Proposition 6 *Given a set of objects* $X = \{x_1, \ldots, x_N\}$ *and a similarity measure* $S(x, y)$ *for all* $x, y \in X$, *the following three methods give the same hierarchical clusters:*

1. *clusters by the single linkage;*
2. *clusters by the transitive closure* S^**;*
3. *clusters as vertices of connected components of fuzzy graph with vertices* X *and edges* $X \times X$ *with membership values* $S(x, y)$.

The connected components of a fuzzy graph is the essential part in this proposition, which means the family of those connected components of all α-cuts of the fuzzy graph. Since connected components grow with decreasing α, those sets of vertices form hierarchical clusters. The proof of this proposition is given in [26] and omitted here, but the idea of the proof is to reduce both the transitive closure and the single linkage clusters to the connected components. Thus fuzzy graph is fundamental in this proposition.

The significance of fuzzy relation and its transitive closure is the algebraic expression of a method of agglomerative clustering in contrast to the general understanding that a method of clustering is essentially a proposal of an algorithm.

Seemingly no new results are included in this theorem. However, Miyamoto [31] showed that ideas in other methods of DBSCAN [11] and Wishart's mode analysis [41] are captured into the above results of equivalence. Concretely, the transitive closure $[S \wedge (\boldsymbol{a}\boldsymbol{a}^\top)]^*$ is proposed in [31], where $\boldsymbol{a}$ is a fuzzy set of X; $\boldsymbol{a}$ is the abstraction of dense points in [41] and core points in [11].

Inductive Property of Hierarchical Clustering

Agglomerative hierarchical clustering in general is non-inductive, as in the assumption that a given X is not in a metric space. When space $\boldsymbol{R}^p$ is given and $X \subset \boldsymbol{R}^p$ is, e.g., with a Euclidean metric, the single linkage can be regarded as an inductive method [33] where the nearest neighbor allocation is used. Roughly, a point x in the Euclidean space can be allocated to the cluster i if a point nearest to x exists in that cluster.

A question arises whether or not this result can be extended to the complete linkage and the average linkage: the furthest neighbor allocation and the average distance allocation can be used, respectively, in these methods. No good answer exists to this question, since it is doubtful that such furthest and/or average allocation methods are as useful as nearest neighbor allocation in the single linkage. Note also that an algebraic expression like the transitive closure is unavailable for the complete linkage or the average linkage.

7 Conclusion

We studied fuzzy clustering and its significance. The method of fuzzy c-means is known to have robustness, and robustness property has been discussed from a theoretical viewpoint using a natural fuzzy classifier, which is derived from substituting an object symbol by a variable. Such function of classification is useful in considering theoretical properties of a clustering method and leads to the concept of inductive clustering, while the original idea of clustering is non-inductive. Kernel fuzzy c-means have been considered in which both non-inductive and inductive algorithms are derived.

Entropy methods including K–L information fuzzy c-means are also discussed which are more closely related to the Gaussian mixture of distributions. They are less robust when compared their theoretical properties with those of the fuzzy c-means including the Gustafson-Kessel method and its extension.

Hierarchical fuzzy clustering was also considered where the transitive closure of a symmetric fuzzy relation is proved to be equivalent to the single linkage method. Thus the transitive closure is an algebraic expression of the well-known agglomerative hierarchical algorithm. Although the result appears purely theoretical, the equivalence leads to a new method of hierarchical clustering [31]. An α-cut of the transitive closure will produce a crisp classifier for the whole space if the problem is given in an Euclidean space, but a fuzzy classifier is difficult to be obtained.

We omitted derivations of solutions of which readers should refer to [3, 16, 30].

An important issue which we omitted here is cluster validity whereby the number of clusters can be decided, which is discussed in [3, 7]. Another topic of recent interest is semi-supervised classification [5, 32, 43] including constrained clustering [1, 37, 40]. Fuzzy classifiers will be useful also in semi-supervised classification, which will be studied in near future.

References

1. S. Basu, I. Davidson, K.L. Wagstaff, *Constrained Clustering*, CRC Press, Boca Raton, 2009.
2. J.C. Bezdek, *Fuzzy Mathematics in Pattern Classification*, Ph.D. Thesis, Cornell Univ., Ithaca, NY, 1973.
3. J.C. Bezdek, *Pattern Recognition with Fuzzy Objective Function Algorithms*, Plenum Press, 1981.
4. J.C. Bezdek, J. Keller, R. Krishnapuram, N.R. Pal, *Fuzzy Models and Algorithms for Pattern Recognition and Image Processing*, Kluwer, Boston, 1999.
5. O. Chapelle, B. Schölkopf, A. Zien, eds., *Semi-Supervised Learning*, MIT Press, Cambridge, Massachusetts, 2006.
6. R.N. Davé, R. Krishnapuram, Robust clustering methods: a unified view, *IEEE Trans. on Fuzzy Systems*, Vol. 5, pp. 270–293, 1997.
7. D. Dumitrescu, B. Lazzerini, L.C. Jain, *Fuzzy Sets and Their Application to Clustering and Training*, CRC Press, Boca Raton, Florida, 2000.
8. R.O. Duda, P.E. Hart, *Pattern Classification and Scene Analysis*, John Wiley & Sons, 1973.

9. J.C. Dunn, A fuzzy relative of the ISODATA process and its use in detecting compact well-separated clusters, *J. of Cybernetics*, Vol. 3, pp. 32–57, 1974.
10. J.C. Dunn, Well-separated clusters and optimal fuzzy partitions, *J. of Cybernetics*, Vol. 4, pp. 95–104, 1974.
11. M. Ester, H.-P. Kriegel, J. Sander, X.W. Xu, A density-based algorithm for discovering clusters in large spatial databases with noise, Proc. of 2nd Intern. Conf. on Knowledge Discovery and Data Mining (KDD-96), AAAI Press, pp. 226–231, 1996.
12. B.S. Everitt, *Cluster Analysis, 3rd Ed.*, Arnold, London, 1993.
13. M. Girolami, Mercer kernel based clustering in feature space, *IEEE Trans. on Neural Networks*, Vol. 13, No. 3, pp. 780–784, 2002.
14. E.E. Gustafson, W.C. Kessel, Fuzzy clustering with a fuzzy covariance matrix, IEEE CDC, San Diego, California, pp. 761–766, 1979.
15. R.J. Hathaway, J.C. Bezdek, Switching regression models and fuzzy clustering, *IEEE Trans. on Fuzzy Systems*, Vol. 1, No. 3, pp. 195–204, 1993.
16. F. Höppner, F. Klawonn, R. Kruse, T. Runkler, *Fuzzy Cluster Analysis*, Jhon Wiley & Sons, 1999
17. H. Ichihashi, K. Honda, N. Tani, Gaussian mixture PDF approximation and fuzzy c-means clustering with entropy regularization, Proc. of Fourth Asian Fuzzy Systems Symposium, Vol. 1, pp. 217–221, 2000.
18. H. Ichihashi, K. Miyagishi, K. Honda, Fuzzy c-means clustering with regularization by K-L information, Proc. of 10th IEEE International Conference on Fuzzy Systems, Vol. 2, pp. 924–927, 2001.
19. A.K. Jain, R.C. Dubes, *Algorithms for Clustering Data*, Prentice Hall, Englewood Cliffs, NJ, 1988.
20. L. Kaufman, P.J. Rousseeuw, *Finding Groups in Data: An Introduction to Cluster Analysis*, Wiley, New York, 1990.
21. T. Kohonen, *Self-Organizing Maps, 2nd Ed.*, Springer, Berlin, 1997.
22. R. Krishnapuram, J. M. Keller, A possibilistic approach to clustering, *IEEE Trans. on Fuzzy Systems*, Vol. 1, pp. 98–110, 1993.
23. R.-P. Li and M. Mukaidono, A maximum entropy approach to fuzzy clustering, Proc. of the 4th IEEE Intern. Conf. on Fuzzy Systems (FUZZ-IEEE/IFES'95), Yokohama, Japan, March 20–24, 1995, pp. 2227–2232, 1995.
24. J.B. MacQueen, Some methods of classification and analysis of multivariate observations, Proc. of 5th Berkeley Symposium on Math. Stat. and Prob., pp. 281–297, 1967.
25. G. McLachlan, D. Peel, *Finite Mixture Models*, Wiley, New York, 2000.
26. S. Miyamoto, *Fuzzy Sets in Information Retrieval and Cluster Analysis*, Kluwer, Dordrecht, 1990.
27. S. Miyamoto, M. Mukaidono, Fuzzy *c*-means as a regularization and maximum entropy approach, Proc. of the 7th International Fuzzy Systems Association World Congress (IFSA'97), June 25–30, 1997, Prague, Czech, Vol. II, pp. 86–92, 1997.
28. S. Miyamoto, *Introduction to Cluster Analysis*, Morikita-Shuppan, Tokyo, 1999 (in Japanese).
29. S. Miyamoto, D. Suizu, Fuzzy *c*-means clustering using kernel functions in support vector machines, *Journal of Advanced Computational Intelligence and Intelligent Informatics*, Vol. 7, No. 1, pp. 25–30, 2003.
30. S. Miyamoto, H. Ichihashi, K. Honda, *Algorithms for Fuzzy Clustering*, Springer, Berlin, 2008.
31. S. Miyamoto, Statistical and non-statistical models in clustering: an introduction and recent topics, A. Okada, D. Vicari, G. Ragozini, Eds., Analysis and Modelling of Complex Data in Behavioural and Social Sciences, JCS-CLADAG 12, Anacapri, Italy, Sept. 3–4, 2012, Cleup, Padova, ISBN 978-88-6129-916-0, pp. 3–6 (Web and USB Proc.) 2012.
32. S. Miyamoto, An Overview of Hierarchical and Non-hierarchical Algorithms of Clustering for Semi-supervised Classification, V. Torra et al. (Eds.): MDAI 2012, LNAI 7647, pp. 1–10, 2012.
33. S. Miyamoto, Inductive and Non-inductive Methods of Clustering, Proc. of 2012 IEEE International Conference on Granular Computing, Aug. 11–12, Hangzhou, China, pp. 12–17, 2012.

34. R.A. Redner, H.F. Walker, Mixture densities, maximum likelihood and the EM algorithm, *SIAM Review*, Vol. 26, No. 2, pp. 195–239, 1984.
35. K. Rose, E. Gurewitz, and G. Fox, "A deterministic annealing approach to clustering," *Pattern Recognition Letters*, Vol. 11, pp. 589–594, 1990.
36. B. Schölkopf, A.J. Smola, *Learning with Kernels*, the MIT Press, 2002.
37. N. Shental, A. Bar-Hillel, T. Hertz, D. Weinshall, Computing Gaussian mixture models with EM using equivalence constraints, In: Advances in Neural Information Processing Systems, Vol. 16, 2004.
38. V.N. Vapnik, *Statistical Learning Theory*, Wiley, New York, 1998.
39. V.N. Vapnik, *The Nature of Statistical Learning Theory: 2nd Ed.*, Springer, New York, 2000.
40. N. Wang, X. Li, X. Luo, Semi-supervised Kernel-based Fuzzy c-Means with Pairwise Constraints, Proc. of WCCI 2008, pp. 1099–1103, 2008.
41. Wishart, D.: Mode analysis: a generalization of nearest neighbour which reduces chaining effects, In: A.J. Cole, ed., Numerical Taxonomy, Proc. Colloq., in Numerical Taxonomy, Univ. of St. Andrews, pp. 283–311, 1968.
42. L.A. Zadeh, Similarity relations and fuzzy orderings, *Information Sciences*, Vol. 3, pp. 177–200, 1971.
43. X. Zhu, A.B. Goldberg, *Introduction to Semi-Supervised Learning*, Morgan and Claypool, 2009.

Fuzzy Clustering/Co-clustering and Probabilistic Mixture Models-Induced Algorithms

Katsuhiro Honda

Abstract While fuzzy c-means (FCM) and its variants have become popular tools in many application fields, their fuzzy partition natures were often discussed only from the empirical viewpoints without theoretical insight. This chapter reviews some fuzzy clustering models induced by probabilistic mixture concepts and discusses the effects of introduction of adjustable fuzziness penalties into statistical models. First, the entropy regularization-based FCM proposed by Miyamoto et al. is revisited from the Gaussian mixtures viewpoint and the fuzzification mechanism is compared with the standard FCM. Second, the regularization concept is discussed in fuzzy co-clustering context and a multinomial mixtures-induced clustering model is reviewed. Some illustrative examples demonstrate the characteristics of fuzzy clustering algorithms with adjustable fuzziness penalties, and the interpretability of object partition is shown to be improved. Finally, a possible future direction of fuzzy clustering research is discussed.

Keywords Fuzzy clustering · Co-clustering · Probabilistic mixture models · Regularization

1 Introduction

Cluster analysis or clustering [1] is a basic technique, which is often utilized in a primary step of analyzing unlabeled data with the goal of summarizing structural information. Besides simple processes of hierarchical algorithms, such as Single-Link and Complete-Link, non-hierarchical algorithms became popular in real world applications because of lower computational efforts. k-means [2] is the most famous

K. Honda (✉)
Graduate School of Engineering, Osaka Prefecture University,
1-1 Gakuen-cho, Nakaku, Sakai 599-8531, Osaka, Japan
e-mail: honda@cs.osakafu-u.ac.jp
URL: http://www.cs.osakafu-u.ac.jp/hi/

V. Torra et al. (eds.), *Fuzzy Sets, Rough Sets, Multisets and Clustering*,
Studies in Computational Intelligence 671, DOI 10.1007/978-3-319-47557-8_3

non-hierarchical clustering algorithm and has been modified for improving its utilities. Fuzzy c-means (FCM) [3] adopted fuzzy partition [4] supported by a fuzzy set-induced membership concept, where fuzzy partition is realized by introducing non-linear nature into k-means objective function with a weighting exponent. Miyamoto and Mukaidono [5, 6] proposed another fuzzification model based on an entropy regularization concept, where additional non-linear penalty term is combined with k-means objective function. This regularization concept was also achieved with a quadric penalty term [7].

k-means-type clustering also has another interpretation from the viewpoint of probabilistic mixture concepts. Hathaway [8] discussed that the pseudo-log-likelihood function of Gaussian mixture models (GMMs) [9, 10] can be decomposed into the hard k-means objective function and the K-L information-induced penalty term for soft partition. This concept supports the validity of the entropy regularized FCM objective function and implies a close connection with FCM clustering and probabilistic mixture models. Then, fuzzy counterparts of several probabilistic mixture models have been proposed based on K-L information regularization, where the degree of fuzziness of probabilistic partitions is tuned with adjustable penalty weights [11, 12].

FCM-type clustering was also extended to fuzzy co-clustering, where the goal is to extract pair-wise clusters of objects and items from their cooccurrence information such as document-keyword frequencies in document analysis and customer-product purchase frequencies in market analysis. Besides the entropy-based and quadric-based regularization models [13, 14], a fuzzy counterpart of multinomial mixture models [15], which is also implemented with an adjustable penalty weight, was proposed [16].

In this chapter, the above models are summarized and their characteristic features are demonstrated through several numerical experiments.

2 Fuzzy c-Means and Gaussian Mixture Models

2.1 *Fuzzification Schemes in FCM Clustering*

Assume that we have p-dimensional observation on n objects such as $\boldsymbol{x}_i = (x_{i1}, \dots, x_{ip})^\top, i = 1, \dots, n$. The goal of k-means-type clustering is to partition the n objects into C clusters with their representative centroids $\boldsymbol{b}_c$, where intra-cluster objects are as mutually similar as possible but inter-cluster objects are dissimilar. Generally, k-means scheme starts with a random centroid assignment (or a random object partition) and alternatively optimizes centroid assignment and object partition until convergence. Object partition can be represented with several membership models adopting different constraints. Let $u_{ci}, i = 1, \dots, n, c = 1, \dots, C$ be the membership degree of object i to cluster c.

Hard c-partition [2] brings crisp exclusive object assignment under the following constraints:

$$u_{ci} \in \{0, 1\}, \forall i, c, \qquad \sum_{c=1}^{C} u_{ci} = 1, \forall i, \qquad \sum_{i=1}^{n} u_{ci} > 0, \forall c. \tag{1}$$

Fuzzy c-partition [4] relaxes the crisp constraint but still holds the exclusive constraint such that:

$$u_{ci} \in [0, 1], \forall i, c, \qquad \sum_{c=1}^{C} u_{ci} = 1, \forall i, \qquad \sum_{i=1}^{n} u_{ci} > 0, \forall c. \tag{2}$$

Possibilistic c-partition [17] gives much more flexible object assignment without exclusive nature among clusters such that:

$$u_{ci} \in [0, 1], \forall i, c, \qquad \sum_{i=1}^{n} u_{ci} > 0, \forall c. \tag{3}$$

Fuzzy c-means (FCM) achieves fuzzy c-partition by modifying the k-means objective function. The standard model proposed by Bezdek [3], which is called as sFCM in this chapter, introduced an additional weighting exponent m such that:

$$L_{sfcm} = \sum_{c=1}^{C} \sum_{i=1}^{n} u_{ci}^{m} ||\boldsymbol{x}_i - \boldsymbol{b}_c||^2, \tag{4}$$

where u_{ci} follows Eq. (2). m ($m > 1$) tunes the degree of fuzziness. A larger m brings very fuzzy partition and $m \to \infty$ implies $u_{ci} \to 1/C$, $\forall i, c$. On the other hand, $m \to 1$ reduces to hard c-partition caused by linear programming nature of Eq. (4) with respect to u_{ci}. In this sense, fuzzy characteristics of c-partition is dependent to the nonlinearity of the k-means objective function.

Miyamoto and Mukaidono [5] introduced nonlinear nature based on another concept. Regularization is a basic approach in the formulation of ill-posed problems and a typical regularization is realized by adding a regularizing function. Entropy-based FCM, which is called eFCM in this chapter, combines the k-means objective function with an entropy-like penalty as follows:

$$L_{efcm} = \sum_{c=1}^{C} \sum_{i=1}^{n} u_{ci} ||\boldsymbol{x}_i - \boldsymbol{b}_c||^2 + \lambda \sum_{c=1}^{C} \sum_{i=1}^{n} u_{ci} \log u_{ci}, \tag{5}$$

where λ is the fuzzification penalty and tunes the degree of fuzziness. A larger λ brings very fuzzy partition and $\lambda \to \infty$ implies $u_{ci} \to 1/C, \forall i, c$ [6].

2.2 *Interpretation of FCM-type Clustering from Probabilistic Mixtures Viewpoint*

Caused by the exclusive partition constraint of Eq. (2), fuzzy c-partition has close relation to probabilistic mixture models, where fuzzy membership u_{ci} is identified with the generative probability of object i from the cth component distribution. Assume that n objects are drawn from one of C independent Gaussian distributions, each of which is a Gaussian component $g_c(\boldsymbol{x}_i|\boldsymbol{b}_c, \Sigma_c)$ with mean $\boldsymbol{b}_c$ and covariance Σ_c. Considering mixing weight α_c, the probability of $\boldsymbol{x}_i$ is represented as:

$$P(\boldsymbol{x}_i) = \sum_{c=1}^{C} \alpha_c g_c(\boldsymbol{x}_i|\boldsymbol{b}_c, \Sigma_c). \tag{6}$$

The maximum likelihood estimator for model parameters are given by maximizing the following log-likelihood function:

$$\begin{aligned} L_{gmm} &= \sum_{i=1}^{n} \log(P(\boldsymbol{x}_i)) \\ &= \sum_{i=1}^{n} \log\left(\sum_{c=1}^{C} \alpha_c g_c(\boldsymbol{x}_i|\boldsymbol{b}_c, \Sigma_c)\right) \\ &= \sum_{i=1}^{n} \log\left(\sum_{c=1}^{C} u_{ci} \frac{\alpha_c g_c(\boldsymbol{x}_i|\boldsymbol{b}_c, \Sigma_c)}{u_{ci}}\right). \end{aligned} \tag{7}$$

Supported by Jensen's inequality [18], the optimal solution can be derived by maximizing the following pseudo-log-likelihood function:

$$\begin{aligned} L_{gmm'} &= \sum_{i=1}^{n}\sum_{c=1}^{C} u_{ci} \log\left(\frac{\alpha_c g_c(\boldsymbol{x}_i|\boldsymbol{b}_c, \Sigma_c)}{u_{ci}}\right) \\ &= \sum_{i=1}^{n}\sum_{c=1}^{C} u_{ci} \log\left(g_c(\boldsymbol{x}_i|\boldsymbol{b}_c, \Sigma_c)\right) + \sum_{i=1}^{n}\sum_{c=1}^{C} u_{ci} \log \frac{\alpha_c}{u_{ci}}, \end{aligned} \tag{8}$$

where u_{ci} is the posterior probability of component c given object i, and u_{ci} and component parameters $(\alpha_c, \boldsymbol{b}_c, \Sigma_c)$ are iteratively optimized through the EM algorithm [19].

Hathaway [8] gave another interpretation on Eq. (8), where the first term is identified with the conventional k-means objective function while the second one is regarded as a penalty for achieving soft partition.

For example, in the case of $\alpha_c = 1/C$ and $\Sigma_c = \sigma_c I, \forall c$, Eq. (8) is reduced to:

$$L_{gmm'} = -\frac{1}{2}\sum_{i=1}^{n}\sum_{c=1}^{C} u_{ci}\frac{||\boldsymbol{x}_i - \boldsymbol{b}_c||^2}{\sigma_c} - \frac{1}{2}\sum_{i=1}^{n}\sum_{c=1}^{C} u_{ci}\log|\sigma_c| \\ -\sum_{i=1}^{n}\sum_{c=1}^{C} u_{ci}\log u_{ci} - \frac{np}{2}\log 2\pi - n\log C, \quad (9)$$

and implies that the crisp k-means objective function is adopted with an additional entropy penalty term.

This interpretation brings another viewpoint on eFCM because Eq. (9) is reduced to the eFCM objective function of Eq. (5) by identifying σ_c/C with fuzzification penalty weight λ, which is a pre-fixed model parameter. The fuzziness degree of the eFCM model can be compared with its statistical counterpart, where $\lambda = \sigma_c/C$.

2.3 A Fuzzy Counterpart of Full-Parameter GMMs

Following the above consideration, a fuzzy counterpart of full-parameter GMMs was proposed by adopting K-L information-based regularization. The FCM objective function to be maximized was extended to fuzzy c-means with K-L information regularization (KLFCM) [11] as:

$$L_{klfcm} = \sum_{i=1}^{n}\sum_{c=1}^{C} u_{ci}(\boldsymbol{x}_i - \boldsymbol{b}_c)^\top \Sigma_c^{-1}(\boldsymbol{x}_i - \boldsymbol{b}_c) \\ -\lambda\sum_{i=1}^{n}\sum_{c=1}^{C} u_{ci}\log\frac{\alpha_c}{u_{ci}} + \sum_{c=1}^{C} u_{ci}\log|\Sigma_c|, \quad (10)$$

where λ is an adjustable weight for tuning the degree of partition fuzziness. When $\lambda = 1$, the model is reduced to GMMs with full parameters of means $\boldsymbol{b}_c$, covariances Σ_c and mixing coefficients α_c for all c. The larger the value of λ, the fuzzier the object partition.

Considering the necessary conditions for the optimality of the objective function, the updating rules for model parameters to be used in the alternative optimization algorithm are given as follows:

$$\alpha_c = \frac{1}{n}\sum_{i=1}^{n} u_{ci}, \quad (11)$$

$$\boldsymbol{b}_c = \frac{\sum_{i=1}^{n} u_{ci}\boldsymbol{x}_i}{\sum_{i=1}^{n} u_{ci}}, \quad (12)$$

$$\Sigma_c = \frac{1}{\sum_{i=1}^{n} u_{ci}} \sum_{i=1}^{n} (\boldsymbol{x}_i - \boldsymbol{b}_c)(\boldsymbol{x}_i - \boldsymbol{b}_c)^\top, \tag{13}$$

$$u_{ci} = \frac{\alpha_c \exp(-d_{ci})}{\sum_{\ell=1}^{C} \alpha_\ell \exp(-d_{\ell i})}, \tag{14}$$

where $d_{ci} = (\boldsymbol{x}_i - \boldsymbol{b}_c)^\top \Sigma_c (\boldsymbol{x}_i - \boldsymbol{b}_c)$.

2.4 Numerical Experiment

In this subsection, an illustrative example is shown to demonstrate the characteristics of FCM variant induced by probabilistic mixtures concepts. An artificial 2-D data set is composed of 200 objects forming four spherical clusters with 50 objects each. In order to demonstrate the effects of tuning the fuzziness degree in GMMs-induced clustering model, the KLFCM algorithm ($C = 4$) was applied with different fuzziness degrees of $\lambda \in \{0.5, 1.0, 2.0\}$.

Figure 1 compares the derived fuzzy partitions. Circles are data plots and gray diamonds are the estimated prototypes. Grayscale image depicts the fuzzy classification function, which is the maximum membership value of four clusters in each location. The result of GMMs (Fig. 1b), which is also achieved by KLFCM with $\lambda = 1.0$, implies a part of the center cluster was exploited by other outer clusters because fuzzy partition often distorts ambiguous boundaries caused by non-Gaussian component densities. In this situation, crisper partition with $\lambda = 0.5$ (Fig. 1a) rather than GMMs is suitable for clarifying cluster boundaries because the boundaries of visual four clusters are clearly and fairly depicted by fine but distinct lines. Then, using a slightly crisper model, we can enjoy the benefits of both crisp k-means partition and fuzzy membership assignment with a linearly separable data set.

On the other hand, in fuzzier case with $\lambda = 2.0$ (Fig. 1c), the center visual cluster was severely shared by multiple clusters and almost undetectable although its cluster center was fairly located in its centroid. Then, using a fuzzier model, outer clusters might be emphasized while inner clusters can be concealed by being shared by multiple outer clusters.

The above result simply demonstrates that the adjustable penalty can contribute to improving the interpretability of cluster partition and crisper or fuzzier partitions rather than GMMs are sometimes more useful for intuitive data summarization. A slightly crisper model is suitable for linearly separable data sets while a fuzzier model may work well in handling severely overlapping clusters.

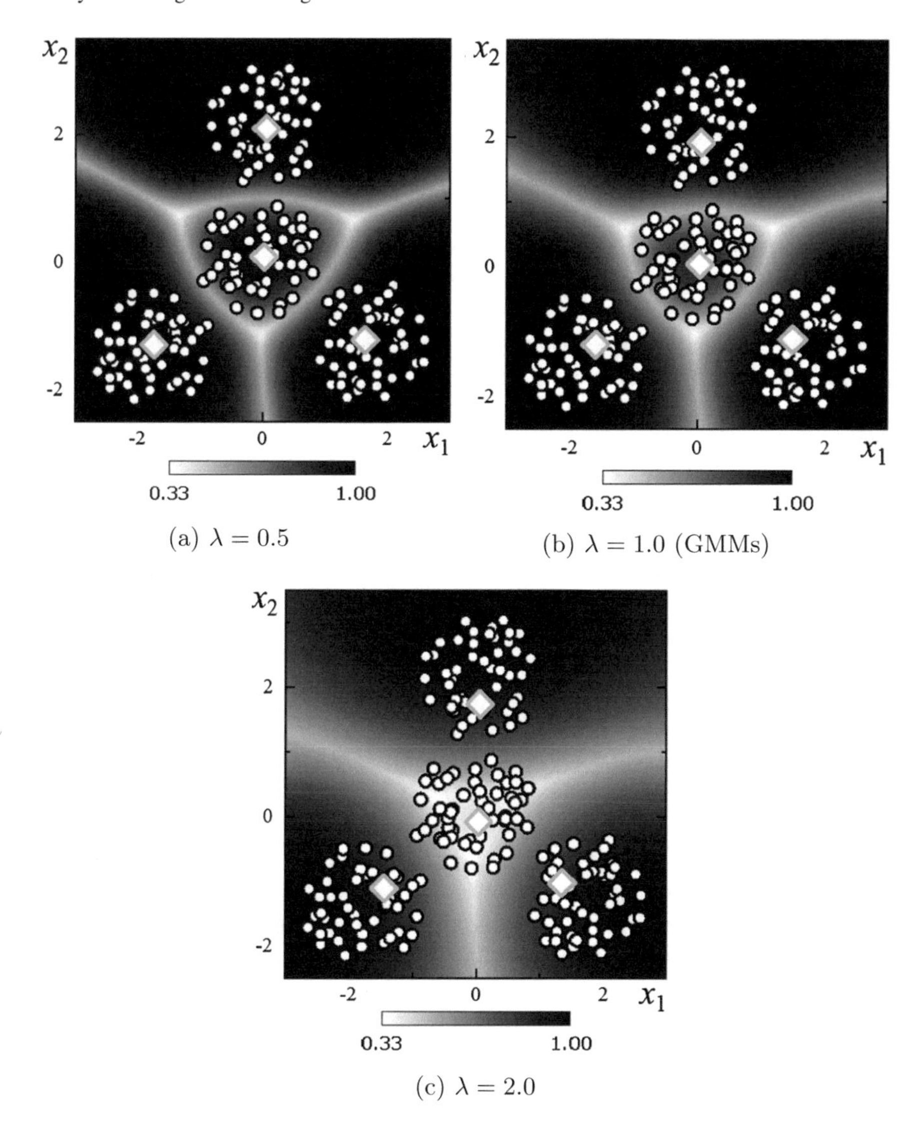

Fig. 1 Comparison of cluster partitions by KLFCM with different fuzziness degrees: Grayscale image depicts the fuzzy classification function, which is the maximum membership value of four clusters in each location

3 Fuzzy Co-clustering and Multinomial Mixture Models

3.1 *Fuzzy Co-clustering Based on Cluster Aggregation Criterion*

Cooccurrence information analysis became more important especially in web-based services. Assume that we have cooccurrence information among n objects and p items such as an $n \times p$ matrix $R = \{r_{ij}\}$, $i = 1, \ldots, n$, $j = 1, \ldots, p$. For example, r_{ij} can be the frequency of keyword (item) j in document (object) i in document-keyword analysis. The goal of co-clustering is to extract pair-wise clusters of familiar objects and items so that we can find object clusters in conjunction with their typical items.

In fuzzy co-clustering, co-cluster structures are represented by using two types of fuzzy memberships: object memberships and item memberships. Object memberships u_{ci}, $i = 1, \ldots, n$, $c = 1, \ldots, C$ have the same feature with the conventional ones of FCM while item memberships w_{cj}, $j = 1, \ldots, p$, $c = 1, \ldots, C$ gives additional information on the typicality of each item in each co-cluster. They are estimated so that mutually familiar object-item pairs with large r_{ij} should have large memberships in a same cluster, and the aggregation degree to be maximized in C clusters is measured by $\sum_{c=1}^{C} \sum_{i=1}^{n} \sum_{j=1}^{p} u_{ci} w_{cj} r_{ij}$. Here, we should note that a different type of constraint must be forced to w_{cj} from $\sum_{c=1}^{C} u_{ci} = 1$ for object memberships because the aggregation is trivially maximized by assigning all objects and items to a solo cluster. Then, w_{cj} are estimated under the constraint of $\sum_{j=1}^{p} w_{cj} = 1$.

Since the aggregation measure is a linear function with respect to both u_{ci} and w_{cj}, fuzzy co-clustering can be realized by introducing non-linear natures in the same manner with the fuzzification schemes in FCM. Fuzzy clustering for categorical multivariate data (FCCM) [13] adopted the entropy regularization scheme [5] for membership fuzzification as follows:

$$L_{fccm} = \sum_{c=1}^{C} \sum_{i=1}^{n} \sum_{j=1}^{p} u_{ci} w_{cj} r_{ij} - \lambda_u \sum_{c=1}^{C} \sum_{i=1}^{n} u_{ci} \log u_{ci} - \lambda_w \sum_{c=1}^{C} \sum_{j=1}^{p} w_{cj} \log w_{cj}, \quad (15)$$

where λ_u and λ_w are independent fuzzification penalty weights. Fuzzy co-clustering of documents and keywords (Fuzzy CoDoK) [14] introduced the quadric regularization scheme [7] as follows:

$$L_{codok} = \sum_{c=1}^{C} \sum_{i=1}^{n} \sum_{j=1}^{p} u_{ci} w_{cj} r_{ij} + \lambda_u \sum_{c=1}^{C} \sum_{i=1}^{n} u_{ci}^2 + \lambda_w \sum_{c=1}^{C} \sum_{j=1}^{p} w_{cj}^2. \quad (16)$$

Such techniques as standard fuzzification approach with penalty weighting exponents is also possible in the aggregation measure [20, 21].

In contrast to some FCM variants with regularized concepts, however, these fuzzy co-clustering models have no comparative models of probabilistic mixtures. So, we have no guideline for evaluating the degree of fuzziness in tuning the penalty weights, and it is often difficult to select appropriate values for two penalty weights.

3.2 Fuzzy Co-clustering Induced by Multinomial Mixtures Concept

A classical model for statistical cooccurrence information analysis is multinomial mixture models (MMMs) [15], in which each probabilistic component is given by a multinomial distribution. If objects are assume to be drawn from one of C distributions, in which the probability of item observation j in component c is w_{cj}, the joint distribution of cooccurrence feature vector $\boldsymbol{r}_i = (r_{i1}, \dots, r_{ip})^\top$ on object i is given as:

$$P(\boldsymbol{r}_i) = \sum_{c=1}^{C} \alpha_c p_c(\boldsymbol{r}_i), \tag{17}$$

where α_c is the a priori probability (mixing coefficient) of component c. $p_c(\boldsymbol{r}_i)$ is the component density in component c as:

$$p_c(\boldsymbol{r}_i) = \frac{T_i!}{r_{i1}! \dots r_{ip}!} \prod_{j=1}^{p} (w_{cj})^{r_{ij}}, \tag{18}$$

where $T_i = \sum_{j=1}^{p} r_{ij}$.

In a same manner with GMMs, maximum likelihood estimators can be estimated from the following pseudo-log-likelihood function:

$$L_{mmms} = \sum_{c=1}^{C} \sum_{i=1}^{n} \sum_{j=1}^{p} u_{ci} r_{ij} \log w_{cj} + \sum_{c=1}^{C} \sum_{i=1}^{n} u_{ci} \log \frac{\alpha_c}{u_{ci}}, \tag{19}$$

where u_{ci} is the posterior probability of component c given object i.

Comparing Eq. (19) with the objective function of fuzzy co-clustering such as FCCM [13], we have another interpretation from the fuzzy co-clustering viewpoint. The first term is in a similar form with the aggregation measure but includes $\log w_{cj}$ instead of w_{cj}. So, the first term has linear nature with respect to u_{ci} but is non-linearized by log function with respect to w_{cj}. Then, the additional K-L information-like penalty is responsible only for soft partition of u_{ci} while soft partition of w_{cj} is achieved by log function-induced non-linearity.

Here, the above co-clustering interpretation of MMMs brings an adjustable model of fuzzy memberships in MMMs-induced fuzzy co-clustering, where degree of partition fuzziness can be tuned by adjusting the non-linear degree of pseudo-log-likelihood function. In [16], fuzziness degrees of both u_{ci} and w_{cj} were discussed, but adjustment of object partition is only considered here for simplicity.

Equation (19) implies that the maximization problem reduces hard object partition without K-L information-like penalty and the degree of fuzziness can be tuned with adjustable penalty weight. Then, Honda et al. proposed fuzzy co-clustering induced by multinomial mixture concept (FCCMM) [16] with the following objective function:

$$L'_{mmms} = \sum_{c=1}^{C} \sum_{i=1}^{n} \sum_{j=1}^{p} u_{ci} r_{ij} \log w_{cj} + \lambda \sum_{c=1}^{C} \sum_{i=1}^{n} u_{ci} \log \frac{\alpha_c}{u_{ci}}, \tag{20}$$

where λ tunes the degree of fuzziness of object partition. When $\lambda = 1$, the model is reduced to the conventional MMMs. The larger the value of λ, the fuzzier the object partition.

Considering the necessary conditions for the optimality of the objective function, the updating rules for model parameters to be used in the alternative optimization algorithm are given as follows:

$$\alpha_c = \frac{1}{n} \sum_{i=1}^{n} u_{ci}, \tag{21}$$

$$u_{ci} = \frac{\alpha_c \prod_{j=1}^{p} (w_{cj})^{r_{ij}/\lambda}}{\sum_{\ell=1}^{C} \alpha_\ell \prod_{j=1}^{p} (w_{\ell j})^{r_{ij}/\lambda}}. \tag{22}$$

$$w_{cj} = \frac{\displaystyle\sum_{i=1}^{n} r_{ij} u_{ci}}{\displaystyle\sum_{\ell=1}^{p} \sum_{i=1}^{n} r_{i\ell} u_{ci}}. \tag{23}$$

3.3 *Numerical Experiment*

In this subsection, an illustrative example is shown to demonstrate the characteristics of fuzzy co-clustering induced by a multinomial mixtures concept. The experiment was performed with two benchmark cooccurrence information data sets, which are available from LINQS webpage of Statistical Relational Learning Group UMD (http://linqs.cs.umd.edu/projects//index.shtml). The CiteSeer data set consists of a $3{,}312 \times 3{,}703$ cooccurrence matrix with the cooccurrence information among 3,312 objects (scientific publications: $n = 3{,}312$) and 3,703 items (words: $p = 3{,}703$), where $r_{ij} \in \{0, 1\}$ indicate the absence/presence of words and each object is classified into one of six classes. The Cora data set consists of a $2{,}708 \times 1{,}433$ matrix with similar elements, where each object is classified into one of seven classes.

In order to demonstrate the effects of tuning the fuzziness degrees, the FCCMM algorithm was applied to the two data sets with different fuzziness degrees of $\lambda \in \{0.5, 1.0, 2.0\}$, where cluster numbers was set as the corresponding class numbers and the initialization was done in a supervised manner, i.e., the initial object memberships are given following the actual class information. Tables 1 and 2 compare the cross-tabulation of maximum membership classification, in which the correct matching of class-cluster are shown in bold. As in the same with the KLFCM algorithm, the result of MMMs, which is also achieved by FCCMM with $\lambda = 1.0$, implies a part of objects was exploited by other clusters. In this situation, crisper partition with $\lambda = 0.5$ rather than MMMs is suitable for clarifying cluster structures. On the other hand, in fuzzier case with $\lambda = 2.0$, mis-assignment became more severe. It is because the fuzzier object partition also caused fuzzier item membership assignment and severe mis-assignment of boundary objects.

The above result simply demonstrates again that the adjustable penalty can contribute to improving the interpretability of cluster partition and crisper or fuzzier partitions rather than MMMs are sometimes more useful. This situation is quite similar to the FCM result shown in the previous section. When each class is expected to have distinct difference in item occurrences from other classes, a slightly crisper partition can clarify the sharp boundaries among classes in co-clustering tasks as well.

Table 1 Comparison of cross tabulation of class-cluster matching (Citeseer)

Cluster		$\lambda = 0.5$						$\lambda = 1.0$ (MMMs)						$\lambda = 2.0$					
		1	2	3	4	5	6	1	2	3	4	5	6	1	2	3	4	5	6
Class	1	**549**	9	5	4	14	15	**534**	15	7	5	14	21	**514**	24	7	6	15	30
	2	11	**203**	8	5	18	4	13	**190**	11	8	22	5	16	**160**	15	9	38	11
	3	12	16	**613**	33	14	13	18	21	**587**	41	20	14	30	27	**552**	55	21	16
	4	17	4	34	**564**	36	13	18	4	43	**542**	45	16	22	4	51	**517**	54	20
	5	16	20	13	27	**507**	7	26	31	16	29	**481**	7	24	53	17	46	**439**	11
	6	17	5	8	17	6	**455**	24	8	9	18	8	**441**	36	10	10	21	15	**416**

Table 2 Comparison of cross tabulation of class-cluster matching (Cora)

Cluster		$\lambda = 0.5$							$\lambda = 1.0$ (MMMs)							$\lambda = 2.0$						
		1	2	3	4	5	6	7	1	2	3	4	5	6	7	1	2	3	4	5	6	7
Class	1	**263**	6	2	2	1	10	14	**248**	8	4	3	3	13	19	**218**	10	6	3	10	15	36
	2	8	**378**	7	3	12	6	4	10	**367**	9	3	14	8	7	7	**359**	16	3	16	6	11
	3	20	22	**631**	66	19	11	49	23	24	**605**	68	22	13	63	34	27	**593**	72	17	7	68
	4	9	2	16	**365**	10	5	19	12	2	22	**342**	11	6	31	30	3	27	**313**	14	3	36
	5	3	6	5	0	**198**	2	3	5	10	10	2	**184**	2	4	11	13	17	5	**167**	0	4
	6	9	0	2	1	3	**156**	9	12	0	2	3	3	**147**	13	26	2	6	2	3	**123**	18
	7	18	1	12	14	6	12	**288**	19	7	16	18	9	15	**267**	23	14	33	18	8	16	**239**

4 Conclusions

In this chapter, several fuzzy clustering models induced from probabilistic mixtures concepts were reviewed with experimental demonstrations of their characteristics. Besides the original FCM clustering with fuzzification heuristics only, several FCM variants can be analyzed under some supports of statistical theories.

KLFCM and FCCMM were constructed by introducing adjustable penalty weights into pseudo-log-likelihood function of GMMs and MMMs, respectively, where the linear measures were non-linearized by additional K-L information-like penalty terms in conjunction with their adjustable contribution. Although the fuzziness degrees of basic fuzzy clustering models were validated heuristically from only the empirical viewpoints [22] except for stability analysis [23], probabilistic models-induced algorithms can tune the fuzziness degrees under the guideline of probabilistic counterparts.

In several illustrative examples, the interpretability of cluster structures given by probabilistic mixture models were demonstrated to be improved by slightly crisper settings. However, it should be also noted that hard or crisper models are often more sensitive to initial partitions rather than fuzzier models and often suffer from the issue of local optimality. Then, the robust feature of fuzzier models can also be utilized for estimating proper crisper models under the support of deterministic annealing [24, 25], where fuzzification penalty weights are identified with the temperature parameters in annealing process. Starting from very fuzzy situation, where we have few local solutions, the partition fuzziness is gradually decreased until the intended fuzziness degrees without dropped into local solutions.

Recently, various improvements of probabilistic mixture models [26–29] have also been emerging in many societies such as statistics, neuro-sciences and information-based induction sciences. Fuzzy clustering theories will become more and more powerful tools in various data mining tasks and will investigate many research directions.

Acknowledgements This work was supported in part by the Ministry of Education, Culture, Sports, Science and Technology, Japan, under Grant-in-Aid for Scientific Research (JSPS KAKENHI Grant Number JP26330281).

References

1. M. R. Anderberg, *Cluster Analysis for Applications*, Academic Press, 1981.
2. J. B. MacQueen, "Some methods of classification and analysis of multivariate observations," *Proc. of 5th Berkeley Symposium on Math. Stat. and Prob.*, pp. 281–297, 1967.
3. J. C. Bezdek, *Pattern Recognition with Fuzzy Objective Function Algorithms*, Plenum Press, 1981.
4. E. H. Ruspini, "A new approach to clustering," *Information and Control*, vol. 15, no. 1, pp. 22–32, July 1969.
5. S. Miyamoto and M. Mukaidono, "Fuzzy *c*-means as a regularization and maximum entropy approach," *Proc. of the 7th Int. Fuzzy Syst. Assoc. World Cong.*, vol. 2, pp. 86–92, 1997.

6. S. Miyamoto, H. Ichihashi and K. Honda, *Algorithms for Fuzzy Clustering*, Springer, 2008.
7. S. Miyamoto and K. Umayahara, "Fuzzy clustering by quadratic regularization," *Proc. 1998 IEEE Int. Conf. Fuzzy Systems and IEEE World Congr. Computational Intelligence*, vol. 2, pp. 1394–1399, 1998.
8. R. J. Hathaway, "Another interpretation of the EM algorithm for mixture distributions," *Statistics & Probability Letters*, vol. 4, pp. 53–56, 1986.
9. R. O. Duda and P. E. Hart, *Pattern Classification and Scene Analysis*, John Wiley & Sons, 1973.
10. C. M. Bishop, *Neural Networks for Pattern Recognition*, Clarendon Press, 1995.
11. H. Ichihashi, K. Miyagishi and K. Honda, "Fuzzy c-means clustering with regularization by K-L information," *Proc. of 10th IEEE International Conference on Fuzzy Systems*, vol. 2, pp. 924–927, 2001.
12. K. Honda and H. Ichihashi, "Regularized linear fuzzy clustering and probabilistic PCA mixture models," *IEEE Trans. Fuzzy Systems*, vol. 13, no. 4, pp. 508–516, 2005.
13. C.-H. Oh, K. Honda and H. Ichihashi, "Fuzzy clustering for categorical multivariate data," *Proc. of Joint 9th IFSA World Congress and 20th NAFIPS International Conference*, pp. 2154–2159, 2001.
14. K. Kummamuru, A. Dhawale and R. Krishnapuram, "Fuzzy co-clustering of documents and keywords," *Proc. 2003 IEEE Int'l Conf. Fuzzy Systems*, vol. 2, pp. 772–777, 2003.
15. L. Rigouste, O. Cappé and F. Yvon, "Inference and evaluation of the multinomial mixture model for text clustering," *Information Processing and Management*, vol. 43, no. 5, pp. 1260–1280, 2007.
16. K. Honda, S. Oshio and A. Notsu, "Fuzzy co-clustering induced by multinomial mixture models," *Journal of Advanced Computational Intelligence and Intelligent Informatics*, vol. 19, no. 6, pp. 717–726, 2015.
17. R. Krishnapuram and J. M. Keller, "A possibilistic approach to clustering," *IEEE Trans. on Fuzzy Systems*, vol. 1, pp. 98–110, 1993.
18. T. Needham, "A visual explanation of Jensen's inequality," *American Mathematical Monthly*, vol. 100, no. 8, pp. 768–771, 1993.
19. A. P. Dempster, N. M. Laird, and D. B. Rubin, "Maximum likelihood from incomplete data via the EM algorithm," *J. of the Royal Statistical Society*, Series B, vol. 39, pp. 1–38, 1977.
20. Y. Kanzawa, "Fuzzy co-clustering algorithms based on fuzzy relational clustering and TIBA imputation," *Journal of Advanced Computational Intelligence and Intelligent Informatics*, vol. 18, no. 2, pp. 182–189, 2014.
21. Y. Kanzawa, "On Bezdek-type fuzzy clustering for categorical multivariate data," *Proc. of Joint 7th Int. Conf. Soft Computing and Intelligent Systems and 15th Int. Sympo. Advanced Intelligent Systems*, pp. 694–699, 2014.
22. N. R. Pal and J. C. Bezdek, "On cluster validity for the fuzzy c-mean model," *IEEE Trans. Fuzzy Systems*, vol. 3, pp. 370–379, Aug. 1995.
23. J. Yu, Q. Cheng and H. Huang, "Analysis of the weighting exponent in the FCM," *IEEE Transactions on Systems, Man, and Cybernetics, Part B: Cybernetics*, vol. 34, no. 1, pp. 634–639, 2004.
24. K. Rose, E. Gurewitz and G. Fox, "A deterministic annealing approach to clustering," *Pattern Recognition Letters*, vol. 11, pp. 589–594, 1990.
25. S. Oshio, K. Honda, S. Ubukata and A. Notsu, "Deterministic annealing framework in MMMs-induced fuzzy co-clustering and its applicability," *International Journal of Computer Science and Network Security*, vol. 16, no. 1, pp. 43–50, 2016.
26. K. Sjölander, K. Karplus, M. Brown, R. Hughey, A. Krogh, I. Saira Mian and D. Haussler, "Dirichlet mixtures: a method for improved detection of weak but significant protein sequence homology," *Computer Applications in the Biosciences*, vol. 12, no. 4, pp. 327–345, 1996.
27. X. Ye, Y.-K. Yu and S. F. Altschul, "Compositional adjustment of Dirichlet mixture priors," *Journal of Computational Biology*, vol. 17, no. 12, pp. 1607–1620, 2010.
28. M. E. Tipping and C. M. Bishop, "Mixtures of probabilistic principal component analysers," *Neural Computation*, vol. 11, no. 2, pp. 443–482, 1999.

29. J. Takeuchi, T. Kawabata, and A. R. Barron, “Properties of Jeffreys mixture for markov sources,” *IEEE Transactions on Information Theory*, vol. 59, no. 1, pp. 438–457, January 2013.

Semi-supervised Fuzzy c-Means Algorithms by Revising Dissimilarity/Kernel Matrices

Yuchi Kanzawa

Abstract Semi-supervised clustering uses partially labeled data, as often occurs in practical clustering, to obtain a better clustering result. One approach uses *hard* constraints which specify data that must and cannot be within the same cluster. In this chapter, we propose another approach to semi-supervised clustering with *soft* pairwise constraints. The clustering method used is fuzzy c-means (FCM), a commonly used fuzzy clustering method. Two previously proposed variants, entropy-regularized relational/kernel fuzzy c-means clustering and indefinite kernel fuzzy c-means clustering algorithm are modified to use the soft constraints. In addition, a method is discussed that propagates pairwise constraints when the given constraints are not sufficient for obtaining the desired clustering result. Using some numerical examples, it is shown that the proposed algorithms obtain better clustering results.

Keywords Semi-supervised clustering · Kernel · Relational clustering · Fuzzy c-means

1 Introduction

Fuzzy c-means (FCM) [1] is a well-known fuzzy clustering method, and many FCM variants have been proposed so far. Here, we refer to the original FCM as the Bezdek-type FCM (bFCM) in order to distinguish it from other variants. Of these variants, the FCM algorithm based on the concept of regularization by entropy was proposed by Miyamoto and Umayahara [2]. This algorithm is called entropy-regularized FCM (eFCM) and is discussed not only for its usefulness but also for its mathematical relations with other techniques.

Another variant is Bezdek-type relational fuzzy c-means (bRFCM) [3], which is used with relational data to quantify the relationship between each pair of objects. In order to deal with non-Euclidean relational data, a modified version of bRFCM,

Y. Kanzawa (✉)
Shibaura Institute of Technology, Tokyo, Japan
e-mail: kanzawa@sic.shibaura-it.ac.jp

V. Torra et al. (eds.), *Fuzzy Sets, Rough Sets, Multisets and Clustering*,
Studies in Computational Intelligence 671, DOI 10.1007/978-3-319-47557-8_4

non-Euclidean bRFCM (NEbRFCM) [4] with β-spreading, has been developed for non-Euclidean relational data. Both bRFCM and NEbRFCM can be naturally applied to data in Euclidean space by calculating the dissimilarity between each datum. The author proposed entropy-regularized relational fuzzy c-means (eRFCM) in [5] and showed that eRFCM does not need the help of β-spreading for any non-Euclidean relational data.

Another two FCM variants called Bezdek-type and entropy-regularized kernel fuzzy c-means (K-bFCM and K-eFCM, respectively) [6, 7] are used with nonlinearly bordered clusters, in which data are transformed into a higher-dimensional space called feature space and clustered in that feature space using some kernel function that calculates the inner product of two data. Correspondingly, for indefinite kernels, the author also proposed indefinite kernel Bezdek-type fuzzy c-means (IK-bFCM) in [8]. bRFCM, NEbRFCM, eRFCM, K-bFCM, K-eFCM, and IK-bFCM with a dissimilarity-based kernel such as a Gaussian kernel all have a common feature in that they use the dissimilarity between data.

Yet another FCM variant is the framework of semi-supervised clustering, in which we start with a supply of unlabeled and labeled data, as often occurs in practical clustering, and some labeled data are used along with the unlabeled data to obtain a better clustering with a higher convergence speed and accuracy. The semi-supervised approach for FCM was first proposed by Pedrycz [9], who assumed that some supervisors for labels are provided, and many similar methods have been proposed since then [10–12]. Another semi-supervised approach for K-means uses two types of pairwise constraints: *must-link*, where two data have to be together in the same cluster and *cannot-link*, where two data must be in different clusters [13]. By relaxing the above mentioned *hard* constraints into *soft* constraints such that two data should be together in the same cluster and that two data should be in different clusters, respectively, some variants of pairwise constraint FCM have been proposed [14, 15] by adding a penalty term for *soft* constraints to the original FCM optimization problem.

In this chapter, we describe another approach to semi-supervised FCM with *soft* pairwise constraints, which completes the author's work in [16] with the help of the author's other works in [5, 8]; this approach is based on the concept that the data in the same cluster may be close to each other and the data in different clusters may be far apart. It is difficult to directly apply this concept to FCM because the FCM optimization problem is described not by dissimilarities between data but by dissimilarities between each datum and the cluster center. Furthermore, we apply NEbRFCM, eRFCM, IK-bFCM, and K-eFCM using a dissimilarity-based kernel because these methods use dissimilarities between data. Therefore, we revise the dissimilarity between data using information about the pairwise constraints and apply NEbRFCM, eRFCM, IK-bFCM, or K-eFCM using the revised dissimilarities. Furthermore, we consider propagating the given pairwise constraints to unconstrained data when the given constraints are not sufficient to obtain the desired clustering result.

The remainder of this chapter is organized follows. In the second section, we define some notation and introduce bRFCM, NEbRFCM, eRFCM, K-bFCM, K-eFCM, and IK-bFCM, which are used for our proposed method. In the third section, we explain our concept and propose four semi-supervised fuzzy c-means algorithms by revising the dissimilarity between data. In the fourth section, we present some numerical examples. In the last section, we conclude this chapter.

2 Preliminaries

For a given data set $x = \{x_k \mid k \in \{1, \dots, N\}\}$, bRFCM, eRFCM and NEbRFCM assume that the dissimilarity data matrix $R \in \mathbb{R}^{N \times N}$ is given. In contrast, K-bFCM, K-eFCM, and IK-bFCM assume that kernel matrix $K \in \mathbb{R}^{N \times N}$ is given. The membership by which x_k belongs to the i-th cluster is denoted by $u_{i,k}$ ($k \in \{1, \dots, N\}$, $i \in \{1, \dots, C\}$) and the set of $u_{i,k}$ is denoted by $u \in \mathbb{R}^{C \times N}$, and called the partition matrix. The constraints for u are

$$u_{i,k} \in [0, 1] \tag{1}$$

and

$$\sum_{i=1}^{C} u_{i,k} = 1. \tag{2}$$

Methods bRFCM [3] and eRFCM [5] are given by Algorithm 1.

Algorithm 1 (*bRFCM* [3] *and eRFCM* [5])

STEP 1. Given dissimilarity data matrix R. Select $m > 1$ for bRFCM, and $\lambda > 0$ for eRFCM. Fix C, and initialize membership u.
STEP 2. Calculate

$$v_i = \left((u_{i,1})^m, \dots, (u_{i,N})^m\right) / \sum_{k=1}^{N} (u_{i,k})^m \tag{3}$$

for bRFCM, and

$$v_i = \left(u_{i,1}, \dots, u_{i,N}\right) / \sum_{k=1}^{N} u_{i,k} \tag{4}$$

for eRFCM.
STEP 3. Calculate

$$d_{i,k} = (Rv_i)_k - v_i^{\mathsf{T}} R v_i. \tag{5}$$

STEP 4. Calculate

$$u_{i,k} = 1/\sum_{j=1}^{C}\left(\frac{d_{i,k}}{d_{j,k}}\right)^{1/(m-1)} \tag{6}$$

for bRFCM, and

$$u_{i,k} = \exp(-\lambda d_{i,k})/\sum_{j=1}^{C}\exp(-\lambda d_{j,k}) \tag{7}$$

for eRFCM.

STEP 5. If (u, d) converge, terminate. Otherwise, return to STEP 2. ■

Matrix R is called *Euclidean* if there exists a set of points $\{y_1, \ldots, y_N\}$ in $\mathbb{R}^{N-1}$ such that $R_{k,\tilde{k}} = \|y_k - y_{\tilde{k}}\|_2^2$. The method of bRFCM may fail for non-*Euclidean* R caused by negative $d_{i,k}$ in STEP 3 of Algorithm 1. For example, in the case with $C = 2, m = 2$, and $\{d_{i,1}\}_{i=1}^2 = \{-1, 2\}$, we have $\{u_{i,1}\}_{i=1}^2 = \{2, -1\}$, which violates the condition given in Eq. (1). To overcome such the limitation, NEbRFCM is given by the following Algorithm [4]:

Algorithm 2 (*NEbRFCM* [4])

STEP 1. Do STEP 1 of Algorithm 1 and initialize $\beta = 0$.
STEP 2. Do STEP 2 of Algorithm 1.
STEP 3. Do STEP 3 of Algorithm 1. If $d_{i,k} < 0$,

$$\Delta\beta = \max\{-2d_{i,k}/\|e_k - v_i\|^2\}, \tag{8}$$

$$d_{i,k} \leftarrow d_{i,k} + \Delta\beta\|e_k - v_i\|^2, \tag{9}$$

$$\beta \leftarrow \beta + \Delta\beta. \tag{10}$$

STEP 4. Do STEP 4 of Algorithm 1.
STEP 5. If (u, d) converge, terminate. Otherwise, return to STEP 2. ■

For a given data set $X = \{x_k \mid k \in \{1, \ldots, N\}\}$, kernel fuzzy clustering assumes that the kernel matrix $K \in \mathbb{R}^{N\times N}$ is given. Let $\mathbb{H}$ be a higher-dimensional feature space, let $\Phi : X \to \mathbb{H}$ be a map from data set X to feature space $\mathbb{H}$, and let $W = \{W_i \in \mathbb{H} \mid i \in \{1, \ldots, C\}\}$ be a set of cluster centers in the feature space.

Methods K-bFCM and K-eFCM are obtained as follows [6, 7].

Algorithm 3 (*K-bFCM and K-eFCM* [6, 7])

STEP 1. Specify the number of clusters C. Select the fuzzifier parameter $m > 1$ for K-bFCM, and $\lambda > 0$ for K-eFCM. Set the initial value of u.
STEP 2. Update cluster centers as

$$W_i = \frac{\left((u_{i,1})^m, \ldots, (u_{i,N})^m\right)^{\mathsf{T}}}{\sum_{k=1}^{N}(u_{i,k})^m} \tag{11}$$

for K-bFCM, and

$$W_i = \frac{\left(u_{i,1}, \ldots, u_{i,N}\right)^{\mathsf{T}}}{\sum_{k=1}^{N} u_{i,k}} \tag{12}$$

for K-eFCM.

STEP 3. Update the dissimilarity between each element in the data set and the cluster center as

$$d_{i,k} = (e_k - W_i)^{\mathsf{T}} K (e_k - W_i). \tag{13}$$

STEP 4. Update the membership as

$$u_{i,k} = \left(\sum_{j=1}^{C}\left(\frac{d_{i,k}}{d_{j,k}}\right)^{\frac{1}{m-1}}\right)^{-1} \tag{14}$$

for K-bFCM, and

$$u_{i,k} = \exp(-\lambda d_{i,k}) / \sum_{j=1}^{C} \exp(-\lambda d_{j,k}) \tag{15}$$

for K-eFCM.

STEP 5. If (u, d, W) converges, terminate this algorithm. Otherwise, return to STEP 2.

K-bFCM is constructed based on the fact that K is positive semidefinite. Even then, K is sometimes introduced without guaranteeing the existence of Φ; in this case, K is not always positive semidefinite. K-bFCM works for an indefinite K when the magnitude of negative eigenvalues is not very large. However K-bFCM fails for indefinite K when the magnitude of negative eigenvalues is extremely large because the memberships cannot be calculated after the dissimilarity between a datum and cluster center has been updated to a negative value. In order to overcome this limitation, the following β-spread transformation of K has been considered [8]:

$$K_\beta = K + \beta E, \tag{16}$$

by which K_β is positive semidifinite for sufficiently large value of $\beta > 0$. K-bFCM with β-spread transformation is given by the following IK-bFCM algorithm.

Algorithm 4 (*IK-bFCM* [8])

STEP 1. Specify the number of clusters C and the fuzzifier parameter m. Set the initial value of u, set $\beta = 0$, and set $K_0 = K$.
STEP 2. Execute STEP 2 of Algorithm 3.
STEP 3. Update $d_{i,k}$ as

$$d_{i,k} = (e_k - W_i)^\mathsf{T} K_\beta (e_k - W_i). \tag{17}$$

STEP 4. If $d_{i,k} < 0$, update $\Delta\beta$, $d_{i,k}$, and β as

$$\Delta\beta = \max\{-d_{i,k}/\|e_k - W_i\|_2^2\}, \tag{18}$$

$$d_{i,k} \leftarrow d_{i,k} + \Delta\beta\|e_k - W_i\|^2, \tag{19}$$

$$\beta \leftarrow \beta + \Delta\beta, \tag{20}$$

$$K_\beta \leftarrow K_\beta + \Delta\beta E. \tag{21}$$

STEP 5. Execute STEP 4 of Algorithm 3.
STEP 6. If the stopping criterion is satisfied, terminate this algorithm. Otherwise, return to STEP 2.

bRFCM, eRFCM, NEbRFCM, K-bFCM, K-eFCM, and IK-bFCM, when used with some kernel matrix such as a Gaussian kernel

$$K_{k,\tilde{k}} = \exp(-\sigma^2\|x_k - x_{\tilde{k}}\|^2), \tag{22}$$

all have the common property that they use the dissimilarity between data.

3 Proposed Method

In this section, we propose four types of semi-supervised fuzzy c-means algorithms by revising the dissimilarity between data following the given *soft* pairwise constraints. The first two approaches apply NEbRFCM and eRFCM to revise the dissimilarity and the others apply IK-bFCM and K-eFCM. First, we describe the concept of the proposed methods and their outlines. In Sects. 3.2 and 3.3, we describe the respective proposed algorithms. Furthermore, we discuss the propagation of the given pairwise constraints to unconstrained data in Sect. 3.4.

3.1 Our Concept and the Outline of the Proposed Algorithm

Clustering is a type of unsupervised learning; however, in some cases, we can have supervision in the form of constraints that specify whether pairs of points should belong to the same cluster or to different clusters. In this setting, a clustering framework was proposed with two types of pairwise constraints [13–15]:

must-link — two data must be or should be together in the same cluster
and
cannot-link — two data must be or should be in different clusters.

Let $\mathcal{M}$ be a set of *must-link* pairs such that $(x_k, x_{\tilde{k}}) \in \mathcal{M}$ implies that x_k and $x_{\tilde{k}}$ must or should be assigned to the same cluster, and $\mathcal{C}$ be a set of *cannot-link* pairs such that $(x_k, x_{\tilde{k}}) \in \mathcal{C}$ implies that x_k and $x_{\tilde{k}}$ must or should be assigned to different clusters. Such pairwise constraints are used in two different manners: *hard* constraints and *soft* constraints. While *hard* constraints must be satisfied in the obtained clustering result [13], *soft* constraints need not be satisfied and are used only as hints in clustering [14, 15]. In this chapter, all pairwise constraints are considered to be *soft*.

The outline of the proposed algorithm is as follows. We can obtain a dissimilarity matrix for the data as well as *soft* pairwise constraints $\mathcal{M}$ and $\mathcal{C}$. We create a new dissimilarity matrix on the basis of the pairwise constraints and their implications. We then supply this new matrix to a dissimilarity-based clustering algorithm such as eRFCM (Algorithm 1), NEbRFCM (Algorithm 2), K-eFCM (Algorithm 3), or IK-bFCM (Algorithm 4) with a dissimilarity-based kernel function.

In our above-mentioned dissimilarity matrix, we would like $(x_k, x_{\tilde{k}}) \in \mathcal{M}$ to be closer to each other than the given dissimilarity, and $(x_k, x_{\tilde{k}}) \in \mathcal{C}$ to be further from each other than the given dissimilarity. In this manner, we revise the original dissimilarity matrix, by decreasing the dissimilarity for $(x_k, x_{\tilde{k}}) \in \mathcal{M}$ and increasing the dissimilarity for $(x_k, x_{\tilde{k}}) \in \mathcal{C}$.

3.2 *Applying Relational Clustering: NEbRFCM and eRFCM*

In this section, we consider applying NEbRFCM and eRFCM to the revised dissimilarity in Sect. 3.1.

Revising the dissimilarity matrix can be, in some cases, interpreted as transforming the data into another space such that the dissimilarity between the transformed data corresponds to the revised dissimilarities. In such cases, the revised dissimilarity matrix is *Euclidean* and we can apply bRFCM (Algorithm 1) to the revised dissimilarity matrix. If the revised dissimilarity matrix is non-*Euclidean*, bRFCM may fail because of the negative $d_{i,k}$ in STEP 3 of Algorithm 2. However, we can apply NEbRFCM instead. Note that eRFCM can also handle non-Euclidean relational data. The above discussion is summarized by Algorithm 5.

Algorithm 5 (*NEbRFCM or eRFCM with a revised dissimilarity matrix*)

STEP 1. Calculate dissimilarity matrix R for given data set x, where $R_{k,\tilde{k}}$ is the dissimilarity between x_k and $x_{\tilde{k}}$.

STEP 2. Revise the dissimilarity matrix R, by decreasing $R_{k,\tilde{k}}$ if $(x_k, x_{\tilde{k}}) \in \mathcal{M}$ and increasing $R_{k,\tilde{k}}$ if $(x_k, x_{\tilde{k}}) \in \mathcal{C}$, to the new dissimilarity matrix $\tilde{R}$.

STEP 3. Apply eRFCM (Algorithm 1) or NEbRFCM (Algorithm 2) to $\tilde{R}$. ■

3.3 Applying Kernel Clustering: IK-bFCM or K-eFCM

In this section, we consider applying kernel clustering algorithms with a dissimilarity-based kernel function to the revised dissimilarity.

A kernel function $K(x, y)$, e.g., the Gaussian in Eq. (22), can be described as

$$K(x, y) = \psi_K(\|x - y\|_2^2) \tag{23}$$

using a mapping $\psi_K : \mathbb{R}_+ \to \mathbb{R}$, where $\psi_{\text{Gaussian}}(z) = \exp(z)$. Let another mapping $\tilde{\psi} : \mathbb{R}_+ \to \mathbb{R}_+$ be the one from the given dissimilarity between data to the revised dissimilarity. Using a selected kernel function and by revising the given dissimilarity, we consider another kernel matrix $\tilde{K}$ such that

$$\tilde{K}_{k,\tilde{k}} = \psi_K(\tilde{\psi}(\|x_k - x_{\tilde{k}}\|_2^2)). \tag{24}$$

If this kernel matrix $\tilde{K}$ is positive semidefinite, we can apply K-bFCM (Algorithm 3) to $\tilde{K}$. If this kernel matrix $\tilde{K}$ is not positive semidefinite, K-bFCM may fail because of negative $d_{i,k}$ in STEP 3 of Algorithm 3. However, we can apply IK-bFCM (Algorithm 4) instead. Note that K-eFCM can be applied not only to a positive semidefinite matrix but also to an indefinite one. The above discussion is summarized by Algorithm 6.

Algorithm 6 (*IK-bFCM or K-eFCM for revised dissimilarity matrix*)

STEP 1. Calculate dissimilarity matrix R for given data set x, where $R_{k,\tilde{k}}$ is the dissimilarity between x_k and $x_{\tilde{k}}$.

STEP 2. Revise the dissimilarity matrix R, by decreasing $R_{k,\tilde{k}}$ if $(x_k, x_{\tilde{k}}) \in \mathcal{M}$ and increasing $R_{k,\tilde{k}}$ if $(x_k, x_{\tilde{k}}) \in \mathcal{C}$, to the new dissimilarity matrix $\tilde{R}$.

STEP 3. Select a dissimilarity-based kernel function ψ_K and construct a kernel matrix $\tilde{K}$ as

$$\tilde{K}_{k,\tilde{k}} = \psi_K(\tilde{R}_{k,\tilde{k}}). \tag{25}$$

STEP 4. Apply K-eFCM (Algorithm 3) or IK-bFCM Algorithm 4 to $\tilde{K}$. ■

3.4 Propagating Pairwise Constraints

In this section, we propose propagating the given pairwise constraints to other pairs of data for cases in which the given constraints are not sufficient for obtaining the desired clustering result.

Following the intuition that if data x_k and $x_{\tilde{k}}$ are close to each other, data close to x_k are also close to $x_{\tilde{k}}$, the given pairwise constraint is expected to propagate to other pairs such that if $(x_k, x_{\tilde{k}}) \in \mathcal{M}$, we would like data close to x_k to be closer to $x_{\tilde{k}}$ than the given dissimilarity. For such a propagation, we adopt a graph-theoretic approach as follows. Consider an undirected graph such that each vertex is x_k and each edge between x_k and $x_{\tilde{k}}$ has a weight $\|x_k - x_{\tilde{k}}\|_2$. Obviously, the minimal route with respect to the sum of its edge weights from x_k to $x_{\tilde{k}}$ is $\|x_k - x_{\tilde{k}}\|_2$, which corresponds to the Euclidean distance between x_k and $x_{\tilde{k}}$. If the weight between x_k and $x_{\tilde{k}}$ is revised following the revised dissimilarity by the given pairwise constraints, the minimal routes from x_k to its neighbors $x_{\tilde{k}}$ are expected to be respectively less than the given weights for the directly connected edges. Such minimal routes can be calculated using the famous Floyd–Warshall algorithm, and the squared minimal routes are applied to Algorithms 5 or 6 as the further revised dissimilarity, reflected by propagating the given pairwise constraints to other pairs of data. The above discussion is summarized by the Algorithm 7.

Algorithm 7 (*Propagating pairwise constraints*)

STEP 1. Consider an undirected graph G each vertex of which is x_k and each edge of which between x_k and $x_{\tilde{k}}$ has a weight $\sqrt{\tilde{R}_{k,\tilde{k}}}$, where $\tilde{R}$ is obtained from STEP 2 of Algorithms 5 or 6.

STEP 2. Apply the Floyd–Warshall algorithm to G and obtain the minimal route $\hat{r}_{k,\tilde{k}}$.

STEP 3. Construct the new revised dissimilarity matrix $\hat{R}$ as

$$\hat{R}_{k,\tilde{k}} = \hat{r}^2_{k,\tilde{k}}. \tag{26}$$

This algorithm is applied between STEP 2 and STEP 3 of Algorithm 5 or 6.

We also propose another approach to propagate the given pairwise constraints to other pairs for Algorithm 6. The kernel matrix obtained in STEP 3 of Algorithm 6 can be interpreted as a revised kernel function value $\tilde{K}(x_k, x_{\tilde{k}})$ of the original $K(x_k, x_{\tilde{k}})$, where

$$\begin{aligned}\tilde{K}(x, y) =& K(x, y) + \sum_{(x_k, x_{\tilde{k}}) \in \mathcal{M}} \alpha_{k,\tilde{k}} \big(\delta(x - x_k, y - x_{\tilde{k}}) + \delta(x - x_{\tilde{k}}, y - x_k)\big) + \\ & \sum_{(x_k, x_{\tilde{k}}) \in \mathcal{C}} \beta_{k,\tilde{k}} \big(\delta(x - x_k, y - x_{\tilde{k}}) + \delta(x - x_{\tilde{k}}, y - x_k)\big),\end{aligned} \tag{27}$$

$$\delta(x, y) = \begin{cases} 1 \ (x = y), \\ 0 \ (\text{otherwise}), \end{cases} \tag{28}$$

$$\alpha_{k,\tilde{k}} = \frac{\psi_K(\tilde{R}_{k,\tilde{k}}) - K(x_k, x_{\tilde{k}})}{2}, \tag{29}$$

$$\beta_{k,\tilde{k}} = -\frac{\psi_K(\tilde{R}_{k,\tilde{k}}) - K(x_k, x_{\tilde{k}})}{2}. \tag{30}$$

This revised kernel function does not influence pairs $(x_k, x_{\tilde{k}}) \notin \mathcal{M} \cup \mathcal{C}$ because function δ is singular in the sense that it is discontinuous at the origin. If δ is regularized as

$$\delta(x, y) = \max\{0, 1 - \omega^{-1}(\|x\|_2^2 + \|y\|_2^2)\} \text{ or} \tag{31}$$

$$\delta(x, y) = \exp(-\omega(\|x\|_2^2 + \|y\|_2^2)), \tag{32}$$

where the parameter $\omega > 0$ specifies the width for the influence of the given pairwise constraints, such a kernel function should influence pairs $(x_k, x_{\tilde{k}}) \notin \mathcal{M} \cup \mathcal{C}$. This chapter only deals with the case of Eq. (32), in which the coefficients $\alpha_{k,\tilde{k}}$ and $\beta_{k,\tilde{k}}$ are determined as

$$\tilde{K}(x_k, x_{\tilde{k}}) = \psi_K(\tilde{R}_{k,\tilde{k}}) \quad ((x_k, x_{\tilde{k}}) \in \mathcal{M} \cup \mathcal{C}). \tag{33}$$

Although such a revised kernel matrix $\tilde{K}$ with elements $\tilde{K}_{k,\tilde{k}} = \tilde{K}(x_k, x_{\tilde{k}})$ is not always positive-semidefinite, we can obtain a positive-semidefinite one in STEP 4 of Algorithm 6. The above discussion is summarized by Algorithm 8.

Algorithm 8 (*Kernel-based propagating constraints*)

STEP 1. For $\mathcal{M}$, $\mathcal{C}$, and revised dissimilarity matrix R in STEP 2 of Algorithm 6, select a dissimilarity-based kernel function ψ_K and set the parameter ω for the width of the influence of the given pairwise constraints.

STEP 2. Construct a kernel matrix $\tilde{K}$ as Eqs. (27), (32), and (33). ■

This algorithm is replaced by STEP 3 of Algorithm 6.

4 Numerical Example

In this section, we show some examples of clustering using our proposed algorithms. In each example, 100 trials for the proposed algorithm with random initial values were tested, and the solution with the minimal objective function value was selected as the final result. The goal is to cluster the data shown in Fig. 1 into two moon-shaped clusters. This data set is constructed of 300 elements in two dimensional Euclidean space.

4.1 Example for Algorithm 5

In this example, we apply Algorithm 5 to the data in Fig. 1.

First, we note that NEbRFCM (Algorithm 2) with $m = 2$ (i.e., neither *must-link* constraints or *cannot-link* constraints are given), fails to cluster correctly, as shown

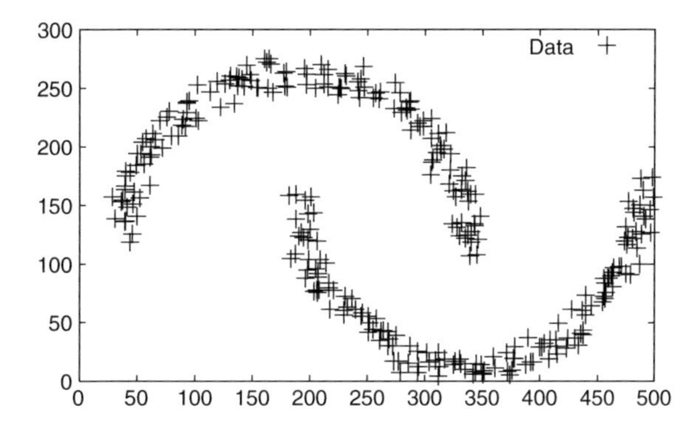

Fig. 1 Data used in the examples

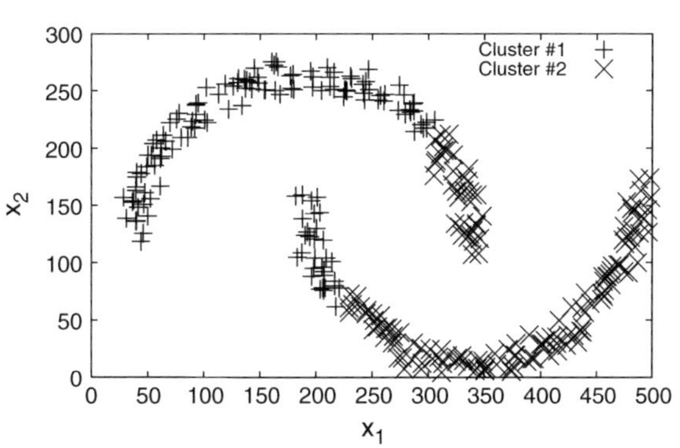

Fig. 2 Clustering results of Algorithms 1 and 2 with $m = 2$

in Fig. 2, where the plus symbol indicates one cluster and the cross symbol indicates the other. Here, the inside edges of the moons are mis-clustered with each other.

Next, we assume that *must-link* and *cannot-link* constraints are given, as shown in Fig. 3, where the black solid lines connect the pairs with *must-link* constraints and the gray dashed lines connect the pairs with *cannot-link* constraints. In Algorithm 5, we revise the dissimilarity matrix at STEP 2 to

$$\tilde{R}_{k,\tilde{k}} = \exp(-\alpha_{k,\tilde{k}} + \beta_{k,\tilde{k}})R_{k,\tilde{k}}, \tag{34}$$

where

$$\alpha_{k,\tilde{k}} = \begin{cases} 10 & ((x_k, x_{\tilde{k}}) \in \mathcal{M}), \\ 0 & ((x_k, x_{\tilde{k}}) \notin \mathcal{M}), \end{cases} \tag{35}$$

$$\beta_{k,\tilde{k}} = \begin{cases} 10 & ((x_k, x_{\tilde{k}}) \in \mathcal{C}), \\ 0 & ((x_k, x_{\tilde{k}}) \notin \mathcal{C}), \end{cases} \tag{36}$$

and we obtain the desired clustering results, as shown in Fig. 4. Thus, this example shows that on the data on which the original algorithm (Algorithm 2) fails, the

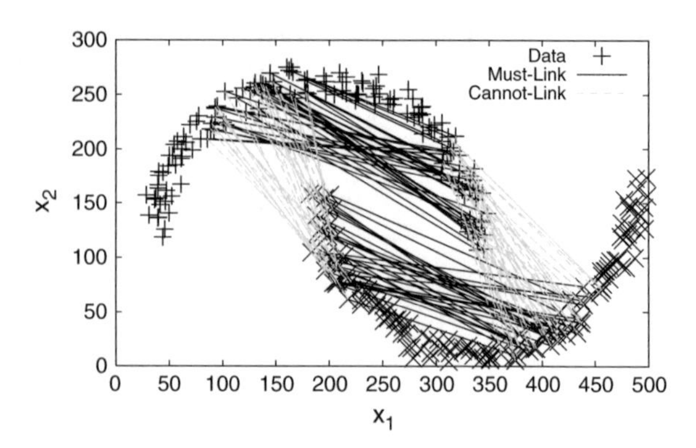

Fig. 3 *Must-links* and *cannot-links* for Algorithm 5

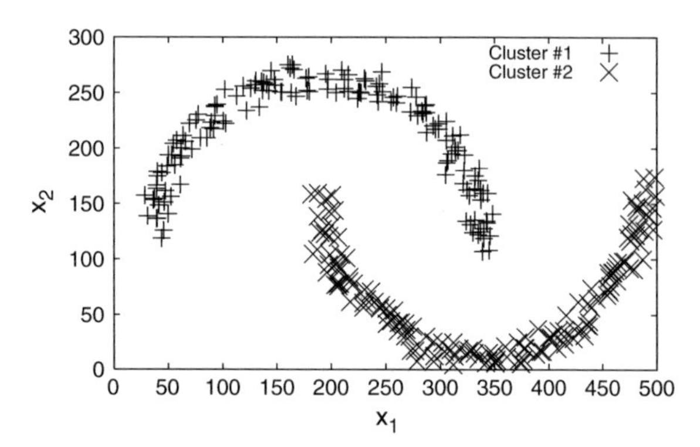

Fig. 4 Clustering results of Algorithm 5

proposed algorithm (Algorithm 5) can achieve a successful clustering result with the help of given pairwise constraints of a data supervisor.

4.2 Example for Algorithm 6

In this example, we apply Algorithm 6 to the data in Fig. 1.

First, we see that K-FCM (Algorithm 3) with $m = 2$ and

$$K_{k,\tilde{k}} = \exp(-0.0002\|x_k - x_{\tilde{k}}\|_2^2), \tag{37}$$

(i.e., neither *must-link* constraints or *cannot-link* constraints are given), fails to cluster correctly, as shown in Fig. 5. In this figure, the right edge of the upper moon is mis-clustered as the other cluster. Although K-bFCM and K-eFCM (Algorithm 3) may cluster correctly with some tuned parameters, we fix the parameters to the ones for which K-bFCM fails because we would like to demonstrate the situation in which the adequate parameters have not been selected for K-bFCM.

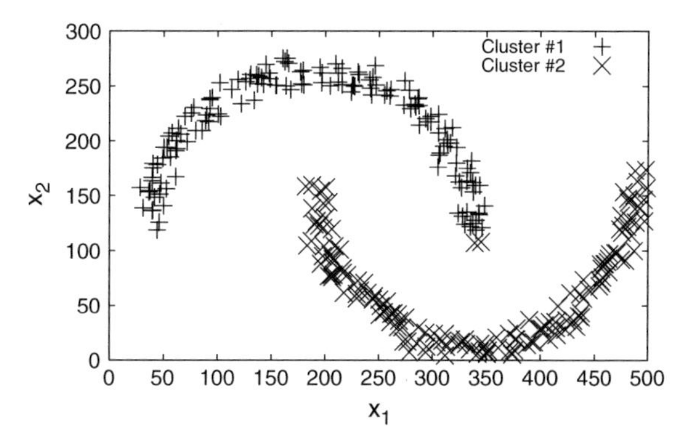

Fig. 5 Clustering results for Algorithm 3 with $\sigma = 0.0002, m = 2$

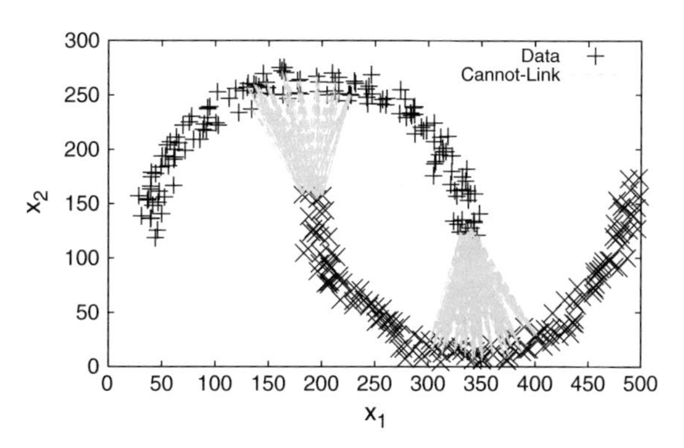

Fig. 6 *Cannot-links* for Algorithm 6

Next, we assume that *cannot-link* constraints are given, as shown in Fig. 6, where the gray dashed lines connect the pairs in *cannot-link* constraints. In Algorithm 6, we revise the dissimilarity matrix at STEP 2 to

$$\tilde{R}_{k,\tilde{k}} = \exp(\beta_{k,\tilde{k}})R_{k,\tilde{k}}, \tag{38}$$

where

$$\beta_{k,\tilde{k}} = \begin{cases} 100 & ((x_k, x_{\tilde{k}}) \in \mathcal{C}), \\ 0 & ((x_k, x_{\tilde{k}}) \notin \mathcal{C}), \end{cases} \tag{39}$$

and we set ψ_K at STEP 3 as

$$\psi_K(z) = \exp(-0.0002z) \tag{40}$$

following Eq. (37). We then obtain the desired clustering results, as shown in Fig. 4 using Algorithm 6. Thus, this example shows that on data that causes Algorithm 3 to fail, the proposed algorithm (Algorithm 6) achieves a successful clustering result with the help of pairwise constraints given by a supervisor. Although we tested Algorithm 6 with some *must-link* constraints, we could not obtain the desired clustering result because all the memberships $u_{i,k}$ converge to the same value 0.5.

4.3 Example for Algorithm 7

In this example, we apply Algorithm 7 to the data in Fig. 1 with fewer *must-link* and *cannot-link* constraints than shown in Fig. 3.

First, we see that Algorithm 5 fails with the given constraints as follows. We assume that *must-link* and *cannot-link* constraints are given as shown in Fig. 7, where the black solid and gray dashed lines indicate *must-link* and *cannot-link* constraints, respectively. In Algorithm 5, we revise the dissimilarity matrix at STEP 2 as using Eqs. (34)–(36), and find that we cannot obtain the desired clustering result. This seems to be because the given constraints are not sufficient to obtain the desired clustering result.

Next, we apply Algorithm 7 between STEP 2 and STEP 3 of Algorithm 5. After we revise the dissimilarity matrix in STEP 2 in Algorithm 5, as per Eqs. (34)–(36), we further revise the dissimilarity matrix using Algorithm 7. We can then obtain the desired clustering result, as shown in Fig. 4. Thus, this example shows that the proposed algorithm (Algorithm 7) is a valid way to achieve successful clustering results when the given constraints are not sufficient for Algorithm 5 to obtain the desired clustering result.

4.4 Example for Algorithm 8

In this example, we apply Algorithm 8 to the data in Fig. 1 with fewer *cannot-link* constraints than shown in Fig. 6.

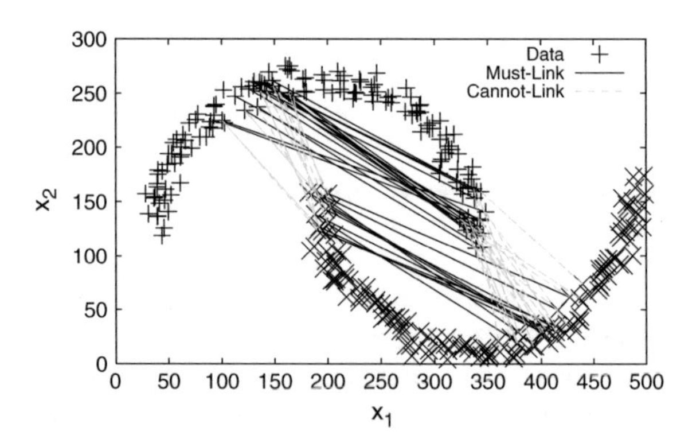

Fig. 7 *Must-links* and *cannot-links* for Algorithm 5 via Algorithm 7

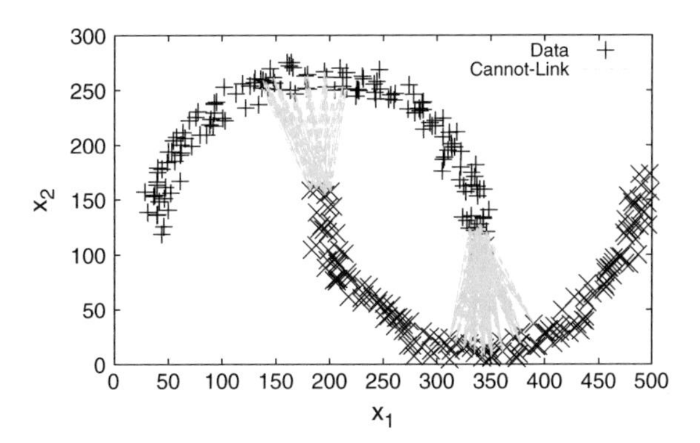

Fig. 8 *Cannot-links* for Algorithm 6 via Algorithm 8

First, we see that Algorithm 6 fails with given constraints as follows. We assume that *cannot-link* constraints are given as shown in Fig. 8, where the gray dashed lines indicate *cannot-link* constraints. In Algorithm 6, we revise the dissimilarity matrix at STEP 3 as per Eqs. (38) and (39) and set ψ_K at STEP 2 as per Eq. (40), but cannot obtain the desired clustering result using Algorithm 6. Again, this seems to be because the given constraints are not sufficient to obtain the desired clustering result.

We next use Algorithm 8 to replace STEP 3 of Algorithm 6. After we set the dissimilarity matrix at STEP 2 in Algorithm 5, as per Eqs. (38) and (39), we set the kernel matrix using Algorithm 8 with $\omega = 0.001$. We can then obtain the desired clustering result as shown in Fig. 4. Thus, this example shows that the proposed algorithm (Algorithm 8) achieves successful clustering results when the given constraints are not sufficient for Algorithm 6 to obtain the desired clustering result.

5 Conclusion

In this chapter, we proposed four approaches for semi-supervised FCM with *soft* pairwise constraints. The first two are apply NEbRFCM and eRFCM to revised dissimilarity matrix by pairwise constraints (Algorithm 5), and the others use IK-bFCM and K-eFCM with a dissimilarity-based kernel function (Algorithm 6). The dissimilarity matrix is revised based on the concept that the data in the same cluster may be close to each other and the data in the different clusters may be far apart. Furthermore, we discussed propagating the given pairwise constraints to unconstrained data when the given constraints are not sufficient to obtain the desired clustering result (Algorithms 7 and 8). We showed some numerical examples for the proposed algorithms. The first two numerical examples show that the proposed algorithms (Algorithms 5 and 6) achieved successful clustering results with the help of pairwise constraints given by a data supervisor on data where the original algorithms (Algorithms 2 and 3) fail. The other two examples show that the proposed algorithms (Algorithms 7 and 8) can achieve successful clustering results when the given constraints are not sufficient for Algorithms 5 and 6. Although the data shown in Fig. 1 were perfectly clustered by all the proposed algorithms, such perfect clustering for other data does not always hold, especially in practical applications. The proposed methods will be applied to several real datasets in our future work.

Our other future tasks are as follows. (1) Because we could not demonstrate a deterministic method to revise the dissimilarity matrix by the given pairwise constraints, we will investigate a reasonable revision to the dissimilarity matrix based on some theoretical background. Through such investigation, we will clarify the reason why the *must-link* constraints obtained the result of $u_{i,k} = 0.5$ for the example in Sect. 4.2, and understand the sensitivity of the revised dissimilarity matrix to this result. (2) Because we considered propagating only *must-link* constraints to other pairs in the graph-theoretic approach, we will also consider propagating *cannot-link* constraints. (3) Because we fixed the number of trials with random initial values to 100 in the experiments, which is without any reasonable discussion, we will also investigate which method achieves the maximal number of trials achieving the minimal objective functions values, along with how many trials are sufficient to obtain the desireble result for each method.

References

1. Bezdek, J.C.: Pattern Recognition with Fuzzy Objective Function Algorithms, Plenum, New York, (1981).
2. Miyamoto, S. and Umayahara, K.: "Methods in Hard and Fuzzy Clustering," in: Liu, Z.-Q. and Miyamoto, S. (eds), Soft Computing and Human-centered Machines, Springer-Verlag Tokyo, (2000).
3. Hathaway, R.J., Davenport, J.W. and Bezdek, J.C.: "Relational Duals of the c-means Clustering Algorithms," Pattern Recognition, Vol. 22, No. 2, pp. 205–212, (1989).
4. Hathaway, R.J. and Bezdek, J.C.: "NERF C-means: Non-Euclidean Relational Fuzzy Clustering," Pattern Recognition, Vol. 27, No. 3, pp. 429–437, (1994).
5. Kanzawa, Y.: "Entropy-Regularized Fuzzy Clustering for Non-Euclidean Relational Data and Indefinite Kernel Data," JACIII, Vol. 16, No. 7, pp. 784–792, (2012).
6. Miyamoto, S. and Suizu, D.: "Fuzzy c-Means Clustering Using Kernel Functions in Support Vector Machines," JACIII, Vol. 7, No. 1, pp. 25–30, (2003).
7. Miyamoto, S., Kawasaki, Y., and Sawazaki, K.: "An Explicit Mapping for Kernel Data Analysis and Application to Text Analysis," Proc. IFSA-EUSFLAT 2009, pp. 618–623, (2009).
8. Kanzawa, Y., Endo, Y., and Miyamoto, S.: "Indefinite Kernel Fuzzy c-Means Clustering Algorithms," Lecture Notes in Computer Science, Vol. 6408, pp. 116–128, (2010).
9. Bouchachia, A. and Pedrycz, W.: "Data Clustering with Partial Supervision," Data Mining and Knowledge Discovery, Vol. 12, pp. 47–78, (2006).
10. Yamazaki, M., Miyamoto, S. and Lee, I.-J.: "Semi-supervised Clustering with Two Types of Additional Functions," Proc. 24th Fuzzy System Symposium, 2E2-01, (2009).
11. Yamashiro, M., Endo, Y., Hamasuna, Y. and Miyamoto, S.: "A Study on Semi-supervised Fuzzy c-Means," Proc. 24th Fuzzy System Symposium, 2E3-04, (2009).
12. Kanzawa, Y., Endo, Y. and Miyamoto, S.: "A Semi-Supervised Entropy Regularized Fuzzy c-Means," Proc. 2009 International Symposium on Nonlinear Theory and Its Applications, pp. 564–567, (2009).

13. Wagstaff, K., Cardie, C., Rogers, S. and Schroedl, S.: "Constrained K-means Clustering with Background Knowledge," Proc. Eighteenth International Conference on Machine-Learning, pp. 577–584, (2001).
14. Grira, N., Crucianu, M. and Boujemaa, N.: "Semi-supervised Image Database Categorization using Pairwise Constraints," Proc. 2005 IEEE International Conference on Image Processing, Vol. 3, pp. 1228–1231, (2005).
15. Kanzawa, Y., Endo, Y. and Miyamoto, S: "Some Pairwise Constrained Semi-Supervised Fuzzy c-Means Clustering," LNAI, Vol. 5681, pp. 268–281, (2009).
16. Kanzawa, Y., Endo, Y., and Miyamoto, S.: "Semi-Supervised Fuzzy c-Means Algorithm by Revising Dissimilarity Between Data," JACIII, Vol. 15, No. 1, pp. 95–101, (2011).

Various Types of Objective-Based Rough Clustering

Yasunori Endo and Naohiko Kinoshita

Abstract Conventional clustering algorithms classify a set of objects into some clusters with clear boundaries, that is, one object must belong to one cluster. However, many objects belong to more than one cluster in real world since the boundaries of clusters generally overlap with each other. Fuzzy set representation of clusters makes it possible for each object to belong to more than one cluster. On the other hand, it is pointed out that the fuzzy degree is sometimes regarded as too descriptive for interpreting clustering results. Instead of fuzzy representation, rough set one could deal with such cases. Clustering based on rough set could provide a solution that is less restrictive than conventional clustering and less descriptive than fuzzy clustering. Therefore, Lingras et al. (Lingras and Peters, Wiley Interdiscip Rev: Data Min Knowl Discov 1(1):64–72, 1207–1216, 2011, [1] and Lingras and West, J Intell Inf Syst 23(1):5–16, 2004, [2]) proposed a clustering method based on rough set, rough K-means (RKM). RKM is almost only one algorithm inspired by KM and some assumptions of RKM are very natural, however it is not useful from the viewpoint that the algorithm is not based on any objective functions. Outputs of non-hierarchical clustering algorithms strongly depend on initial values and the "better" output among many outputs from different initial values should be chosen by comparing the value of the objective function of the output with each other. Therefore the objective function plays very important role in clustering algorithms. From the standpoint, we have proposed some rough clustering algorithms based on objective functions. This paper shows such rough clustering algorithms which is based on optimization of an objective function.

Y. Endo (✉)
Faculty of Engineering, Information and Systems, University of Tsukuba,
1-1-1, Tennnodai, Ibaraki, Tsukuba 305-8573, Japan
e-mail: endo@risk.tsukuba.ac.jp

N. Kinoshita
JSPS, University of Tsukuba, 1-1-1, Tennnodai,
Ibaraki, Tsukuba 305-8573, Japan
e-mail: s1430167@u.tsukuba.ac.jp

V. Torra et al. (eds.), *Fuzzy Sets, Rough Sets, Multisets and Clustering*,
Studies in Computational Intelligence 671, DOI 10.1007/978-3-319-47557-8_5

1 Introduction

As the importance of data analysis increases, clustering techniques have been more focused [3] and more clustering algorithms have been proposed.

Conventional clustering algorithms partition a set of objects into some clusters with clear boundaries. In other words, one object must belong to one cluster. k-means (KM) [4], also called hard c-means (HCM), is representative one.

However, the boundaries may not be clear in practice and quite a few objects should belong to more than one cluster. Fuzzy set representation of clusters makes it possible for each object to belong to more than one cluster and the degree of belongingness of an object to each cluster is represented as a value in an unit interval [0, 1]. Fuzzy c-means (FCM) [5, 6] achieves the representation by introducing a fuzzification parameter into KM.

On the other hand, it is pointed out that the fuzzy degree sometimes may be too descriptive for interpreting clustering results [1]. In such cases, rough set representation is considered as an useful and powerful tool [7, 8]. The basic concept of the representation is based on two definitions of lower and upper approximations of a set. The lower approximation means that "an object surely belongs to the set" and the upper one means that "an object possibly belongs to the set". Clustering based on rough set could provide a solution that is less restrictive than conventional clustering and less descriptive than fuzzy clustering [1, 9], and therefore, the rough set based clustering has attracted increasing interest of researchers [1, 2, 10–14].

Rough k-means (RKM) proposed by Lingras et al. [1, 2] is one of initial rough set based clustering. In RKM, the degree of belongingness and cluster centers are calculated by iterative process like KM or FCM. However, RKM has a problem that the algorithm is not constructed based on optimization of an objective function. Here, we call clustering based on optimization of an objective function "objective-based clustering". In other words, calculation outputs of the objective-based clustering make the objective function minimize.

Many non-hierarchical clustering algorithms such as KM and FCM are objective-based clustering. Calculation outputs of such algorithms strongly depend on initial values. Hence, we need some indicator when we choose the "better" outputs among many outputs from different initial values. The objective functions play very important role as the indicator, that is, we can choose the "better" outputs by comparing the value of the objective function of the output with each other.

RKM is one of the most representative algorithms inspired by KM and some assumptions of RKM are very natural, however it is not useful from the viewpoint that the algorithm is not based on any objective functions because we do not have any indicator to choose "better" outputs. Some rough set based clustering algorithms based on an objective function are proposed [12], however these may be a bit complicated and not easy to expand the theoretical discussion.

We have proposed some objective-based rough clustering methods. This paper shows objective functions and algorithms of the methods, type-I rough c-means, type-II rough c-means, and rough non metric model. In each method, we show rough hard clustering and rough fuzzy one.

2 Rough Sets

2.1 Concept of Rough Sets

Let U be the universe and $R \subseteq U \times U$ be an equivalence relation on U. R is also called indiscernibility relation. The pair $X = (U, R)$ is called an approximation space. If $x, y \in U$ and $(x, y) \in R$, we say that x and y are indistinguishable in X.

Equivalence classes of the relation R is called elementary sets in X. The set of all elementary sets is denoted by U/R. The empty set is also elementary in every X.

Every finite union of elementary sets in X is called a composed set in X.

Since it is impossible to distinguish the elements in the same equivalence class, we may not be able to get a precise representation for an arbitrary subset $A \subset U$. Instead, any A can be represented by its lower and upper bounds. The upper bound $\overline{A}$ is the least composed set in X that contains A, called the best upper approximation or, in short, the upper approximation. The lower bound $\underline{A}$ is the greatest composed set in X that is included in A, called the best lower approximation or, briefly, the lower approximation. The set $\mathrm{Bnd}(A) = \overline{A} - \underline{A}$ is called the boundary of A in X.

The pair $(\underline{A}, \overline{A})$ is the representation of an ordinary set A in the approximation space X, or simply the rough set of A. The elements in the lower approximation of A definitely belong to A, while elements in the upper bound of A may or may not belong to A.

2.2 Conditions of Rough Clustering

Let a set of objects and a set of equivalence classes by an equivalence relation R be $U = \{x_k \mid x_k = (x_{k1}, \ldots, x_{kp})^T \in \Re^p,\ k = 1, \ldots, n\}$ and $U/R = \{A_i \mid i = 1, \ldots, c\}$, respectively. $v_i = (v_{i1}, \ldots, v_{ip})^T \in \Re^p$ $(i = 1, \ldots, c)$ means a cluster center of a cluster A_i. We notice that $A_i \neq \emptyset$ for any i. That is, $\underline{A} = \emptyset$ means that $\mathrm{Bnd}(A) \neq \emptyset$. Similarly, $\mathrm{Bnd}(A) = \emptyset$ means that $\underline{A} \neq \emptyset$.

Lingras et al., who proposed rough K-means (RKM) [1, 2], put the following conditions. Their conditions are very natural from the viewpoint of the definition of rough sets.

(C1) An object x can be part of at most one lower bound.
(C2) If $x \in \underline{A}_i$, $x \in \overline{A}_i$.

(C3) An object x is not part of any lower bound if and only if x belongs to two or more upper bounds.

Note that the above conditions are not necessarily independent or complete.

3 Type-I Rough c-Means

3.1 Type-I Rough Hard c-Means

In this section, we describe type-I rough c-means (RCM-I). In order to distinguish the later mentioned rough fuzzy c-means, we also write type-I rough hard c-means (RHCM-I).

3.1.1 Objective Function

For any objects $x_k = (x_{k1}, \dots, x_{kp})^T \in \Re^p$ $(k = 1, \dots, n)$, ν_{ki} and u_{ki} $(i = 1, \dots, c)$ mean belongingness of an object x_k to a lower approximation of A_i and a boundary of A_i, respectively. Partition matrices of ν_{ki} and u_{ki} are denoted by $N = \{\nu_{ki}\}$ and $U = \{u_{ki}\}$, respectively. We define an objective function of RCM-I as follows:

$$J_{\text{RCM-I}}(N, U, V) = \sum_{k=1}^{n} \sum_{l=1}^{n} \sum_{i=1}^{c} \Big(\nu_{ki} u_{li} (\underline{w} d_{ki} + \overline{w} d_{li}) + (\nu_{ki} \nu_{li} + u_{ki} u_{li}) D_{kl} \Big), \quad (1)$$

where

$$d_{ki} = \|x_k - v_i\|^2, \quad D_{kl} = \|x_k - x_l\|^2.$$

For any k, constraints are as follows:

$$\underline{w} + \overline{w} = 1,$$
$$\nu_{ki} \in \{0, 1\}, \quad u_{ki} \in \{0, 1\},$$
$$\sum_{i=1}^{c} \nu_{ki} \in \{0, 1\}, \quad \sum_{i=1}^{c} u_{ki} \neq 1,$$
$$\sum_{i=1}^{c} \nu_{ki} = 1 \Longleftrightarrow \sum_{i=1}^{c} u_{ki} = 0.$$

The last term of (1) is a regularized term. If the term does not exist, it results in trivial solutions of $\nu_{ki} = 0$ and $u_{ki} = 0$. From the above constraints, we can derive the following relation for any k:

$$\sum_{i=1}^{c} \nu_{ki} = 0 \Longleftrightarrow \sum_{i=1}^{c} u_{ki} \geq 2$$

It is obvious that these relations are equivalent to (C1)–(C3) in Sect. 2.2.

3.1.2 Derivation of Optimal Solutions and Algorithm

We'll obtain an optimal solution to v_i with fixing ν_{ki} and u_{ki}. Here, we introduce the following function:

$$J^i_{\text{RCM-I}}(V) = \sum_{k=1}^{n} \sum_{l=1}^{n} \left(\nu_{ki} u_{li} (\underline{w} d_{ki} + \overline{w} d_{li}) \right). \tag{2}$$

Since ν_{ki} and u_{ki} are fixed, v_i which minimizes $J^i_{\text{RCM-I}}$ is an optimal solution which also minimizes $J_{\text{RCM-I}}$. Now, we have to consider the following two cases:

1. $\underline{A}_i \neq \emptyset$ and $\text{Bnd}(A_i) \neq \emptyset$, that is, $|\underline{A}_i| \cdot |\text{Bnd}(A_i)| \neq 0$.
2. $\underline{A}_i = \emptyset$ or $\text{Bnd}(A_i) = \emptyset$, that is, $\nu_{ki} = 0$ or $u_{ki} = 0$ for any k.

If $\underline{A}_i \neq \emptyset$ and $\text{Bnd}(A_i) \neq \emptyset$, from partially differentiating (2) by v_i,

$$\frac{\partial J^i_{\text{RCM-I}}}{\partial v_i} = \sum_{k=1}^{n} \sum_{l=1}^{n} \nu_{ki} u_{li} \left(\underline{w}(x_k - v_i) + \overline{w}(x_l - v_i) \right).$$

From $\frac{\partial J^i_{\text{RCM-I}}}{\partial v_i} = 0$,

$$\sum_{k=1}^{n} \sum_{l=1}^{n} \nu_{ki} u_{li} v_i = \sum_{k=1}^{n} \sum_{l=1}^{n} \nu_{ki} u_{li} (\underline{w} x_k + \overline{w} x_l),$$

then, we get

$$\sum_{k=1}^{n} \nu_{ki} \sum_{l=1}^{n} u_{li} v_i = \underline{w} \sum_{l=1}^{n} u_{li} \sum_{k=1}^{n} \nu_{ki} x_k + \overline{w} \sum_{k=1}^{n} \nu_{ki} \sum_{l=1}^{n} u_{li} x_l. \tag{3}$$

We here notice the following relations:

$$|\underline{A}_i| = \sum_{k=1}^{n} \nu_{ki}, \quad |\text{Bnd}(A_i)| = \sum_{l=1}^{n} u_{li}.$$

Then, (3) can be rewritten as follows:

$$|\underline{A}_i| \cdot |\mathrm{Bnd}(A_i)| v_i = \underline{w} |\mathrm{Bnd}(A_i)| \sum_{x_k \in \underline{A}_i} x_k + \overline{w} |\underline{A}_i| \sum_{x_k \in \mathrm{Bnd}(A_i)} x_k.$$

Since $|A_i| \cdot |\mathrm{Bnd}(A_i)| \neq 0$,

$$v_i = \underline{w} \frac{\sum_{x_k \in \underline{A}_i} x_k}{|\underline{A}_i|} + \overline{w} \frac{\sum_{x_k \in \mathrm{Bnd}(A_i)} x_k}{|\mathrm{Bnd}(A_i)|}.$$

On the other hand, if $\underline{A}_i = \emptyset$ or $\mathrm{Bnd}(A_i) = \emptyset$, $v_{ki} = 0$ or $u_{ki} = 0$ for any k. In the both cases, $J^i_{\mathrm{RCM\text{-}I}}$ becomes the minimum value 0 in spite of v_i. Therefore, we can determine v_i as follows:

$$v_i = \frac{\sum_{x \in \overline{A}_i} x}{|\overline{A}_i|}.$$

From the above discussion, the optimal solution to v_i is (4).

$$v_i = \begin{cases} \underline{w} \dfrac{\sum_{x \in \underline{A}_i} x}{|\underline{A}_i|} + \overline{w} \dfrac{\sum_{x \in \mathrm{Bnd}(A_i)} x}{|\mathrm{Bnd}(A_i)|}, & (\underline{A}_i \neq \emptyset \wedge \mathrm{Bnd}(A_i) \neq \emptyset) \\ \dfrac{\sum_{x \in \overline{A}_i} x}{|\overline{A}_i|}. & (\text{otherwise}) \end{cases} \tag{4}$$

Optimal solutions to v_{ki} and u_{ki} can be obtained by comparing the following two cases:

1. x_k belongs to the lower approximation $\underline{A}_{p_k}$ of which the cluster center v_i is nearest to x_k. Here,

$$p_k = \arg\min_i d_{ki}.$$

In this case, the value of the term for x_k of the objective function can be calculated as follows:

$$\begin{aligned} J^v_k &= \sum_{l=1, l \neq k}^{n} \left(v_{kp_k} u_{lp_k} (\underline{w} d_{kp_k} + \overline{w} d_{lp_k}) + (v_{kp_k} v_{lp_k} + u_{kp_k} u_{lp_k}) D_{kl} \right) \\ &= \sum_{l=1, l \neq k}^{n} \left(v_{kp_k} u_{lp_k} (\underline{w} d_{kp_k} + \overline{w} d_{lp_k}) + v_{kp_k} v_{lp_k} D_{kl} \right). \end{aligned}$$

2. x_k belongs to the upper approximation of two clusters $\overline{A}_{p_k}$ and $\overline{A}_{q_k}$ of which the cluster centers v_{p_k} and v_{q_k} are the first and second nearest to x_k. Here,

$$q_k = \arg\min_{i \neq p_k} d_{ki}.$$

In this case, the value of the terms for x_k of the objective function can be calculated as follows:

$$J_k^u = \sum_{l=1,l\neq k}^{n} \sum_{i=p_k,q_k}^{c} \Big(\nu_{li}u_{ki}(\underline{w}d_{li} + \overline{w}d_{ki}) + (\nu_{ki}\nu_{li} + u_{ki}u_{li})D_{kl}\Big)$$

$$= \sum_{l=1,l\neq k}^{n} \sum_{i=p_k,q_k}^{c} \Big(\nu_{li}u_{ki}(\underline{w}d_{ki} + \overline{w}d_{ki}) + u_{ki}u_{li}D_{kl}\Big).$$

In comparison with J_k^ν and J_k^u, we determine ν_{ki} and u_{ki} as follows:

$$\nu_{ki} = \begin{cases} 1, & (J_k^\nu < J_k^u \wedge i = p_k) \\ 0. & \text{(otherwise)} \end{cases}$$

$$u_{ki} = \begin{cases} 1, & \Big(J_k^\nu \geq J_k^u \wedge (i = p_k \vee i = q_k)\Big) \\ 0. & \text{(otherwise)} \end{cases}$$

Here, we construct RCM-I algorithm using optimal solutions to N, V, and U which are derived in the above. In practice, the optimal solutions are calculated through iterative optimization. We show RCM-I algorithm as Algorithm 1.

Algorithm 1 RCM-I

RCM-I.1 Set initial values and calculate initial cluster centers.
RCM-I.2 Update partition matrices N and U by fixing V.
RCM-I.3 Update cluster centers V by fixing N and U.
RCM-I.4 If a stopping criterion satisfies, finish the algorithm. Otherwise, back to **RCM-I.2**.

We can consider the Sequential RCM-I (SRCM-I) in which cluster centers are re-calculated every time cluster partition changes as Algorithm 2.

Algorithm 2 SRCM-I

SRCM-I.1 Set initial values and calculate initial cluster centers.
SRCM-I.2 Iterate **SRCM-I2**, **SRCM-I3**, and **SRCM-I4** for each x_k $(k = 1, \ldots, n)$.
SRCM-I.3 Update partition matrices N and U by fixing V.
SRCM-I.4 Update cluster centers V by fixing N and U.
SRCM-I.5 If a stopping criterion satisfies for each x_k $(k = 1, \ldots, n)$, finish the algorithm. Otherwise, back to **SRCM-I.2**.

3.2 Type-I Rough Fuzzy c-Means

We have two ways to fuzzify RCM-I by introducing fuzzy-set representation.

The first is to introduce the fuzzification parameter m, and the second is to introduce the entropy term. These ways are known to be very useful. We call the method using the first way type-I rough fuzzy c-means (RFCM-I). and that using the second way entropy-regularized type-I rough fuzzy c-means (ERFCM-I), as mentioned above. In this paper, we describe RFCM-I.

In RFCM-I, degrees of belongingness to $\mathrm{Bnd}(A_i)$ are only fuzzified.

3.2.1 Objective Function

The objective function of RFCM-I is defined as follows:

$$J_{\text{RFCM-I}}(N, U, V) = \sum_{k=1}^{n} \sum_{l=1}^{n} \sum_{i=1}^{c} \left(u_{ki}^m \nu_{li} (\underline{\omega} d_{li} + \overline{\omega} d_{ki}) + (\nu_{ki} \nu_{li} + u_{ki}^m u_{li}^m) D_{kl} \right). \tag{5}$$

Constraints are as follows:

$$\underline{\omega} + \overline{\omega} = 1,$$
$$\nu_{ki} \in \{0, 1\}, \quad u_{li} \in [0, 1], \forall k, i$$
$$\sum_{i=1}^{c} (\nu_{ki} + u_{ki}) = 1, \ \forall k$$

3.2.2 Derivation of Optimal Solutions and Algorithm

To get an optimal solution to v_i, we partially differentiate (5) with respect to v_i, getting

$$v_i = \begin{cases} \dfrac{\sum_{x_k \in \underline{A}_i} x_k}{|\underline{A}_i|}, & (\mathrm{Bnd}(A_i) = \emptyset) \\ \dfrac{\sum_{k=1}^{n} u_{ki}^m x_k}{\sum_{k=1}^{n} u_{ki}^m}, & (\underline{A}_i = \emptyset) \\ \underline{\omega} \times \dfrac{\sum_{x_k \in \underline{A}_i} x_k}{|\underline{A}_i|} + \overline{\omega} \times \dfrac{\sum_{k=1}^{n} u_{ki}^m x_k}{\sum_{k=1}^{n} u_{ki}^m}. & (\text{otherwise}) \end{cases}$$

We must consider the following two cases to derive optimal solutions to N and U:

1. x_k belongs to $\underline{A}_{p_k}$.

2. x_k belongs to $\mathrm{Bnd}(A_i)$. $\forall i$

If x_k belongs to $\underline{A}_{p_k}$, optimal solutions and the objective function are represented as follows:

$$\nu_{ki} = 1, \quad u_{ki} = 0,$$

$$\underline{J}^k_{\text{RFCM-I}} = \sum_{l=1, l \neq k}^{n} (u^m_{lp_k}(\underline{\omega} d_{kp_k} + \overline{\omega} d_{lp_k}) + 2\nu_{lp_k} D_{kl}). \tag{6}$$

If x_k belongs to $\mathrm{Bnd}(A_i)$, optimal solutions and the objective function are represented as follows:

$$\nu_{ki} = 0, \quad u_{ki} = \frac{\left(\frac{1}{\alpha_i}\right)^{\frac{1}{m-1}}}{\sum_{j=1}^{c} \left(\frac{1}{\alpha_j}\right)^{\frac{1}{m-1}}},$$

$$\overline{J}^k_{\text{RFCM-I}} = \sum_{i=1}^{c} \sum_{l=1, l \neq k}^{n} (u^m_{ki} \nu_{li}(\underline{\omega} d_{li} + \overline{\omega} d_{ki}) + 2u^m_{ki} u^m_{li} D_{kl}). \tag{7}$$

Here,

$$\alpha_i = \sum_{l=1, l \neq k}^{n} (\nu_{li}(\underline{\omega} d_{li} + \overline{\omega} d_{ki}) + 2u^m_{li} D_{kl})$$

We calculate the optimal solution to u_{ki} by using the Lagrange multiplier. From (5), the Lagrange function of RFCM-I is defined as follows:

$$L_{\text{RFCM-I}} = \sum_{k=1}^{n} \sum_{l=1}^{n} \sum_{i=1}^{c} \big(u^m_{ki} \nu_{li}(\underline{\omega} d_{li} + \overline{\omega} d_{ki})$$
$$+(\nu_{ki} \nu_{li} + u^m_{ki} u^m_{li}) D_{kl}\big) - \sum_{k=1}^{n} \lambda_k \left(\sum_{i=1}^{c} u_{ki} - 1 \right).$$

Comparing (6) and (7), the optimal solutions to N and U are as follows:

$$\nu_{ki} = \begin{cases} 1, & (J^\nu_k < J^u_k \wedge i = p_k) \\ 0, & (\text{otherwise}) \end{cases}$$

$$u_{ki} = \begin{cases} 0, & (J^\nu_k < J^u_k \wedge i = p_k) \\ \dfrac{\left(\frac{1}{\alpha_i}\right)^{\frac{1}{m-1}}}{\sum_{j=1}^{c} \left(\frac{1}{\alpha_j}\right)^{\frac{1}{m-1}}} & (\text{otherwise}) \end{cases}.$$

Last, we describe the algorithm of RFCM-I.

Algorithm 3 RFCM-I

RFCM-I.1 Set initial approximations and calculate initial cluster centers.
RFCM-I.2 Update lower approximations and boundaries.
RFCM-I.3 Update cluster centers.
RFCM-I.4 If the convergence criterion is satisfied, finish. Otherwise back to **RFCM-I.2.**

4 Type-II Rough c-Means

We propose another method: type-II rough c-means (RCM-II) or type-II rough hard c-means (RHCM-II) to solve Lingras's problems. The objective function of RCM-II is simpler than RCM-II.

4.1 Type-II Rough Hard c-Means

4.1.1 Objective Function

Let $N = (\nu_{ki})_{1 \le k \le n,\ 1 \le i \le c}$ and $U = (u_{ki})_{1 \le k \le n,\ 1 \le i \le c}$ be degrees of belongingness of x_k to $\underline{A}_i$ and $\mathrm{Bnd}(A_i)$. Let V be a set of cluster centers. The objective function of RCM-II is defined as follows:

$$J_{\text{RCM-II}}(N, U, V) = \sum_{k=1}^{n} \sum_{i=1}^{c} (\nu_{ki}\underline{w} + u_{ki}\overline{w}) \|x_k - v_i\|^2. \tag{8}$$

$$\underline{w} + \overline{w} = 1,\ \underline{w} > 0,\ \overline{w} > 0.\ (1 \le k \le n,\ 1 \le i \le c)$$

Constraints are as follows:

$$\nu_{ki}, u_{ki} \in \{0, 1\}, \quad \forall k, i$$

$$\sum_{i=1}^{c} \nu_{ki} \in \{0, 1\}, \quad \sum_{i=1}^{c} u_{ki} \ne 1, \quad \forall k$$

$$\sum_{i=1}^{c} \nu_{ki} = 1 \iff \sum_{i=1}^{c} u_{ki} = 0. \quad \forall k$$

From these constraints, the following restriction holds true:

$$\sum_{i=1}^{c} \nu_{ki} = 0 \iff \sum_{i=1}^{c} u_{ki} > 1. \quad \forall k$$

These constraints are clearly equivalent to **(C1)–(C3)**. $J_{\text{RCM-II}}$ is minimized under these constraints.

4.1.2 Derivation of Optimal Solutions and Algorithm

We partially differentiate (8) with respect to v_i. We get

$$v_i = \frac{\sum_{k=1}^{n}(\underline{w}\nu_{ki} + \overline{w}u_{ki})x_k}{\sum_{k=1}^{n}(\underline{w}\nu_{ki} + \overline{w}u_{ki})}. \tag{9}$$

We must consider the following two cases to derive optimal solutions to N and U:

1. x_k belongs to $\underline{A}_{p_k}$.
2. x_k belongs to $\text{Bnd}(A_{p_k})$ and $\text{Bnd}(A_{q_k})$.

Here,

$$p_k = \min_i \|x_k - v_i\|^2,$$
$$q_k = \min_{i \neq p_k} \|x_k - v_i\|^2.$$

If x_k belongs to $\underline{A}_{p_k}$, we get the value of the objective function as follows:

$$\underline{J}^k_{\text{RCM-II}} = \underline{w}\|x_k - v_{p_k}\|^2. \tag{10}$$

If x_k belongs to $\text{Bnd}(A_{p_k})$ and $\text{Bnd}(A_{q_k})$, we get the value of the objective function as follows:

$$\overline{J}^k_{\text{RCM-II}} = \sum_{i=p_k,q_k} \overline{w}\|x_k - v_i\|^2. \tag{11}$$

Comparing (10) and (11), we derive the optimal solution to N and U as follows:

$$\nu_{ki} = \begin{cases} 1, & (\underline{J}^k_{\text{RCM-II}} < \overline{J}^k_{\text{RCM-II}} \wedge i = p_k) \\ 0. & (\text{otherwise}) \end{cases}$$

$$u_{ki} = \begin{cases} 1, & (\underline{J}^k_{\text{RCM-II}} \geq \overline{J}^k_{\text{RCM-II}} \wedge (i = p_k \vee i = q_k)) \\ 0. & (\text{otherwise}) \end{cases}$$

We describe the RCM-II algorithm as follows:

Algorithm 4 RCM-II

RCM-II.1 Set initial cluster centers.
RCM-II.2 Update the belongingness of x_k to $\underline{A}_i$ and $\mathrm{Bnd}(A_i)$ by using the procedure in Sect. 4.1.2
RCM-II.3 Update v_i by (9).
RCM-II.4 If the convergence criterion is satisfied, finish. Otherwise back to **RCM-II.2.**

4.2 *Type-II Rough Fuzzy c-Means*

4.2.1 Objective Function

Here, we propose another method: type-II rough fuzzy c-means (RFCM-II) to solve Lingras's problems. RFCM-II is an extended method using the concept of fuzzy theory. In RFCM-II, degrees of belongingness to $\mathrm{Bnd}(A_i)$ are only fuzzified. The objective function of RFCM-II is defined as follows:

$$J_{\text{RFCM-II}}(N, U, V) = \sum_{k=1}^{n} \sum_{i=1}^{c} (\underline{w}\nu_{ki} + \overline{w}u_{ki}^{m}) \|x_k - v_i\|^2. \tag{12}$$

$$\underline{w} + \overline{w} = 1,\ \underline{w} > 0,\ \overline{w} > 0.\ (1 \leq k \leq n,\ 1 \leq i \leq c)$$

Constraints are as follows:

$$\nu_{ki}, u_{ki} \geq 0, \quad \forall k, i$$

$$\sum_{i=1}^{c} (\nu_{ki} + u_{ki}) = 1, \quad \forall k$$

$J_{\text{RFCM-II}}$ is minimized under these constraints.

4.2.2 Derivation of Optimal Solutions and Algorithm

First, we derive an optimal solution of the cluster center. Similar to Sect. 4.1.2, we get

$$v_i = \frac{\sum_{k=1}^{n} (\underline{w}\nu_{ki} + \overline{w}u_{ki}^{m}) x_k}{\sum_{k=1}^{n} (\underline{w}\nu_{ki} + \overline{w}u_{ki}^{m})}. \tag{13}$$

Next, we derive optimal solutions of lower approximation and boundary. Similar to Sect. 4.1.2, we must consider the following two cases to derive optimal solutions to N and U:

1. x_k belongs to $\underline{A}_{p_k}$.
2. x_k belongs to $\mathrm{Bnd}(A_i)$ $\forall i$.

If x_k belongs to $\underline{A}_{p_k}$, we get the value of the objective function as follows:

$$\underline{J}^k_{\text{RFCM-II}} = \underline{w}\nu_{ki}\|x_k - v_{p_k}\|^2. \tag{14}$$

If x_k belongs to $\mathrm{Bnd}(A_i)$, we get the value of the objective function as follows:

$$\overline{J}^k_{\text{RFCM-II}} = \sum_{i=1}^{c} \overline{w}u^m_{ki}\|x_k - v_i\|^2. \tag{15}$$

Comparing (14) and (15), we derive the optimal solution to N and U as follows:

$$\nu_{ki} = \begin{cases} 1, & (\underline{J}^k_{\text{RFCM-II}} < \overline{J}^k_{\text{RFCM-II}} \wedge i = p_k) \\ 0. & (\text{otherwise}) \end{cases}$$

$$u_{ki} = \begin{cases} \dfrac{\left(\frac{1}{\|x_k - v_i\|^2}\right)^{\frac{1}{m-1}}}{\sum_{j=1}^{c}\left(\frac{1}{\|x_k - v_j\|^2}\right)^{\frac{1}{m-1}}}, & (\underline{J}^k_{\text{RFCM-II}} \geq \overline{J}^k_{\text{RFCM-II}}) \\ 0. & (\text{otherwise}) \end{cases}$$

We calculate the optimal solution to u_{ki} by using the Lagrange multiplier method. The Lagrange function of RFCM-II is defined as follows:

$$L_{\text{RFCM-II}} = \sum_{k=1}^{n}\sum_{i=1}^{c} \left(\underline{w}\nu_{ki} + \overline{w}u^m_{ki}\right)\|x_k - v_i\|^2 - \sum_{k=1}^{n}\lambda_k\left(\sum_{i=1}^{c} u_{ki} - 1\right).$$

From the above discussion, we describe the RFCM-II algorithm.

Algorithm 5 RFCM-II

RFCM-II.1 Set initial cluster centers.
RFCM-II.2 Update the belongingness of x_k to $\underline{A}_i$ and $\mathrm{Bnd}(A_i)$ by using the procedure in Sect. 4.2.2.
RFCM-II.3 Update v_i by (13).
RFCM-II.4 If the convergence criterion is satisfied, finish. Otherwise back to **RCM-II.2.**

5 Rough Non Metric Model

5.1 Rough Hard Non Metric Model

5.1.1 Objective Function

To construct a new relational clustering algorithm based on rough sets, rough non metric model (RNM) or rough hard non metric model (RHNM), we define the following objective function based on Non Metric Model by Roubens [15]:

$$J_{\text{RNM}}(N, U) = \underline{w} \sum_{i=1}^{c} \sum_{k=1}^{n} \sum_{t=1}^{n} \nu_{ki} \nu_{ti} D_{kt} + \overline{w} \sum_{i=1}^{c} \sum_{k=1}^{n} \sum_{t=1}^{n} u_{ki} u_{ti} D_{kt}. \tag{16}$$

Here $\underline{w} + \overline{w} = 1$ and $\underline{w} \in (0, 1)$. If $\underline{w}$ is close to 0, almost all objects belong to the lower approximation. If $\underline{w}$ is close to 1, however, almost all objects belong to the upper approximation. $\underline{w}$ (or $\overline{w}$) therefore controls belongingness and it plays a very important role in our proposed methods. D_{kt} means a dissimilarity between x_k and x_t. One of the examples is a Euclidean norm:

$$D_{kt} = \|x_k - x_t\|^2.$$

We consider the following conditions for ν_{ki} and u_{ki}:

$$\nu_{ki} \in \{0, 1\}, \qquad u_{ki} \in \{0, 1\}.$$

From (C1)–(C3) in Sect. 2.2, we derive the following constraints:

$$\sum_{i=1}^{c} \nu_{ki} \in \{0, 1\}, \qquad \sum_{i=1}^{c} u_{ki} \neq 1,$$

$$\sum_{i=1}^{c} \nu_{ki} = 1 \Longleftrightarrow \sum_{i=1}^{c} u_{ki} = 0.$$

From the above constraints, we derive the following relation for any k:

$$\sum_{i=1}^{c} \nu_{ki} = 0 \Longleftrightarrow \sum_{i=1}^{c} u_{ki} \geq 2.$$

It is obvious that these relations are equivalent to (C1)–(C3) in Sect. 2.2

5.1.2 Derivation of Optimal Solutions and Algorithm

Optimal solutions to v_{ki} and u_{ki} are obtained by comparing the following two cases for each x_k:

1. x_k belongs to the lower approximation $\underline{A}_{p_k}$.
2. x_k belongs to the boundaries of two clusters $\overline{A}_{q_k^1}$ and $\overline{A}_{q_k^2}$.

We describe the details of each case as follows.

In the first case, let us assume that x_k belongs to the lower approximation $\underline{A}_{p_k}$. p_k is derived as follows:

The objective function J is rewritten as follows:

$$J_{\mathrm{RNM}}(N, U) = \underline{w} \sum_{i=1}^{c} \underline{J}_i + \overline{w} \sum_{i=1}^{c} \sum_{l=1}^{n} \sum_{t=1}^{n} u_{li} u_{ti} D_{lt}.$$

Here

$$\underline{J}_i = \left(2 v_{ki} \sum_{t=1}^{n} v_{ti} D_{kt} + \sum_{l=1, l \neq k}^{n} \sum_{t=1, t \neq k}^{n} v_{li} v_{ti} D_{lt} \right).$$

Note that $D_{kk} = 0$ and $D_{kt} = D_{tk}$, therefore

$$p_k = \arg\min_i \sum_{t=1}^{n} v_{ti} D_{kt}. \tag{17}$$

This means the following relations:

$$v_{ki} = \begin{cases} 1, & (i = p_k) \\ 0, & (\text{otherwise}) \end{cases}$$
$$u_{ki} = 0. \quad (\forall i)$$

In this case, the value of the objective function is calculated as follows:

$$\begin{aligned} J_{\mathrm{RNM}}(N, U) &= \underline{w} \left(2 \sum_{t=1}^{n} v_{tp_k} D_{kt} + \sum_{i=1}^{c} \sum_{l=1, l \neq k}^{n} \sum_{t=1, t \neq k}^{n} v_{li} v_{ti} D_{lt} \right) + \overline{w} \sum_{i=1}^{c} \sum_{l=1}^{n} \sum_{t=1}^{n} u_{li} u_{ti} D_{lt} \\ &= 2 J_k^{v} + J_c. \end{aligned}$$

Here

$$J_k^v = \sum_{t=1}^{n}\left(\underline{w}v_{tp_k} + \sum_{i=1}^{c}\overline{w}u_{ki}u_{ti}\right)D_{kt} = \underline{w}\sum_{t=1}^{n} v_{tp_k}D_{kt}, \tag{18}$$

$$J_c = \underline{w}\sum_{i=1}^{c}\sum_{l=1,l\neq k}^{n}\sum_{t=1,t\neq k}^{n} v_{li}v_{ti}D_{lt} + \overline{w}\sum_{i=1}^{c}\sum_{l=1,l\neq k}^{n}\sum_{t=1,t\neq k}^{n} u_{li}u_{ti}D_{lt}.$$

In the second case, let us assume that x_k belongs to the boundaries of two clusters $\overline{A}_{q_k^1}$ and $\overline{A}_{q_k^2}$. q_k^1 and q_k^2 are derived as follows:
The objective function J is rewritten as follows:

$$J_{\mathrm{RNM}}(N,U) = \overline{w}\sum_{i=1}^{c}\overline{J}_i + \underline{w}\sum_{i=1}^{c}\sum_{l=1}^{n}\sum_{t=1}^{n} v_{ki}v_{ti}D_{lt}.$$

Here

$$\overline{J}_i = \left(2u_{ki}\sum_{t=1}^{n} u_{ti}D_{kt} + \sum_{l=1,l\neq k}^{n}\sum_{t=1,t\neq k}^{n} u_{li}u_{ti}D_{lt}\right).$$

Therefore

$$q_k^1 = \arg\min_i \sum_{t=1}^{n} u_{ti}D_{kt}, \tag{19}$$

$$q_k^2 = \arg\min_{i,i\neq q_k^1} \sum_{t=1}^{n} u_{ti}D_{kt}. \tag{20}$$

This means the following relations:

$$v_{ki} = 0, \quad (\forall i)$$

$$u_{ki} = \begin{cases} 1, & (i = q_k^1 \vee i = q_k^2) \\ 0. & (\text{otherwise}) \end{cases}$$

In this case, the value of the objective function is calculated as follows:

$$\begin{aligned} J_{\mathrm{RNM}}(N,U) = \underline{w}\sum_{i=1}^{c}\sum_{l=1}^{n}\sum_{t=1}^{n} v_{li}v_{ti}D_{lt} + \overline{w}\Bigg(2\sum_{t=1}^{n}(u_{tq_k^1} + u_{tq_k^2})D_{kt} \\ + \sum_{i=1}^{c}\sum_{l=1,l\neq k}^{n}\sum_{t=1,t\neq k}^{n} u_{li}u_{ti}D_{lt}\Bigg) \\ = 2J_k^u + J_c. \end{aligned}$$

Here

$$J_k^u = \sum_{t=1}^{n}\left(\sum_{i=1}^{c} \underline{w} \nu_{ki} \nu_{ti} + \overline{w}(u_{tq_k^1} + u_{tq_k^2})\right) D_{kt}$$
$$= \overline{w} \sum_{t=1}^{n} (u_{tq_k^1} + u_{tq_k^2}) D_{kt}. \tag{21}$$

In comparison with J_k^ν and J_k^u, we determine ν_{ki} and u_{ki} as follows:

$$\nu_{ki} = \begin{cases} 1, & (J_k^\nu < J_k^u \wedge i = p_k) \\ 0, & \text{(otherwise)} \end{cases}$$
$$u_{ki} = \begin{cases} 1, & \left(J_k^\nu \geq J_k^u \wedge (i = q_k^1 \vee i = q_k^2)\right) \\ 0. & \text{(otherwise)} \end{cases}$$

From the above discussion, we show the RNM algorithm as Algorithm 6. The proposed algorithm is constructed based on iterative optimization.

Algorithm 6 RNM

RNM.1 The iteration number $L = 0$. Give D_{kt} and set initial values of $\nu_{ki}^{(0)}$ and $u_{ki}^{(0)}$.
RNM.2 Update $\nu_{ki}^{(L+1)}$ and $u_{ki}^{(L+1)}$ as follows:

$$\nu_{ki}^{(L+1)} = \begin{cases} 1, & (J_k^\nu < J_k^u \wedge i = p_k) \\ 0, & \text{(otherwise)} \end{cases}$$
$$u_{ki}^{(L+1)} = \begin{cases} 1, & \left(J_k^\nu \geq J_k^u \wedge (i = q_k^1 \vee i = q_k^2)\right) \\ 0. & \text{(otherwise)} \end{cases}$$

$p_k, q_k^1, q_k^2, J_k^\nu$, and J_k^u are calculated by (17), (19), (20), (18), and (21), respectively, with $\nu_{ti}^{(L)}$ and $u_{ti}^{(L)}$.
RNM.3 If the stop criterion is satisfied, finish. Otherwise $L := L + 1$ and back to **RNM.2**.

5.2 *Rough Fuzzy Non Metric Model*

In the previous section, we proposed the RNM algorithm. In the algorithm, an object x_k belongs to just two boundaries if x_k does not belong to any lower approximation, since $u_{ki} \in \{0, 1\}$ and the objective function (16) is linear for u_{ki}. In this section, we therefore propose the RFNM algorithm to make x_k belong to more than one boundary if x_k does not belong to any lower approximation.

We have two ways to fuzzify RNM. The first is to introduce the fuzzification parameter m, and the second is to introduce the entropy term. These ways are known to be very useful. We call the method using the first way rough fuzzy non metric model (RFNM) and that using the second way entropy-regularized rough fuzzy non metric model (ERFNM), as mentioned above. In this paper, we describe RFNM.

5.2.1 Objective Function

We consider the following objective function of RFNM:

$$J_{\mathrm{RFNM}}(N, U) = \underline{\omega} \sum_{i=1}^{c} \sum_{k=1}^{n} \sum_{t=1}^{n} \nu_{ki} \nu_{ti} D_{kt} + \overline{\omega} \sum_{i=1}^{c} \sum_{k=1}^{n} \sum_{t=1}^{n} u_{ki}^{m} u_{ti}^{m} D_{kt}. \tag{22}$$

Here $\underline{w} + \overline{w} = 1$. D_{kt} means a dissimilarity between x_k and x_t. The last entropy term means fuzzification of u_{ki} and makes the objective function nonlinear for u_{ki}. Hence, the value of the optimal solution on u_{ki} that minimizes the objective function (22) is in [0, 1).

We assume the following conditions for ν_{ki} and u_{ki}:

$$\nu_{ki} \in \{0, 1\}, \qquad u_{ki} \in [0, 1).$$

From (C1)–(C3) in Sect. 2.2, we derive the following constraints:

$$\sum_{i=1}^{c} \nu_{ki} \in \{0, 1\}, \tag{23}$$

$$\sum_{i=1}^{c} u_{ki} \in \{0, 1\}, \tag{24}$$

$$\sum_{i=1}^{c} \nu_{ki} = 1 \Longleftrightarrow \sum_{i=1}^{c} u_{ki} = 0. \tag{25}$$

From the above constraints, we derive the following relation for any k:

$$\sum_{i=1}^{c} \nu_{ki} = 0 \Longleftrightarrow \sum_{i=1}^{c} u_{ki} = 1. \tag{26}$$

It is obvious that these relations are equivalent to (C1)–(C3) in Sect. 2.2.

5.2.2 Derivation of Optimal Solutions and Algorithm

Same as RNM, optimal solutions to ν_{ki} and u_{ki} are obtained by comparing two cases for each x_k:

1. x_k belongs to the lower approximation $\underline{A}_{p_k}$.
2. x_k belongs to the boundaries of two clusters $\overline{A}_{q_k^1}$ and $\overline{A}_{q_k^2}$.

In the first case, let us assume that x_k belongs to the lower approximation $\underline{A}_{p_k}$. p_k is derived as follows:

The objective function J is rewritten as follows:

$$J_{\text{RFNM}}(N, U) = \underline{\omega} \sum_{i=1}^{c} \underline{J}_i + \overline{\omega} \sum_{i=1}^{c} \sum_{l=1}^{n} \sum_{t=1}^{n} u_{li}^m u_{ti}^m D_{lt}.$$

Here

$$\underline{J}_i = 2\nu_{ki} \sum_{t=1}^{n} \nu_{ti} D_{kt} + \sum_{l=1, l \neq k}^{n} \sum_{t=1, t \neq k}^{n} \nu_{li} \nu_{ti} D_{lt}.$$

Note that $D_{kk} = 0$ and $D_{kt} = D_{tk}$. Therefore

$$p_k = \arg\min_i \sum_{t=1}^{n} \nu_{ti} D_{kt}. \tag{27}$$

This means the following relations:

$$\nu_{ki} = \begin{cases} 1 \ (i = p_k) \\ 0 \ (\text{otherwise}) \end{cases}$$
$$u_{ki} = 0$$

In this case, the value of the objective function is calculated as follows:

$$\begin{aligned} J_{\text{RFNM}}(N, U) &= \underline{\omega} \left(2 \sum_{t=1}^{n} \nu_{tp_k} D_{kt} + \sum_{i=1}^{c} \sum_{l=1, l \neq k}^{n} \sum_{t=1, t \neq k}^{n} \nu_{li} \nu_{ti} D_{lt} \right) + \overline{\omega} \sum_{i=1}^{c} \sum_{l=1}^{n} \sum_{t=1}^{n} u_{li}^m u_{ti}^m D_{lt} \\ &= 2J_k^{\nu} + J_c. \end{aligned}$$

Here

$$J_k^\nu = \sum_{t=1}^{n}\left(\underline{\omega}\nu_{tp_k} + \sum_{i=1}^{c}\overline{\omega}u_{ki}^m u_{ti}^m\right)D_{kt} = \underline{\omega}\sum_{t=1}^{n}\nu_{tp_k}D_{kt}, \tag{28}$$

$$J_c = \underline{\omega}\sum_{i=1}^{c}\sum_{l=1,l\neq k}^{n}\sum_{t=1,t\neq k}^{n}\nu_{li}\nu_{ti}D_{lt} + \overline{\omega}\sum_{i=1}^{c}\sum_{l=1,l\neq k}^{n}\sum_{t=1,t\neq k}^{n}u_{li}^m u_{ti}^m D_{lt}.$$

In the second case, let us assume that x_k belongs to the boundaries of more than one cluster. The objective function J is convex for u_{ki}, hence we derive an optimal solution to u_{ki} using a Lagrange multiplier.

Here we introduce the following Lagrange function with the constraint (26):

$$L = J + \sum_{k=1}^{n}\eta_k\sum_{i=1}^{c}(u_{ki} - 1).$$

We partially differentiate L by u_{ki} and get the following equation:

$$\begin{aligned}\frac{\partial L}{\partial u_{ki}} &= 2m\overline{\omega}u_{ki}^{m-1}\left(\sum_{t=1,t\neq k} u_{ti}^m D_{kt} + u_{ki}^m D_{kk}\right) + \eta_k \\ &= 2m\overline{\omega}u_{ki}^{m-1}\sum_{t=1,t\neq k} u_{ti}^m D_{kt} + \eta_k\end{aligned}$$

$\frac{\partial L}{\partial u_{ki}} = 0$, we obtain the following relation:

$$u_{ki} = \frac{(-\eta_k)^{\frac{1}{m-1}}}{D_{ki}^{\frac{1}{m-1}}}, \tag{29}$$

where

$$D_{ki} = 2m\overline{\omega}\sum_{t=1,t\neq k}^{n} u_{ti}^m D_{kt} = 2m\overline{\omega}\sum_{t=1}^{n} u_{ti}^m D_{kt}. \tag{30}$$

From the constraint (26) and the above Eq. (29), we get the following equation:

$$\sum_{i=1}^{c}u_{ki} = \sum_{i=1}^{c}\frac{(-\eta_k)^{\frac{1}{m-1}}}{D_{ki}^{\frac{1}{m-1}}} = 1,$$

$$(-\eta_k)^{\frac{1}{m-1}} = 1/\sum_{j=1}^{c}\frac{1}{D_{kj}^{\frac{1}{m-1}}}.$$

We then obtain the following optimal solution:

$$u_{ki} = \frac{\left(\frac{1}{D_{ki}}\right)^{\frac{1}{m-1}}}{\sum_{j=1}^{c}\left(\frac{1}{D_{kj}}\right)^{\frac{1}{m-1}}}$$

This means the following relations:

$$v_{ki} = 0, (\forall i)$$

$$u_{ki} = \frac{\left(\frac{1}{D_{ki}}\right)^{\frac{1}{m-1}}}{\sum_{j=1}^{c}\left(\frac{1}{D_{kj}}\right)^{\frac{1}{m-1}}}.(\forall i)$$

In this case, the value of the objective function is calculated as follows:

$$\begin{aligned} J_{\mathrm{RFNM}}(N, U) &= \underline{\omega}\sum_{i=1}^{c}\sum_{l=1}^{n}\sum_{t=1}^{n} v_{li}v_{ti}D_{lt} + \overline{\omega}\sum_{i=1}^{c}\left(2mu_{ki}^{m-1}\sum_{t=1}^{n}u_{ti}^{m}D_{kt}\right. \\ &\quad \left. + \sum_{l=1,l\neq k}^{n}\sum_{t=1,t\neq k}^{n} u_{li}^{m}u_{ti}^{m}D_{kt}\right) \\ &= 2J_k^u + J_c. \end{aligned}$$

Here

$$\begin{aligned} J_k^u &= \sum_{t=1}^{n}\left(\sum_{i=1}^{c}\underline{\omega}v_{ki}v_{ti} + \sum_{i=1}^{c}\overline{\omega}mu_{ki}^{m-1}u_{ti}^{m}\right)D_{kt} \\ &= \sum_{i=1}^{c}mu_{ki}^{m-1}\sum_{t=1}^{n}\overline{\omega}u_{ti}^{m}D_{kt}. \end{aligned} \tag{31}$$

In comparison with J_k^v andJ_k^u, we determine v_{ki} and u_{ki} as follows:

$$v_{ki} = \begin{cases} 1 \ (J_k^v < J_k^u \wedge i = p_k) \\ 0 \ (\text{otherwise}) \end{cases}$$

$$u_{ki} = \begin{cases} \dfrac{\left(\frac{1}{D_{ki}}\right)^{\frac{1}{m-1}}}{\sum_{j=1}^{c}\left(\frac{1}{D_{kj}}\right)^{\frac{1}{m-1}}} & (J_k^v \geq J_k^u) \\ 0 & (\text{otherwise}) \end{cases}$$

From the above discussion, we show the RFNM algorithm as Algorithm 7. The proposed algorithm is also constructed based on iterative optimization.

Algorithm 7 RFNM

RFNM.1 The iteration number $L = 0$. Give D_{kt} and set initial values of$v_{ki}^{(0)}$ and $u_{ki}^{(0)}$.
RFNM.2 Update $v_{ki}^{(L+1)}$ and $u_{ki}^{(L+1)}$ as follows:

$$v_{ki}^{(L+1)} = \begin{cases} 1, & (J_k^v < J_k^u \wedge i = p_k) \\ 0, & \text{(otherwise)} \end{cases}$$

$$u_{ki}^{(L+1)} = \begin{cases} \dfrac{\left(\frac{1}{D_{ki}}\right)^{\frac{1}{m-1}}}{\sum_{j=1}^{c}\left(\frac{1}{D_{kj}}\right)^{\frac{1}{m-1}}} & \left(J_k^v \geq J_k^u\right), \\ 0. & \text{(otherwise)} \end{cases}$$

p_k, D_{ki}, J_k^v, and J_k^u are calculated by (27), (30), (28), and (31).
RFNM.3 If the stop criterion is satisfied, finish. Otherwise $L := L + 1$ and go back to **RFNM2**.

6 Conclusion

This paper showed various types of objective functions of objective-based rough clustering and their algorithm.

As mentioned above, many non-hierarchical clustering algorithms are based on optimization of some objective function. The reason is that we could choose the "better" output among many outputs from different initial values by comparing the value of the objective function of the output with each other. Lingras's algorithm is almost only one algorithm with rough set representation inspired by KM, however it is not useful from the viewpoint that the algorithm is not based on any objective functions. Therefore, our proposed algorithms could be expected to be more useful in the field of rough clustering.

In objective-based clustering methods, the concept of classification function is very important. The classification function gives us belongingness of unknown datum to each cluster. It is impossible to derive the classification functions of our algorithms in this paper analytically, hence we can not show the functions explicitly. However, as we have seen, the value of the belongingness numerically. In future works, we will develop these discussion.

References

1. P. Lingras and G. Peters. Rough clustering. *Wiley Interdisciplinary Reviews: Data Mining and Knowledge Discovery, Vol. 1, Issue 1, pp. 64–72*, pages 1207–1216, 2011.
2. P. Lingras and C. West. Interval set clustering of web users with rough k-means. *Journal of Intelligent Information Systems, Vol. 23, No. 1, pp. 5–16*, 2004.

3. R. O. Duda and P. E. Hart. *Pattern Classification and Scene Analysis*. John Wiley & Sons, New York, second edition, 1973.
4. J. B. MacQueen. Some methods for classification and analysis of multivariate observations. *Proceedings of 5-th Berkeley Symposium on Mathematical Statistics and Probability, Berkeley, University of California Press, Vol. 1, pp. 281–297*, 1967.
5. J. C. Dunn. A fuzzy relative of the isodata process and its use in detecting compact well-separated clusters. *Journal of Cybernetics, Vol. 3, pp. 32–57*, 1973.
6. J. C. Bezdek. Pattern recognition with fuzzy objective function algorithms. *Plenum Press, New York*, 1981.
7. Z. Pawlak. Rough sets. *International Journal of Computer and Information Sciences, Vol. 11, No. 5, pp. 341–356*, 1982.
8. M. Inuiguchi. Generalizations of rough sets: From crisp to fuzzy cases. *Proceedings of Rough Sets and Current Trends in Computing, pp. 26–37*, 2004.
9. Z. Pawlak. Rough classification. *International Journal of Man-Machine Studies, Vol. 20, pp. 469–483*, 1984.
10. S. Hirano and S. Tsumoto. An indiscernibility-based clustering method with iterative refinement of equivalence relations. *Journal of Advanced Computational Intelligence and Intelligent Informatics, Vol. 7, No. 2, pp. 169–177*, 2003.
11. S. Mitra, H. Banka, and W. Pedrycz. Rough-fuzzy collaborative clustering. *IEEE Transactions on Systems Man, and Cybernetics, Part B, Cybernetics, Vol. 36, No. 5, pp. 795–805*, 2006.
12. P. Maji and S. K. Pal. Rough set based generalized fuzzy c-means algorithm and quantitative indices. *IEEE Transactions on System, Man and Cybernetics, Part B, Cybernetics, Vol. 37, No. 6, pp. 1529–1540*, 2007.
13. G. Peters. Rough clustering and regression analysis. *Proceedings RSKT'07, LNAI 2007, Vol. 4481, pp. 292–299*, 2007.
14. S. Mitra and B. Barman. Rough-fuzzy clustering: An application to medical imagery. *Rough Set and Knowledge Technology, LNCS 2008, Vol. 5009, pp. 300–307*, 2008.
15. M. Roubens. Pattern classification problems and fuzzy sets, *Fuzzy Sets and Systems, Vol. 1, pp. 239–253*, 1978.

On Some Clustering Algorithms Based on Tolerance

Yukihiro Hamasuna and Yasunori Endo

Abstract A large number of clustering algorithms have been proposed to handle target data and deal with various real-world problems such as uncertain data mining, semi-supervised learning and so on. We focus above two topics and introduce two concepts to construct significant clustering algorithms. We propose tolerance and penalty-vector concepts for handling uncertain data. We also propose clusterwise tolerance concept for semi-supervised learning. These concepts are quite similar approach in the viewpoint of handling objects to be flexible to each clustering topics. We construct two clustering algorithms FCMT and FCMQ for handling uncertain data. We also construct two clustering algorithms FCMCT and SSFCMCT for semi-supervised learning. We consider that those concepts have a potential to resolve conventional and brand new clustering topics in various ways.

1 Introduction

Clustering is one of the common data analysis methods that divides a set of objects into groups called clusters. Objects classified in the same cluster are considered similar, while those in different clusters are considered dissimilar. Hard c-means (HCM) which is also known as k-means [13] and fuzzy c-means (FCM) [2, 15] are the most well-known clustering methods. A large number of clustering algorithms have been proposed to handle target data and deal with various real-world problems. Uncertain data mining [1, 12] and semi-supervised learning [3] which are discussed in this article are typical clustering topics. These two topics are seems to be quite

Y. Hamasuna (✉)
School of Science and Engineering, Kindai University,
3-4-1 Kowakae, Higashi-Osaka 577-8502, Japan
e-mail: yhama@info.kindai.ac.jp

Y. Endo
Faculty of Engineering, Information and Systems, University of Tsukuba,
1-1-1 Tennodai, Tsukuba, Ibaraki 305-8573, Japan
e-mail: endo@risk.tsukuba.ac.jp

V. Torra et al. (eds.), *Fuzzy Sets, Rough Sets, Multisets and Clustering*,
Studies in Computational Intelligence 671, DOI 10.1007/978-3-319-47557-8_6

different from the viewpoint of their purpose. However, they are constructed in a similar way or framework, that is, representation of data and clustering algorithms based on it. For uncertain data mining, significant representation of data to handle inherent uncertainty such as incompleteness or errors and clustering algorithms are required [1, 12]. Also, a significant framework to handle a prior knowledge of data or experts' knowledge is required in semi-supervised learning [3]. To discuss and consider these two different topics in the same framework, we have proposed the concept of tolerance [5] and several clustering algorithms based on it [6, 8, 11].

Objects handled in data analysis methods contain inherent uncertainty. In recent years, the objects measured by sensor network and social medias are stored in massive and complex databases. These objects also contain inherent uncertainty, because the accuracy of these objects is depended on communication environment. Handling uncertain data is still one of the important topics for many years. Several methods for handling uncertain data have been proposed and applied to real-world problems to extract useful value [1, 12]. We have proposed two new concepts of tolerance [5] and penalty vector [8] to handle uncertain data in clustering. These concepts are quite simple and enable us to handle uncertain data within the optimization framework. We have proposed the hard c-means based methods [9], fuzzy c-means based ones [5, 6, 8], and c-regression models based one [7]. We introduce fuzzy c-means based methods as one of the clustering algorithms for uncertain data. The one is fuzzy c-means for data with tolerance [5, 6, 17] and the other is fuzzy c-means using quadratic penalty-vector regularization [8].

In addition to above, semi-supervised learning has also been studied in many research fields and applied to various real-world problems [3]. In the field of clustering, must-link and cannot-link which are referred to as pairwise constraints are frequently used in order to improve clustering performances [18]. The pairwise constraints are also introduced into fuzzy clustering model [11, 16] and hierarchical clustering [4, 14] by several form. The pairwise constraints are divided roughly into two groups. One is hard constraints based methods, and the other is soft ones. In the case of hard constraints, pairwise constraints are always satisfied, while they are not always satisfied in the case of soft constraints. In the semi-supervised clustering, pairwise constraints are used as prior knowledge about which data should be in the same or different cluster. It is difficult to introduce pairwise constraints in the Euclidean-space because of the squared Euclidean-norm that is used as a dissimilarity. We introduce the concept of clusterwise tolerance as a extension of the concept of tolerance and clustering algorithms for data with clusterwise tolerance [10] After that, we propose semi-supervised fuzzy c-means clustering for data with clusterwise tolerance to handle pairwise constraints in the same optimization framework [11].

This paper is organized as follows: In Sect. 2, we introduce symbols and fuzzy c-means. In Sect. 3, we propose the concept of tolerance, penalty vector regularization, and clustering algorithms for uncertain data. In Sect. 4, we construct clusterwise tolerance based clustering algorithms and semi-supervised clustering method based on clusterwise tolerance. In Sect. 5, we conclude this paper.

2 Fuzzy c-Means

A set of objects to be clustered is given and denoted by $X = \{x_1, \ldots, x_n\}$ in which x_k $(k = 1, \ldots, n)$ is an object. In most cases, each object x_k is a vector in p-dimensional Euclidean space $\Re^p$, that is, an object $x_k \in \Re^p$. A cluster is denoted by G_i and a collection of clusters is given by $\mathcal{G} = \{G_1, \ldots, G_c\}$. A cluster center of G_i is denoted by $v_i \in \Re^p$ and a set of v_i is given by $V = \{v_1, \ldots, v_c\}$. A membership degree of x_k belonging to G_i and a partition matrix is denoted as u_{ki} and $U = (u_{ki})_{1 \le k \le n,\ 1 \le i \le c}$.

The method of fuzzy c-means clustering (FCM) [2, 15] is an extension of k-means [13] and one of the well known clustering algorithms. These clustering algorithms divide a set of objects into clusters by optimizing an objective function under the constraint on membership degree u_{ki}.

The following two objective functions J_s and J_e are typical examples of FCM.

$$J_s(U, V) = \sum_{k=1}^{n} \sum_{i=1}^{c} (u_{ki})^m d_{ki},$$

$$J_e(U, V) = \sum_{k=1}^{n} \sum_{i=1}^{c} \{u_{ki} d_{ki} + \lambda u_{ki} \log u_{ki}\}.$$

$m > 1$ and $\lambda > 0$ are fuzzification parameters. d_{ki} is the dissimilarity between an object x_k and cluster center v_i. The squared L_2-norm $d_{ki} = \|x_k - v_i\|^2$ is a typical dissimilarity in several clustering algorithms.

J_s is an objective function for the standard fuzzy c-means [2] and J_e is an objective function for the entropy based fuzzy c-means clustering (eFCM) [15].

The constraint on membership degree u_{ki} for FCM is as follows:

$$\mathcal{U}_f = \left\{ (u_{ki}) : u_{ki} \in [0, 1],\ \sum_{i=1}^{c} u_{ki} = 1,\ {}^{\forall} k \right\}. \tag{1}$$

The optimal solutions for u_{ki} and v_i are derived from objective function J_s or J_e by considering the constraint (1). The algorithms of sFCM and eFCM are summarized in Algorithms 1 and 2, respectively.

A number of repetitions, convergence of each variables, or convergence of objective function are used as convergence criterion in **sFCM 4** and **eFCM 4**.

Algorithm 1: sFCM

sFCM 1 Set initial cluster centers and parameter m.
sFCM 2 Calculate $u_{ki} \in U$ by using following equation:

$$u_{ki} = \frac{\left(\frac{1}{d_{ki}}\right)^{\frac{1}{m-1}}}{\sum_{l=1}^{c}\left(\frac{1}{d_{kl}}\right)^{\frac{1}{m-1}}}$$

sFCM 3 Calculate $v_i \in V$ by using following equation:

$$v_i = \frac{\sum_{k=1}^{n}(u_{ki})^m x_k}{\sum_{k=1}^{n}(u_{ki})^m}$$

sFCM 4 If convergence criterion is satisfied, stop. Otherwise go back to **sFCM 2**.

Algorithm 2: eFCM

eFCM 1 Set initial cluster centers and parameter λ.
eFCM 2 Calculate $u_{ki} \in U$ by using following equation:

$$u_{ki} = \frac{\exp\left(-\frac{d_{ki}}{\lambda}\right)}{\sum_{l=1}^{c}\exp\left(-\frac{d_{kl}}{\lambda}\right)}$$

eFCM 3 Calculate $v_i \in V$ by using following equation:

$$v_i = \frac{\sum_{k=1}^{n} u_{ki} x_k}{\sum_{k=1}^{n} u_{ki}}$$

eFCM 4 If convergence criterion is satisfied, stop. Otherwise go back to **eFCM 2**.

3 Clustering Algorithms for Uncertain Data

We introduce two clustering algorithms for uncertain data based on fuzzy c-means. The one is fuzzy c-means for data with tolerance [5, 6, 17] and the other is fuzzy c-means using quadratic penalty-vector regularization [8].

3.1 Fuzzy c-Means for Data with Tolerance

In most clustering algorithms, each object is regarded as one point and classified into clusters. The uncertain data should be often represented in other form such as interval or probability density function instead of a point in $\Re^p$. The concept of tolerance has been proposed to handle the range of uncertainty as the tolerance and define the tolerance vector inside the tolerance [5]. A tolerance $\kappa_k = (\kappa_{k1}, \ldots, \kappa_{kp})^T \in \Re^p$ is the admissible range of each object. A set of tolerance vector is denoted as $E = \{\varepsilon_1, \ldots, \varepsilon_n\}$ in which $\varepsilon_k \in \Re^p$ is a tolerance vector. A tolerance vector is the vector within the tolerance. In the conventional studies, a data is represented as x_k. On the other hand, data with tolerance is represented as $x_k + \varepsilon_k$ by using this concept.

A constraint on tolerance vector is as follows:

$$\left(\varepsilon_{kj}\right)^2 \leq \left(\kappa_{kj}\right)^2, \ (\kappa_{kj} \geq 0), {}^\forall k, \ j. \tag{2}$$

From these formulation, uncertain data is handled as data with tolerance in clustering procedures. Figure 1 is an illustrative example of ε_k and κ_k in $\Re^2$.

By describing uncertain data as data with tolerance, the dissimilarity squared Euclidean distance between the data with tolerance $x_k + \varepsilon_k$ and the cluster center v_i is denoted as follows:

$$d_{ki} = \|x_k + \varepsilon_k - v_i\|^2 = \sum_{j=1}^{p} \left(x_{kj} + \varepsilon_{kj} - v_{ij}\right)^2.$$

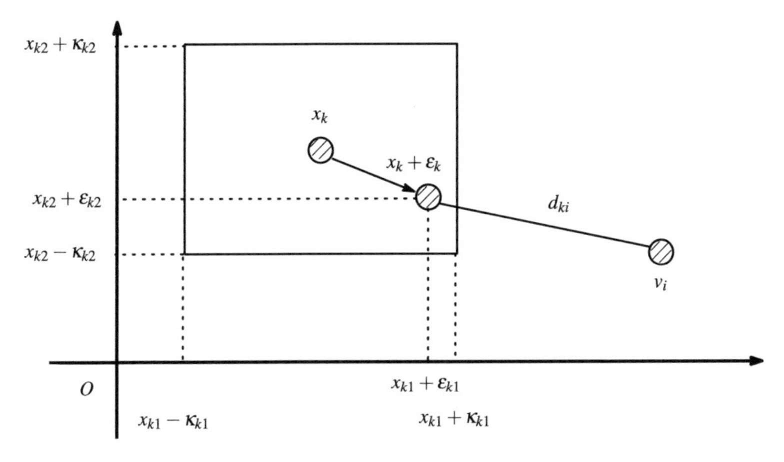

Fig. 1 An illustrative example of ε_k and κ_k in $\Re^2$

First, we introduce standard fuzzy c-means for data with tolerance (sFCMT) [5, 6]. The objective function of sFCMT $J_{st}(U, E, V)$ is as follows:

$$J_{st}(U, E, V) = \sum_{k=1}^{n} \sum_{i=1}^{c} (u_{ki})^m \|x_k + \varepsilon_k - v_i\|^2.$$

The constraints on u_{ki} and ε_{kj} are the same as (1) and (2). The optimal solutions for u_{ki}, v_i, and ε_k are derived from the convexity of objective function J_{st}, the constraint (1) and (2), and Karush–Kuhn–Tucker conditions. The algorithms of sFCMT is summarized in Algorithm 3.

Algorithm 3: sFCMT

---

sFCMT 1 Set initial cluster centers and parameters m and κ_k.
sFCMT 2 Calculate $u_{ki} \in U$ by using following equation:

$$u_{ki} = \frac{\left(\frac{1}{d_{ki}}\right)^{\frac{1}{m-1}}}{\sum_{l=1}^{c} \left(\frac{1}{d_{kl}}\right)^{\frac{1}{m-1}}}$$

sFCMT 3 Calculate $v_i \in V$ by using following equation:

$$v_i = \frac{\sum_{k=1}^{n} (u_{ki})^m (x_k + \varepsilon_k)}{\sum_{k=1}^{n} (u_{ki})^m}$$

sFCMT 4 Calculate $\varepsilon_k \in E$ by using following equation:

$$\varepsilon_{kj} = -\alpha_{kj} \sum_{i=1}^{c} (u_{ki})^m \left(x_{kj} - v_{ij}\right)$$

$$\alpha_{kj} = \min \left\{ \frac{\kappa_{kj}}{\left|\sum_{i=1}^{c} (u_{ki})^m \left(x_{kj} - v_{ij}\right)\right|}, \frac{1}{\sum_{i=1}^{c} (u_{ki})^m} \right\}$$

sFCMT 5 If convergence criterion is satisfied, stop. Otherwise go back to **sFCMT 2**.

FCMT can handle uncertain data analytically in the optimization problem for clustering by handling uncertain data as data with tolerance. Entropy based FCMT is also constructed by the same procedures. It is, however, omitted here, see reference [5, 6] for details.

3.2 *Fuzzy c-Means Using Quadratic Penalty-Vector Regularization*

Next, we introduce standard fuzzy c-means clustering using quadratic penalty-vector regularization (sFCMQ) [8].

κ_k is the maximum tolerance range which means the range of data uncertainty in sFCMT. sFCMT, however, can not handle the uncertain data in case with unknown range. Moreover, the range of uncertainty is not obtained in many cases. Algorithms fail to classify uncertain data in those cases. We introduce penalty vector regularization to overcome such problems [8]. A set of penalty vector is denoted as $\Delta = \{\delta_1, \ldots, \delta_n\}$ in which $\delta_k \in \Re^p$ is a penalty vector. The uncertainty of data is described as the constraint (2) in sFCMT, while the regularization term is considered in sFCMQ.

The objective function of sFCMQ $J_{sq}(U, \Delta, V)$ is as follows:

$$J_{sq}(U, \Delta, V) = \sum_{k=1}^{n} \sum_{i=1}^{c} (u_{ki})^m \|x_k + \delta_k - v_i\|^2 + \sum_{k=1}^{n} \delta_k^T W_k \delta_k.$$

The quadratic regularization term is described as follows:

$$\sum_{k=1}^{n} \delta_k^T W_k \delta_k = \sum_{l=1}^{p} \sum_{j=1}^{p} w_{klj} \delta_{kl} \delta_{kj},$$

where,

$$W_k = \begin{pmatrix} w_{k11} & \cdots & w_{k1p} \\ \vdots & \ddots & \vdots \\ w_{kp1} & \cdots & w_{kpp} \end{pmatrix}$$

is called a penalty matrix and assumed to be a symmetrical and a positive definite matrix. $w_{klj} (w_{klj} \geq 0)$ is a given penalty coefficient.

The constraints on u_{ki} is the same as (1). The optimal solutions for u_{ki}, v_i, and δ_k are derived from the convexity of objective function J_{sq} under the constraint (1). The algorithm of sFCMQ is summarized in Algorithm 4.

Algorithm 4: sFCMQ

sFCMQ 1 Set initial cluster centers and parameters m and W_k.
sFCMQ 2 Calculate $u_{ki} \in U$ by using following equation:

$$u_{ki} = \frac{\left(\frac{1}{d_{ki}}\right)^{\frac{1}{m-1}}}{\sum_{l=1}^{c}\left(\frac{1}{d_{kl}}\right)^{\frac{1}{m-1}}}$$

sFCMQ 3 Calculate $v_i \in V$ by using following equation:

$$v_i = \frac{\sum_{k=1}^{n}(u_{ki})^m(x_k + \delta_k)}{\sum_{k=1}^{n}(u_{ki})^m}$$

sFCMQ 4 Calculate $\delta_k \in \Delta$ by using following equation:

$$\delta_k = -\left(\sum_{i=1}^{c} u_{ki}^m\, I + W_k\right)\left(\sum_{i=1}^{c} u_{ki}^m (x_k - v_i)\right) \quad (I \text{ is a unit matrix.})$$

sFCMQ 5 If convergence criterion is satisfied, stop. Otherwise go back to **sFCMQ 2**.

Entropy based FCMQ is also constructed by the same procedures. It is, however, omitted here, see Ref. [8] for details.

4 Clustering Algorithms Using Clusterwise Tolerance

We introduce the concept of clusterwise tolerance and fuzzy c-means based on it [10]. After that, we propose semi-supervised fuzzy c-means based on the clusterwise tolerance and pairwise constraints.

4.1 *Fuzzy c-Means Using Clusterwise Tolerance*

A set of clusterwise tolerance vector is defined as $\Gamma = \{\gamma_{11}, \ldots, \gamma_{nc}\}$ in which γ_{ki} is a clusterwise tolerance vector. A clusterwise tolerance vector $\gamma_{ki} \in \Re^p$ is the p-dimensional vector with real components. A clusterwise tolerance κ_{ki} means the admissible range of γ_{ki}. In the conventional studies, an object is represented as x_k. On the other hand, data with clusterwise tolerance is represented as $x_k + \gamma_{ki}$ by using this concept.

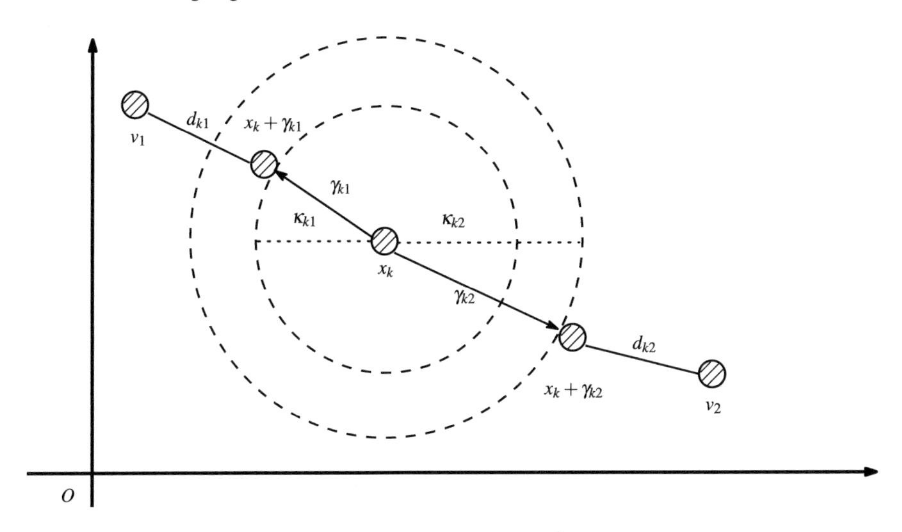

Fig. 2 An illustrative example of clusterwise tolerance in $\Re^2$

A constraint on clusterwise tolerance vector is as follows:

$$\|\gamma_{ki}\|^2 \leq (\kappa_{ki})^2 \quad (\kappa_{ki} \geq 0), {}^\forall k,\ i. \tag{3}$$

Figure 2 shows a clusterwise tolerance in $\Re^2$.

The dissimilarity between the data with clusterwise tolerance $x_k + \gamma_{ki}$ and the cluster center v_i is denoted as follows:

$$d_{ki} = \|x_k + \gamma_{ki} - v_i\|^2 = \sum_{j=1}^{p} \left(x_{kj} + \gamma_{kij} - v_{ij}\right)^2. \tag{4}$$

First, we introduce standard fuzzy c-means for data with clusterwise tolerance (sFCMCT) [10]. The objective function of sFCMCT $J_{sct}(U, \Gamma, V)$ is as follows:

$$J_{sct}(U, \Gamma, V) = \sum_{k=1}^{n} \sum_{i=1}^{c} (u_{ki})^m \|x_k + \gamma_{ki} - v_i\|^2.$$

The constraints on u_{ki} and γ_{ki} are the same as (1) and (3). The optimal solutions for u_{ki}, v_i, and γ_{ki} are derived from the convexity of objective function J_{sct}, the constraints, and Karush–Kuhn–Tucker conditions. The algorithm of sFCMCT is summarized in Algorithm 5.

Algorithm 5: sFCMCT

sFCMCT 1 Set initial cluster centers and parameters m and κ_{ki}.
sFCMCT 2 Calculate $u_{ki} \in U$ by using following equation:

$$u_{ki} = \frac{\left(\frac{1}{d_{ki}}\right)^{\frac{1}{m-1}}}{\sum_{l=1}^{c}\left(\frac{1}{d_{kl}}\right)^{\frac{1}{m-1}}}$$

sFCMCT 3 Calculate $v_i \in V$ by using following equation:

$$v_i = \frac{\sum_{k=1}^{n}(u_{ki})^m (x_k + \gamma_{ki})}{\sum_{k=1}^{n}(u_{ki})^m}$$

sFCMCT 4 Calculate $\gamma_{ki} \in \Gamma$ by using following equation:

$$\gamma_{ki} = -\alpha_{ki}(x_k - v_i),$$
$$\alpha_{ki} = \min\left\{\frac{\kappa_{ki}}{\|x_k - v_i\|}, 1\right\}$$

sFCMCT 5 If convergence criterion is satisfied, stop. Otherwise go back to **sFCMCT 2**.

FCMCT can handle a set of objects which consists different sizes or shapes of clusters by determining the suitable value of parameter κ_{ki} Entropy based FCMCT and possibilistic model is also constructed by the same procedures. See reference [10] for details.

4.2 Semi-supervised Fuzzy c-Means by Opposite Criteria

The pairwise constraints are considered as a prior knowledges about which data should be in the same or different cluster [3, 18]. We propose clusterwise tolerance based pairwise constraints from that sense: if $(x_k, x_l) \in ML$, γ_{ki} and γ_{li} are calculated to be in the same cluster, while $(x_q, x_r) \in CL$, γ_{qi} and γ_{ri} are calculated to be in the different cluster.

First, we define a set of must or cannot-linked objects. A set $ML(x_k)$ consists of must-linked objects which are linked with a data x_k, while $CL(x_k)$ consists of cannot-linked ones which are linked with a data x_k.

$$ML(x_k) = \{\xi \mid \xi \in X, (x_k, \xi) \in ML\},$$
$$CL(x_k) = \{\xi \mid \xi \in X, (x_k, \xi) \in CL\}.$$

The proposed clusterwise tolerance based pairwise constraints uses these sets to determine the upper bound of clusterwise tolerance vector $|K(x_k, v_i)|$ which is defined between an object and a cluster center.

A value of $K(x_k, v_i)$ is calculated as the sum of κ_{ki} which are in $ML(x_k)$ and $CL(x_k)$.

$$K(x_k, v_i) = \sum_{x_q \in ML(x_k)} \kappa_{qi} - \sum_{x_r \in CL(x_k)} \kappa_{ri}.$$

If $K(x_k, v_i) > 0$, γ_{ki} is calculated to be near the cluster center v_i, while $K(x_k, v_i) < 0$, γ_{ki} is calculated to be distant the cluster center v_i. In case that $K(x_k, v_i) = 0$, it is ineffective to calculate γ_{ki}. From above, we can handle a prior knowledge as the vector in L_2-space by using clusterwise tolerance concept.

Next, we propose semi-supervised standard fuzzy c-means for data with clusterwise tolerance by opposite criteria (SSsFCMCT). The objective function of SSsFCMCT is as follows:

$$J_{ssct}(U, \Gamma, V) = \sum_{k=1}^{n} \sum_{i=1}^{c} (u_{ki})^m \|x_k + \gamma_{ki} - v_i\|^2.$$

The dissimilarity is described as (4) as well as sFCMCT. The constraint on u_{ki} remains the same as (1). The constraint on γ_{ki} (3) is rewritten as follows:

$$\|\gamma_{ki}\|^2 \leq (K(x_k, v_i))^2, \forall k, i. \tag{5}$$

The optimal solutions for u_{ki}, v_i, and γ_{ki} are derived from the convexity of objective function J_{ssct}, the constraints, and Karush–Kuhn–Tucker conditions. We consider three cases for deriving the optimal solution for γ_{ki}. First, we consider the case of $K(x_k, v_i) = 0$. In this case, we can get trivial optimal solution $\gamma_{ki} = 0$, because the constraint (5) is not considered. Second, we consider the case of $K(x_k, v_i) > 0$. In this case, we can get the optimal solution for γ_{ki} by minimizing objective function J_{ssct}, because x_k is considered to be near the cluster center v_i. Third, we consider the case of $K(x_k, v_i) < 0$. In this case, we can get the optimal solution for δ_{ki} by maximizing objective function J_{ssct}, because x_k is considered to be distant from the cluster center v_i.

The algorithm of SSsFCMT is summarized in Algorithm 6.

Algorithm 6: SSsFCMCT

SSsFCMCT 1 Set initial cluster centers and parameters m and κ_{ki}.
SSsFCMCT 2 Calculate $u_{ki} \in U$ by using following equation:

$$u_{ki} = \frac{\left(\frac{1}{d_{ki}}\right)^{\frac{1}{m-1}}}{\sum_{l=1}^{c}\left(\frac{1}{d_{kl}}\right)^{\frac{1}{m-1}}}$$

SSsFCMCT 3 Calculate $v_i \in V$ by using following equation:

$$v_i = \frac{\sum_{k=1}^{n} (u_{ki})^m x_k + \gamma_{ki}}{\sum_{k=1}^{n} (u_{ki})^m}$$

SSsFCMCT 4 Calculate $\gamma_{ki} \in \Gamma$ by using following equation:

$$\gamma_{ki} = \begin{cases} -\alpha_{ki}(x_k - v_i), & K(x_k, v_i) > 0, \\ 0 & K(x_k, v_i) = 0, \\ \frac{|K(x_k, v_i)|(x_k - v_i)}{\|x_k - v_i\|} & K(x_k, v_i) < 0, \end{cases}$$

$$\alpha_{ki} = \min\left\{\frac{\kappa_{ki}}{\|x_k - v_i\|}, 1\right\}$$

SSsFCMCT 5 If convergence criterion is satisfied, stop. Otherwise go back to **SSsFCMCT 2**.

SSFCMCT handle the pairwise constraints as soft or hard constraints without breaking the L_2-space by using the concept of clusterwise tolerance and determining the suitable value of κ_{ki}. The effectiveness of SSsFCMCT and entropy based model are omitted here, see reference [11] for details of numerical experiments.

5 Conclusions

We introduced two types of new concepts to construct significant clustering algorithms. One is tolerance and penalty-vector concepts for handling uncertain data in clustering procedures. The other is clusterwise tolerance concept for semi-supervised learning. These concepts are quite similar approach in the viewpoint of handling objects to be flexible to the clustering topics. We construct two clustering algorithms FCMT and FCMQ for handling uncertain data. We also construct SSFCMCT for semi-supervised learning based on FCMCT.

These concepts have a potential for not only clustering but also data analysis topics in various ways. We will consider and propose the way to apply those concepts to resolve conventional and brand new clustering topics. We will moreover show the effectiveness of tolerance based methods and the difference between ones and conventional methods.

References

1. Aggarwal C. C., A survey of uncertain data algorithms and applications , *IEEE Trans. on Knowledge and Data Engineering*, Vol. 21, No. 5, pp. 609–623, 2008.
2. Bezdek J. C., 'Pattern Recognition with Fuzzy Objective Function Algorithms', Plenum Press, New York, 1981.
3. Chapelle O., Schoölkopf B., and Zien A., eds., 'Semi-Supervised Learning', MIT Press, 2006.
4. Davidson I. and Ravi S.S., Agglomerative hierarchical clustering with constraints: theoretical and empirical results, *Proc. of 9th European Conference on Principles and Practice of Knowledge Discovery in Databases* (KDD 2005), pp. 59–70, 2005.
5. Endo Y., Murata R., Haruyama H., and Miyamoto S., Fuzzy c-means for data with tolerance', *Proc. of International Symposium on Nonlinear Theory and Its Applications* (Nolta2005), pp. 345–348, 2005.
6. Endo Y., Hasegawa Y., Hamasuna Y., and Miyamoto S., Fuzzy c-means for data with rectangular maximum tolerance range, *Journal of Advanced Computational Intelligence and Intelligent Informatics* (JACIII), Vol. 12, No. 5, pp. 461–466, 2008.
7. Endo Y., Kurihara K., Miyamoto S, and Hamasuna Y., Hard and fuzzy c-regression models for data with tolerance in independent and dependent variables, *Proc. of The 2010 IEEE World Congress on Computational Intelligence* (WCCI2010), pp. 1842–1849, 2010.
8. Endo Y., Hasegawa Y., Hamasuna Y., and Kanzawa Y., Fuzzy c-means clustering for uncertain data using quadratic regularization of penalty vectors, *Journal of Advanced Computational Intelligence and Intelligent Informatics* (JACIII), Vol. 15, No. 1, pp. 76–82, 2011.
9. Hamasuna Y., Endo Y., Hasegawa Y., and Miyamoto S., Two clustering algorithms for data with tolerance based on hard c-means, *2007 IEEE International Conference on Fuzzy Systems* (FUZZ-IEEE2007), pp. 688–691 2007.
10. Hamasuna Y., Endo Y., and Miyamoto S., On tolerant fuzzy c-means clustering and tolerant possibilistic clustering, *Soft Computing*, Vol. 14, No. 5, pp. 487–494, 2010.
11. Hamasuna Y. and Endo Y., On semi-supervised fuzzy c-means clustering for data with clusterwise tolerance by opposite criteria, *Soft Computing*, Vol. 17, No. 1, pp. 71–81, 2013.
12. Hathaway R. J. and Bezdek J. C., Fuzzy c-means clustering of incomplete data, *IEEE Trans. on Systems, Man, and Cybernetics Part B*, Vol. 31, No. 5, pp. 735–744, 2001.
13. Jain A. K., Data clustering: 50 years beyond K-means, *Pattern Recognition Letters*, Vol. 31, No. 8, pp. 651–666, 2010.
14. Klein D., Kamvar S., and Manning C., From instance-level constraints to space-level constraints: making the most of prior knowledge in data clustering, *Proc. of the 19th International Conference on Machine Learning* (ICML 2002), pp. 307–314, 2002.
15. Miyamoto S., Ichihashi H., and Honda K., 'Algorithms for Fuzzy Clustering', Springer, Heidelberg, 2008.
16. Miyamoto S., Yamazaki M., and Terami A., On semi-supervised clustering with pairwise constraints, *Proc. of The 7th International Conference on Modeling Decisions for Artificial Intelligence* (MDAI 2009), pp. 245–254, 2009 (CD-ROM).
17. Torra V., Endo Y., and Miyamoto S., Computationally Intensive Parameter Selection for Clustering Algorithms: The Case of Fuzzy c-Means with Tolerance, *International Journal of Intelligent Systems*, Vol. 26, No. 4, pp. 313–322, 2011.
18. Wagstaff K., Cardie C., Rogers S., and Schroedl S., Constrained k-means clustering with background knowledge', *Proc. of the 18th International Conference on Machine Learning* (ICML2001), pp. 577–584, 2001.

Robust Clustering Algorithms Employing Fuzzy-Possibilistic Product Partition

László Szilágyi

Abstract One of the main challenges in the field of clustering is creating algorithms that are both accurate and robust. The fuzzy-possibilistic product partition c-means clustering algorithm was introduced with the main goal of producing accurate partitions in the presence of outlier data. This chapter presents several clustering algorithms based on the fuzzy-possibilistic product partition, specialized for the detection of clusters having various shapes including spherical and ellipsoidal shells. The advantages of applying the fuzzy-possibilistic product partition are presented in comparison with previous c-means clustering models. Besides being more robust and accurate than previous probabilistic-possibilistic mixture partitions, the product partition is easier to handle due to its reduced number of parameters.

Keywords Fuzzy c-means clustering · Possibilistic c-means clustering · Spherical shell clustering · Elliptic shell clustering · Fuzzy partition

1 Introduction

Robustness in clustering refers to the stability or reproducibility of the achieved partition, and insensitivity to several kinds of noise including extremely outlier data. The fuzzy c-means (FCM) algorithm introduced by Bezdek [1] is a very popular clustering model due to the fine partitions it usually produces and its easily comprehensible alternating optimization (AO) scheme. However, the probabilistic constraints involved in FCM make it sensitive to outlier data. To combat this problem, several solutions have been proposed, which found their way towards relaxing the probabilistic constraint.

An early solution was given by Davé [2], who introduced an extra, specially treated noisy class to attract feature vectors situated far from all normal cluster prototypes. This theory was later extended by Menard et al. [3], while further noise tolerant

L. Szilágyi (✉)
Faculty of Technical and Human Science, Sapientia - Hungarian
Science University of Transylvania, Tîrgu Mureş, Romania
e-mail: lalo@ms.sapientia.ro

V. Torra et al. (eds.), *Fuzzy Sets, Rough Sets, Multisets and Clustering*,
Studies in Computational Intelligence 671, DOI 10.1007/978-3-319-47557-8_7

fuzzy c-means versions were introduced by Chintalapudi and Kam [4], and Alanzado and Miyamoto [5]. Alternately, Krishnapuram and Keller proposed the possibilistic c-means algorithm (PCM) [6], which produces the partition based on statistical rules. This approach seemed a fine solution for the sensitivity to outliers, but it frequently led to coincident clusters [7]. To avoid the latter, Timm et al. [8] set up a repulsive force between all couples of cluster prototypes of PCM, the strength of which decreased with distance. Their method successfully avoided coincident clusters, but failed to accurately treat cases when two clusters are really close to each other. Two versions of fuzzy-possibilistic partition mixtures were proposed by Pal et al. [9, 10], out of which the second one (possibilistic fuzzy c-means (PFCM)) appears to be a reliable clustering model in most scenarios. However, even PFCM can fail in the presence of extreme outliers [11]. The fuzzy-possibilistic product partition (FPPP) and the fuzzy-possibilistic product partition c-means algorithm (FP^{3}CM) [11] were intended to be a solution for this problem. This chapter will revisit and discuss the achievements of FPPP in this matter.

All c-means algorithms mentioned above work with point-type cluster prototypes that are computed as weighted means of the input data. However, frequently emerge situations when the shape of the clusters differs from the default. To cope with such scenarios, several solutions have been proposed. Linear manifolds are usually modeled via adaptive fuzzy c-varieties [12], while spherical ones via the fuzzy c-spherical shell (FCSS) algorithm model by Krishnapuram et al. [13]. Elliptical prototypes were introduced by the adaptive fuzzy c-shells algorithm [14], and the fuzzy c-ellipsoidal shells clustering model [15]. Generalized versions of shell clusters to the quadric case were given by Krishnapuram et al. [16]. Further quadric prototype models include the fuzzy c-quadrics [17] and the fuzzy c-quadric shells [18]. The norm-induced shell prototypes introduced by Bezdek et al. [19] can be adapted to detect ellipses, quadrics, and rectangles as well. Later, Hoeppner extended the palette of detectable shapes with his fuzzy c-rectangular shell models [20]. Out of these shaped cluster models, only FCSS has explicit expressions to compute the cluster prototypes in each iteration of the main AO loop. All others use nonlinear implicit expressions that need to be solved in an iterative way. The latter, besides being more complicated to implement, also represents a higher computational load.

This chapter is dedicated to present the fuzzy-possibilistic product partition c-means algorithm, and its c-spherical shell and c-elliptical shell versions: their definition, optimization scheme, and the clusters they find in various scenarios, emphasizing the advantages they provide in comparison to previous c-means and c-shell clustering models.

2 Preliminaries

All c-means clustering models partition a set of object data $\mathbf{X} = \{\boldsymbol{x}_1, \boldsymbol{x}_2, \ldots, \boldsymbol{x}_n\}$ into a number of c clusters based on the minimization of a quadratic objective function. According to the partitions they employ, there are three fundamental approaches, namely

1. Hard c-means algorithm (HCM) [21] that uses probabilistic crisp partition,
2. Fuzzy c-means algorithm (FCM) [1] that uses probabilistic fuzzy partition,
3. Possibilistic c-means algorithm (PCM) [13] that uses a less constrained fuzzy partition than the one used by FCM, without sticking to probabilistic conditions.

Mixed partitions were introduced in the theory of c-means clustering models to improve certain properties of the fundamental algorithms. Some of these solutions are presented in the following sections.

2.1 The Fuzzy-Possibilistic c-Means Clustering Algorithm

The fuzzy-possibilistic c-means (FPCM) clustering algorithm introduced by Pal et al. [9] uses a linear combination of a probabilistic and a possibilistic partition. The objective function is defined as:

$$J_{\mathrm{FPCM}} = \sum_{i=1}^{c} \sum_{k=1}^{n} [u_{ik}^m + t_{ik}^p] ||\boldsymbol{x}_k - \boldsymbol{v}_i||^2 = \sum_{i=1}^{c} \sum_{k=1}^{n} [u_{ik}^m + t_{ik}^p] d_{ik}^2 \ , \qquad (1)$$

where $\boldsymbol{v}_i$ represents the prototype or centroid value or representative element of cluster i ($i = 1 \ldots c$), $u_{ik} \in [0, 1]$ is the probabilistic fuzzy membership function showing the degree to which input vector $\boldsymbol{x}_k$ belongs to cluster i, $t_{ik} \in [0, 1]$ is the possibilistic fuzzy membership function showing the degree of compatibility of input vector $\boldsymbol{x}_k$ with cluster i, $m > 1$ and $p > 1$ are the probabilistic and possibilistic fuzzyfication exponents, respectively, and $d_{ik} = ||\boldsymbol{x}_k - \boldsymbol{v}_i||$. The above objective function is minimized under the probabilistic and possibilistic constraints, written as

$$\sum_{i=1}^{c} u_{ik} = 1 \quad \forall k = 1 \ldots n \quad \text{and} \quad \sum_{k=1}^{n} t_{ik} = 1 \quad \forall i = 1 \ldots c \ . \qquad (2)$$

Using the zero gradient conditions of the above objective function, we obtain the following optimization formulas for the iterative AO scheme of the algorithm:

$$u_{ik} = \frac{d_{ik}^{-2/(m-1)}}{\sum\limits_{j=1}^{c} d_{jk}^{-2/(m-1)}} \quad \text{and} \quad t_{ik} = \frac{d_{ik}^{-2/(p-1)}}{\sum\limits_{l=1}^{n} d_{il}^{-2/(p-1)}} \qquad \begin{matrix} \forall i = 1 \ldots c \\ \forall k = 1 \ldots n \end{matrix} \ , \qquad (3)$$

$$\boldsymbol{v}_i = \frac{\sum\limits_{k=1}^{n} [u_{ik}^m + t_{ik}^p] \boldsymbol{x}_k}{\sum\limits_{k=1}^{n} [u_{ik}^m + t_{ik}^p]} \qquad \forall i = 1 \ldots c \ . \qquad (4)$$

FPCM has the main advantage of not using the penalty terms η_i, thus making the parameter adjustment easier. However, the possibilistic effect of the algorithm loses its strength as the number of input vectors grows. In case of $n >> c$, FPCM practically reduces to FCM, regardless of the value of the exponent p.

2.2 *The Possibilistic-Fuzzy c-Means Clustering Algorithm*

The possibilistic-fuzzy c-means (PFCM) algorithm, proposed by Pal et al. [10], minimizes the objective function

$$J_{\text{PFCM}} = \sum_{i=1}^{c} \sum_{k=1}^{n} [a u_{ik}^m + b t_{ik}^p] d_{ik}^2 + \sum_{i=1}^{c} \eta_i \sum_{k=1}^{n} (1 - t_{ik})^p \ , \tag{5}$$

constrained by the conventional probabilistic and possibilistic conditions of FCM and PCM, given as

$$\begin{cases} 0 \leq u_{ik} \leq 1 & \forall i = 1 \ldots c \quad \forall k = 1 \ldots n \\ \sum_{i=1}^{c} u_{ik} = 1 & \forall k = 1 \ldots n \end{cases} \ , \tag{6}$$

$$\begin{cases} 0 \leq t_{ik} \leq 1 & \forall i = 1 \ldots c \quad \forall k = 1 \ldots n \\ 0 < \sum_{i=1}^{c} t_{ik} < c & \forall k = 1 \ldots n \end{cases} \ . \tag{7}$$

Tradeoff parameters a and b control the strength of the possibilistic and probabilistic term in the mixed partition. Exponents m and p have the same role as in FPCM, while penalty terms η_i are inherited from the definition of the PCM algorithm.

The minimization formulas for updating the partition are

$$u_{ik} = \frac{d_{ik}^{\frac{-2}{m-1}}}{\sum_{j=1}^{c} d_{jk}^{\frac{-2}{m-1}}} \quad \text{and} \quad t_{ik} = \left[1 + \left(\frac{b d_{ik}^2}{\eta_i} \right)^{\frac{1}{p-1}} \right]^{-1} \qquad \begin{matrix} \forall i = 1 \ldots c \\ \forall k = 1 \ldots n \end{matrix} \ , \tag{8}$$

while cluster prototypes are obtained as:

$$\boldsymbol{v}_i = \frac{\sum_{k=1}^{n} [a u_{ik}^m + b t_{ik}^p] \boldsymbol{x}_k}{\sum_{k=1}^{n} [a u_{ik}^m + b t_{ik}^p]} \qquad \forall i = 1 \ldots c \ . \tag{9}$$

PFCM was found accurate and robust [10], but still sensitive to outlier data to a certain degree [11].

2.3 *Further Hybrid Fuzzy c-Means Clustering Models*

Another mixed c-means clustering model was proposed by Szilágyi et al. [22], which incorporated all three fundamental partition components into a linearly combined partition. The trade-off among the three components was controlled by two parameters α and β, while all other parameters were inherited from FCM and PCM. The algorithm was found runtime efficient due to quick convergence. It was successfully employed in an application involving medical image segmentation.

3 The Fuzzy-Probabilistic Product Partition Fuzzy *c*-Means Clustering Algorithm

In a probabilistic fuzzy partition, any outlier input vector $\boldsymbol{x}_{\text{out}}$ receives high membership values with respect to all clusters, that is, $u_{i,\text{out}} \approx 1/c$, which strongly influence all cluster prototypes. On the other hand, in a possibilistic approach, outlier input vectors receive very low typicality values with respect to all clusters. Motivated by an intuition stemming from the above facts, the fuzzy-possibilistic product partition and the corresponding c-means clustering model was launched as a hypothesis that a cluster prototype update formula of form

$$\boldsymbol{v}_i = \frac{\sum_{k=1}^{n} \mu_{ik}^m \tau_{ik}^p \boldsymbol{x}_k}{\sum_{k=1}^{n} \mu_{ik}^m \tau_{ik}^p} \qquad \forall i = 1 \ldots c \ . \tag{10}$$

where μ_{ik} $(i = 1 \ldots c, k = 1 \ldots n)$ describe a probabilistic fuzzy partition that is not necessarily equivalent with the FCM's one, and τ_{ik}, $(i = 1 \ldots c, k = 1 \ldots n)$ stand for the elements of a possibilistic partition matrix, could be beneficial in terms of partition accuracy and outlier suppression. To obtain such an optimization formula, the objective function of FP^{3}CM was defined as:

$$J_{\text{FP}^3\text{CM}} = \sum_{i=1}^{c} \sum_{k=1}^{n} u_{ik}^m \left[t_{ik}^p d_{ik}^2 + (1 - t_{ik})^p \eta_i \right] \ , \tag{11}$$

constrained by the conventional probabilistic condition given in Eq. (6) and the conventional possibilistic conditions given in Eq. (7). The only parameters of FP^{3}CM are the fuzzy exponent $m > 1$, the possibilistic exponent $p > 1$, and the conventional penalty terms of the possibilistic partition denoted by η_i $(i = 1 \ldots c)$.

The minimization formulas are obtained using zero gradient conditions, aided by Lagrange multipliers in case of the probabilistic term. We need to compute the partial derivatives of the functional:

$$\mathcal{L} = J_{\mathrm{FP^3CM}} + \sum_{k=1}^{n} \lambda_k \left(1 - \sum_{i=1}^{c} u_{ik}\right) , \tag{12}$$

where λ_k ($k = 1 \dots n$) stand for the Lagrange multipliers. The zero crossing of the partial derivatives with respect to t_{ik} ($\forall i = 1 \dots c$, $\forall k = 1 \dots n$) leads to:

$$\frac{\partial \mathcal{L}}{\partial t_{ik}} = 0 \quad \Rightarrow \quad u_{ik}^{m} \left[p t_{ik}^{p-1} d_{ik}^{2} - \eta_i p (1 - t_{ik})^{p-1} \right] = 0.$$

If $u_{ik} = 0$, the value of t_{ik} does not make a difference. Otherwise we get

$$\left(\frac{1 - t_{ik}}{t_{ik}}\right)^{p-1} = \frac{d_{ik}^2}{\eta_i} \quad \Rightarrow \quad \frac{1}{t_{ik}} - 1 = \left(\frac{d_{ik}^2}{\eta_i}\right)^{1/(p-1)} ,$$

which finally leads to a formula that is identical with the prototype update formula of PCM:

$$t_{ik} = \left[1 + \left(\frac{d_{ik}^2}{\eta_i}\right)^{1/(p-1)} \right]^{-1} \qquad \begin{matrix} \forall i = 1 \dots c \\ \forall k = 1 \dots n \end{matrix} . \tag{13}$$

Further on, let us examine the zero crossing of partial derivatives with respect to u_{ik}. For any $i = 1 \dots c$ and any $k = 1 \dots n$ we get

$$\frac{\partial \mathcal{L}}{\partial u_{ik}} = 0 \quad \Rightarrow \quad m u_{ik}^{m-1} \left[t_{ik}^{p} d_{ik}^{2} + \eta_i (1 - t_{ik})^{p} \right] = \lambda_k,$$

which implies

$$u_{ik} = \left(\frac{\lambda_k}{m}\right)^{1/(m-1)} \times \left[t_{ik}^{p} d_{ik}^{2} + \eta_i (1 - t_{ik})^{p} \right]^{-1/(m-1)} . \tag{14}$$

The probabilistic condition says $\sum_{j=1}^{c} u_{jk} = 1$, which by the means of Eq. (14) becomes

$$1 = \left(\frac{\lambda_k}{m}\right)^{1/(m-1)} \times \sum_{j=1}^{c} \left[t_{jk}^{p} d_{jk}^{2} + \eta_j (1 - t_{jk})^{p} \right]^{-1/(m-1)} . \tag{15}$$

Dividing Eq. (14) by (15) term by term, leads to

$$u_{ik} = \frac{[t_{ik}^p d_{ik}^2 + \eta_i (1 - t_{ik})^p]^{-1/(m-1)}}{\sum_{j=1}^{c} [t_{jk}^p d_{jk}^2 + \eta_j (1 - t_{jk})^p]^{-1/(m-1)}} , \tag{16}$$

which holds for any $i = 1 \ldots c$, and any $k = 1 \ldots n$. Finally, let us investigate the zero crossings of the partial derivatives with respect to $\boldsymbol{v}_i$ $(i = 1 \ldots c)$:

$$\frac{\partial \mathcal{L}}{\partial \boldsymbol{v}_i} = 0 \quad \Rightarrow \quad -2 \sum_{k=1}^{n} u_{ik}^m t_{ik}^p (\boldsymbol{x}_k - \boldsymbol{v}_i) = 0 ,$$

which implies

$$\boldsymbol{v}_i \sum_{k=1}^{n} u_{ik}^m t_{ik}^p = \sum_{k=1}^{n} u_{ik}^m t_{ik}^p \boldsymbol{x}_k \quad \Rightarrow \quad \boldsymbol{v}_i = \frac{\sum_{k=1}^{n} u_{ik}^m t_{ik}^p \boldsymbol{x}_k}{\sum_{k=1}^{n} u_{ik}^m t_{ik}^p} , \tag{17}$$

valid for any $i = 1 \ldots c$, having indeed the form we wished in Eq. (10). Let us remark the followings:

1. The possibilistic memberships t_{ik} are established exactly the same way, as in the PCM algorithm. This does not imply that the penalty terms η_i should be set exactly as recommended by Krishnapuram and Keller [6].
2. The probabilistic memberships u_{ik} are somewhat similar to the ones given by FCM, but distances are distorted according to the possibilitic memberships t_{ik} and penalty terms η_i. In fact, u_{ik} values tend to the ones given by FCM as $\eta_i \to +\infty$.
3. Outlier input vectors $\boldsymbol{x}_k$ are indicated by the algorithm with a low value of $\max\{ \sqrt[m+p]{u_{ik}^m t_{ik}^p}, i = 1 \ldots c\}$.
4. The defuzzification of the final partition should be performed according to the following rule: $\boldsymbol{x}_k$ is assigned to cluster with index w_k, where

$$w_k = \arg\max_j \left(u_{jk}^m t_{jk}^p | j = 1 \ldots c \right) . \tag{18}$$

In case of equal η_i values, for any $i = 1 \ldots c$, the rule may be applied as: $w_k = \arg\max_j \left(u_{jk} | j = 1 \ldots c \right) = \arg\min_j \left(d_{jk} | j = 1 \ldots c \right)$.

The AO algorithm of FP^{3}CM is summarized in Algorithm 1.

Algorithm 1: The alternating optimization algorithm of FP^3CM clustering algorithm

Data: Input data $\mathbf{X} = \{\boldsymbol{x}_1, \boldsymbol{x}_2, \ldots, \boldsymbol{x}_n\}$
Result: Final cluster prototypes $\boldsymbol{v}_1, \boldsymbol{v}_2, \ldots, \boldsymbol{v}_c$
Result: Partition matrices $\mathbf{U} = \{u_{ik}\}$ and $\mathbf{T} = \{t_{ik}\}$, with $i = 1 \ldots c$, $k = 1 \ldots n$
Fix the number of clusters c, $2 \leq c \leq n$;
Set fuzzy exponent m and possibilistic exponent p, both greater than 1;
Set possibilistic penalty terms η_i $(i = 1 \ldots c)$;
Initialize cluster prototypes $\boldsymbol{v}_i$ $(i = 1 \ldots c)$;
repeat
    Update possibilistic membership values using Eq. (13);
    Update probabilistic membership values using Eq. (16);
    Update cluster prototypes using Eq. (17);
until *cluster prototypes* $\boldsymbol{v}_i$ $(i = 1 \ldots c)$ *converge*;
Defuzzify of the obtained product partition as indicated in Eq. (18).

4 Spherical Shell Clustering Algorithm Using Fuzzy-Possibilistic Product Partition

As introduced in [23], the fuzzy-possibilistic product partition c-spherical shell clustering (FP^3CSS) algorithm minimizes

$$J_{FP^3CSS} = \sum_{i=1}^{c} \sum_{k=1}^{n} u_{ik}^m \left[t_{ik}^p d_{ik}^2 + (1 - t_{ik})^p \eta_i \right] , \tag{19}$$

where the distance between vector $\boldsymbol{x_k}$ and the prototype of cluster with index i is defined as

$$d_{ik} = \left| ||\boldsymbol{x}_k - \boldsymbol{\theta}_i||^2 - \rho_i^2 \right| , \tag{20}$$

cluster i being represented by a spheroid having its center in $\boldsymbol{\theta}_i$ and radius ρ_i. The above presented objective function is optimized under the conventional probabilistic and possibilistic constraints given in Eqs. (6) and (7). The only parameters of FP^3CSS are the fuzzy exponent $m > 1$, the possibilistic exponent $p > 1$, and the conventional penalty terms of the possibilistic partition denoted by η_i $(i = 1 \ldots c)$.

To obtain a minimization algorithm of the objective function J_{FP^3CSS}, zero gradient conditions and Lagrange multipliers are used. The partition update formulas are obtained exactly in the same format as in case of FP^3CM, indicated in Eqs. (13) and (16), but using the definition of d_{ik} given in Eq. (20).

To deduce an update formula for cluster prototypes, we will turn to the technique proposed by Krishnapuram et al. [13], namely we reformulate the distance d_{ik} in a different form that is equivalent to the previous one given in Eq. (20)

$$d_{ik}^2 = \boldsymbol{\xi}_i^T \begin{bmatrix} \boldsymbol{x_k} \\ 1 \end{bmatrix} \begin{bmatrix} \boldsymbol{x_k} \\ 1 \end{bmatrix}^T \boldsymbol{\xi}_i + 2\boldsymbol{x}_k^T \boldsymbol{x}_k \begin{bmatrix} \boldsymbol{x_k} \\ 1 \end{bmatrix}^T \boldsymbol{\xi}_i + (\boldsymbol{x}_k^T \boldsymbol{x}_k)^2 , \tag{21}$$

where $\boldsymbol{\xi}_i$ describes the prototype of cluster i in the following manner:

$$\boldsymbol{\xi}_i = \begin{bmatrix} -2\boldsymbol{\theta}_i \\ \boldsymbol{\theta}_i^T \boldsymbol{\theta}_i - \rho_i^2 \end{bmatrix}. \tag{22}$$

Now let us investigate the zero crossing of the partial derivative of the objective function with respect to $\boldsymbol{\xi}_i$, $i = 1 \dots c$:

$$\frac{\partial \mathcal{L}}{\partial \boldsymbol{\xi}_i} = 0 \Rightarrow 2 \sum_{k=1}^{n} u_{ik}^m t_{ik}^p \left(\begin{bmatrix} \boldsymbol{x}_k \\ 1 \end{bmatrix} \begin{bmatrix} \boldsymbol{x}_k \\ 1 \end{bmatrix}^T \boldsymbol{\xi}_i + (\boldsymbol{x}_k^T \boldsymbol{x}_k) \begin{bmatrix} \boldsymbol{x}_k \\ 1 \end{bmatrix} \right) = 0,$$

which implies

$$\left(\sum_{k=1}^{n} u_{ik}^m t_{ik}^p \begin{bmatrix} \boldsymbol{x}_k \\ 1 \end{bmatrix} \begin{bmatrix} \boldsymbol{x}_k \\ 1 \end{bmatrix}^T \right) \boldsymbol{\xi}_i = - \left(\sum_{k=1}^{n} u_{ik}^m t_{ik}^p (\boldsymbol{x}_k^T \boldsymbol{x}_k) \begin{bmatrix} \boldsymbol{x}_k \\ 1 \end{bmatrix} \right)$$

and finally leads to

$$\boldsymbol{\xi}_i = - \left(\sum_{k=1}^{n} u_{ik}^m t_{ik}^p \begin{bmatrix} \boldsymbol{x}_k \\ 1 \end{bmatrix} \begin{bmatrix} \boldsymbol{x}_k \\ 1 \end{bmatrix}^T \right)^{-1} \left(\sum_{k=1}^{n} u_{ik}^m t_{ik}^p (\boldsymbol{x}_k^T \boldsymbol{x}_k) \begin{bmatrix} \boldsymbol{x}_k \\ 1 \end{bmatrix} \right), \tag{23}$$

valid for any $i = 1 \dots c$. The center $\boldsymbol{\theta}_i$ and radius ρ_i can be extracted from $\boldsymbol{\xi}_i$ using Eq. (22).

The initialization of cluster prototypes is a key issue. As the cost function may have several local minima, it is important to start the algorithm in the neighborhood of the global optimum. Krishnapuram et al. [13] recommended using the FCM algorithm to produce initial estimates for the shell centers $\boldsymbol{\theta}_{i0}$ and radii ρ_i obtained as:

$$\begin{cases} \boldsymbol{\theta}_{i0} = \boldsymbol{v}_i \\ \rho_{i0} = \dfrac{\sum_{k=1}^{n} u_{ik}^m ||\boldsymbol{x}_k - \boldsymbol{v}_i||}{\sum_{k=1}^{n} u_{ik}^m} \end{cases}, \tag{24}$$

where $\boldsymbol{v}_i$ $(i = 1 \dots c)$ are final FCM cluster prototypes and u_{ik} $(i = 1 \dots c, k = 1 \dots n)$ are final FCM fuzzy membership functions.

The AO algorithm of FP^{3}CSS is summarized in Algorithm 2.

Algorithm 2: The alternating optimization algorithm of FP^{3}CSS clustering algorithm

Data: Input data $\mathbf{X} = \{\boldsymbol{x}_1, \boldsymbol{x}_2, \ldots, \boldsymbol{x}_n\}$
Result: Final cluster prototype descriptors $\boldsymbol{\xi}_1, \boldsymbol{\xi}_2, \ldots, \boldsymbol{\xi}_c$
Result: Partition matrices $\mathbf{U} = \{u_{ik}\}$ and $\mathbf{T} = \{t_{ik}\}$, with $i = 1 \ldots c$, $k = 1 \ldots n$
Fix the number of clusters c, $2 \leq c \leq n/3$;
Set fuzzy exponent m and possibilistic exponent p, both greater than 1;
Set possibilistic penalty terms η_i $(i = 1 \ldots c)$, as recommended by Krishnapuram and Keller in [6];
Initialize cluster prototypes descriptors $\boldsymbol{\xi}_i$ $(i = 1 \ldots c)$ according to Eq. (24);
repeat
  Update distances d_{ik} $(i = 1 \ldots c, k = 1 \ldots n)$ using Eq. (20);
  Update possibilistic membership values using Eq. (13);
  Update probabilistic membership values using Eq. (16);
  Update cluster prototype descriptors $\boldsymbol{\xi}_i$ $(i = 1 \ldots c)$ using Eq. (23);
  Identify cluster prototypes $(\boldsymbol{\theta}_i, \rho_i)$, $i = 1 \ldots c$ according to Eq. (22);
until *cluster prototypes* $\boldsymbol{\xi}_i$ $(i = 1 \ldots c)$ *converge*;
The degree of membership of input vector $\boldsymbol{x}_k$ with respect to cluster i is given by $\sqrt[m+p]{u_{ik}^m t_{ik}^p}$;
Vector $\boldsymbol{x}_k$ is an identified outlier, if $\sum_{i=1}^{c} \sqrt[m+p]{u_{ik}^m t_{ik}^p} < \varepsilon$, where ε is a predefined small constant (e.g. $\varepsilon = 0.01$).

5 Elliptical Shell Clustering Algorithm Using Fuzzy-Possibilistic Product Partition

Ellipsoids in computational geometry are usually described by a center point, one radius value in each dimension, and a rotation vector which defines the orientation of the ellipsoid. Here we define an ellipsoid as a collection of points $\boldsymbol{x}$ that satisfy the equation:

$$(\boldsymbol{x} - \boldsymbol{\theta})^T \mathbf{A} (\boldsymbol{x} - \boldsymbol{\theta}) = \rho^2 \ ,$$

where $\boldsymbol{\theta}$ is the center of the ellipsoid, ρ is a variable that controls the size of the ellipsoid (radii vary proportionally with ρ), and matrix $\mathbf{A}$ is a positive definite matrix that describes the shape and orientation of the ellipsoid. Under such circumstances, the distance of any data point $\boldsymbol{x}_k$ $(k = 1 \ldots n)$ from cluster prototype number i $(i = 1 \ldots c)$, defined by $\boldsymbol{\theta}_i$, ρ_i, and $\mathbf{A}_i$ is computed as:

$$d_{ik} = \zeta_{ik} - \rho_i \ , \tag{25}$$

where

$$\zeta_{ik} = ||\boldsymbol{x}_k - \boldsymbol{\theta}_i||_{\mathbf{A}} = \sqrt{(\boldsymbol{x}_k - \boldsymbol{\theta}_i)^T \mathbf{A} (\boldsymbol{x}_k - \boldsymbol{\theta}_i)} \ . \tag{26}$$

The objective function of the fuzzy-possibilistic product partition elliptic shell (FP^{3}CES) clustering algorithm is:

$$J_{\mathrm{FP3CES}} = \sum_{i=1}^{c} \sum_{k=1}^{n} u_{ik}^{m} [t_{ik}^{p} d_{ik}^{2} + (1 - t_{ik})^{p} \eta_i] \ ,$$

which will be optimized under the probabilistic and possibilistic constraints given in Eqs. (6) and (7), and a further constraint $\det(\mathbf{A}_i) = \rho_i$ fixed, which assures that each possible ellipsoid has a unique description using its own $\boldsymbol{\theta}_i$, ρ_i, and $\mathbf{A}_i$ values [24].

In order to optimize the above objective function, we need to find the optimal cluster prototypes ($\boldsymbol{\theta}_i, \rho_i, \mathbf{A}_i$ for any $i = 1 \dots c$) and optimal fuzzy membership functions u_{ik} and t_{ik}, for any $i = 1 \dots c$ and $k = 1 \dots n$. The optimum is reached via grouped coordinate minimization by alternately optimizing the partition with fixed cluster prototypes, and then optimizing the cluster prototypes keeping the partition fixed. The alternating optimization is stopped when the norm of variation of cluster prototypes during an iteration stays below a predefined constant ε. The alternately applied optimization formulas are obtained from zero crossing of the objective function's partial derivatives, using Lagrange multipliers where necessary. The optimization formulas thus obtained are necessary conditions of finding the objective function's optimum:

- The partition update formulas are obtained exactly in the same format as in case of FP^{3}CM, indicated in Eqs. (13) and (16), but using the definition of d_{ik} given in Eqs. (25) and (26).
- Cluster prototype centers $\boldsymbol{\theta}_i$ and radii ρ_i ($\forall i = 1 \dots c$) need to be extracted via Newton's method from implicit equations:

$$\begin{cases} \sum\limits_{k=1}^{n} u_{ik}^{m} t_{ik}^{p} \frac{d_{ik}}{\zeta_{ik}} (\boldsymbol{x}_k - \boldsymbol{\theta}_i) = \mathbf{0} \\ \sum\limits_{k=1}^{n} u_{ik}^{m} t_{ik}^{p} d_{ik} = 0 \end{cases} . \tag{27}$$

- Ellipse orientation matrices $\mathbf{A}_i$ ($i = 1 \dots c$) are obtained as:

$$\mathbf{A}_i = \sqrt[z]{\rho_i \det(\mathbf{S}_i)} \mathbf{S}_i^{-1} \ , \tag{28}$$

where

$$\mathbf{S}_i = \sum_{k=1}^{n} u_{ik}^{m} t_{ik}^{p} \frac{d_{ik}}{\zeta_{ik}} (\boldsymbol{x}_k - \boldsymbol{\theta}_i)(\boldsymbol{x}_k - \boldsymbol{\theta}_i)^T \ . \tag{29}$$

and z is the number of dimensions.

The FP^{3}CES algorithm is summarized in Algorithm 3.

Algorithm 3: The alternating optimization algorithm of FP^{3}CES clustering algorithm

Data: Input data $\mathbf{X} = \{\boldsymbol{x}_1, \boldsymbol{x}_2, \ldots, \boldsymbol{x}_n\}$
Result: Final cluster prototype descriptors $\boldsymbol{\theta}_i$, ρ_i, $\mathbf{A}_i$, for all $i = 1 \ldots c$
Result: Partition matrices $\mathbf{U} = \{u_{ik}\}$ and $\mathbf{T} = \{t_{ik}\}$, with $i = 1 \ldots c$, $k = 1 \ldots n$
Fix the number of clusters c, $2 \leq c \leq n/3$;
Set fuzzy exponent m and possibilistic exponent p, both greater than 1;
Set possibilistic penalty terms η_i $(i = 1 \ldots c)$, as recommended by Krishnapuram and Keller in [6];
Initialize cluster prototypes to represent the circles described by Eq. (24);
repeat
  Update ζ_{ik} and d_{ik} values $(i = 1 \ldots c, k = 1 \ldots n)$ according to Eqs. (26) and (25), respectively.
  Update possibilistic membership values using Eq. (13);
  Update probabilistic membership values using Eq. (16);
  Update cluster centers $\boldsymbol{\theta}_i$ and radii ρ_i $(i = 1 \ldots c)$ via Newton's method applied to equation system (27);
  Update matrices $\mathbf{S}_i$ $(i = 1 \ldots c)$ using Eq. (29);
  Update matrices $\mathbf{A}_i$ $(i = 1 \ldots c)$ using Eq. (28);
until *cluster prototypes converge*;
The degree of membership of input vector $\boldsymbol{x}_k$ with respect to elliptic cluster i is given by $\sqrt[m+p]{u_{ik}^m t_{ik}^p}$.
$\boldsymbol{x}_k$ is an identified outlier, if $\sum_{i=1}^{c} \sqrt[m+p]{u_{ik}^m t_{ik}^p} < \varepsilon$, where ε is a predefined small constant (e.g. 0.01).

6 Experimental Evaluation

In this section, we will perform some numerical tests to evaluate the behavior of the FPPP partition in various scenarios, examining the robustness and accuracy. Each FPPP-based algorithm is compared with its corresponding probabilistic fuzzy clustering algorithm derived from FCM, and with the additive mixture model derived from PFCM. The pure possibilistic algorithm is excluded from these tests due to its frequently coincident cluster prototypes.

6.1 Evaluation of the c-Means Clustering Algorithms

Two clusters and one outlier input vector. Let us consider two sets of ν data points each, uniformly distributed along unit-radius circles: $\boldsymbol{x}_k = (\cos\frac{2k\pi}{\nu}, 2 + \sin\frac{2k\pi}{\nu})^T$ and $\boldsymbol{x}_{\nu+k} = (\cos\frac{2k\pi}{\nu}, -2 + \sin\frac{2k\pi}{\nu})^T, \forall k = 1 \ldots \nu$. The input data set also includes an outlier, situated at $\boldsymbol{x}_{2\nu+1} = (\delta, 0)^T$, where δ is a parameter with positive real value. We will attempt to classify these $n = 2\nu + 1$ vectors into $c = 2$ clusters,

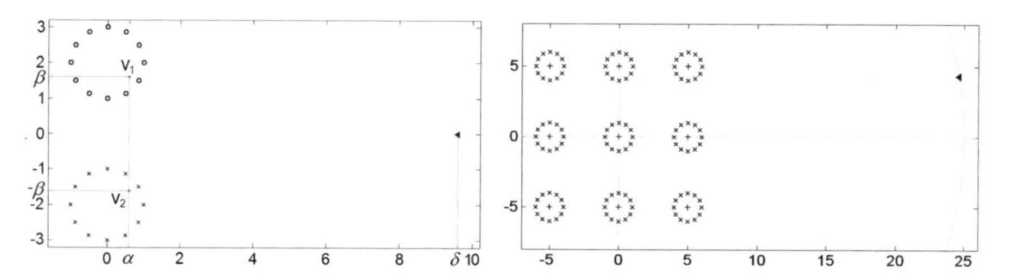

Fig. 1 Two scenarios for the numerical test of robustness: (*left*) two clusters and an outlier, (*right*) nine clusters and an outlier

setting the initial cluster prototypes in the middle of the two circles: $\boldsymbol{v}_1 = (0, 2)^T$ and $\boldsymbol{v}_2 = (0, -2)^T$.

During the iterative optimization of all tested algorithms, the cluster prototypes will be attracted by the outlier vector. As long as the outlier cannot tear off any of the two prototypes, $\boldsymbol{v}_1$ and $\boldsymbol{v}_2$ will behave symmetrically, having their coordinates $\boldsymbol{v}_1 = (\alpha, \beta)^T$ and $\boldsymbol{v}_2 = (\alpha, -\beta)^T$. A graphical representation of the problem is shown in Fig. 1(left). The question is, how α and β will depend on the outlier's position δ in case of all tested algorithms, and how far the outlier vector can go without tearing off one of the cluster prototypes.

Figure 2 presents the outcome of numerical simulations performed on all tested algorithms in various circumstances. In case of all previous algorithms, the further the outlier goes, the stronger it attracts the centroids, and at a certain boundary, one of the prototypes is torn out by the outlier. On the other hand, FP^{3}CM behaves like a gravity system: the further the outlier is situated, the weaker its effect is upon the cluster centroids. No matter how far the outlier is, the obtained partition is correct. The outlier receives such a low membership value to both clusters that it can be easily assigned to the noisy class at defuzzification. Figure 2c shows the behavior of FP^{3}CM in case of various values of possibilistic exponent p, at a constant value of fuzzy exponent $m = 2$. The plots reveal that stronger possibilistic component or lower values of p lead to more efficient rejection of the outlier effect. However, when the outlier is not too far, lower exponent values also cause stronger deviation of the cluster centroids.

Accuracy test with nine regular clusters and an outlier. As it is shown in Fig. 1(right), the input data in this second test consists of 9 sets of vectors uniformly distributed along unit radius circles, situated in the neighborhood of the origin. Initially, the cluster prototypes are placed in the middle of the nine circles. The single outlier vector is placed somewhere along the big circle of radius δ, with its center in the origin. The aim of this study is to establish, which is the limit value for δ where tested algorithms crash in various circumstances.

The obtained limit distances are summarized in Table 1. These values emphasize the fact that previous algorithms may have enhanced the robustness of FCM, they may have enabled the outlier to fall somewhat further (no more than by one order of magnitude) without making the clustering crash. The novel clustering model FP^{3}CM

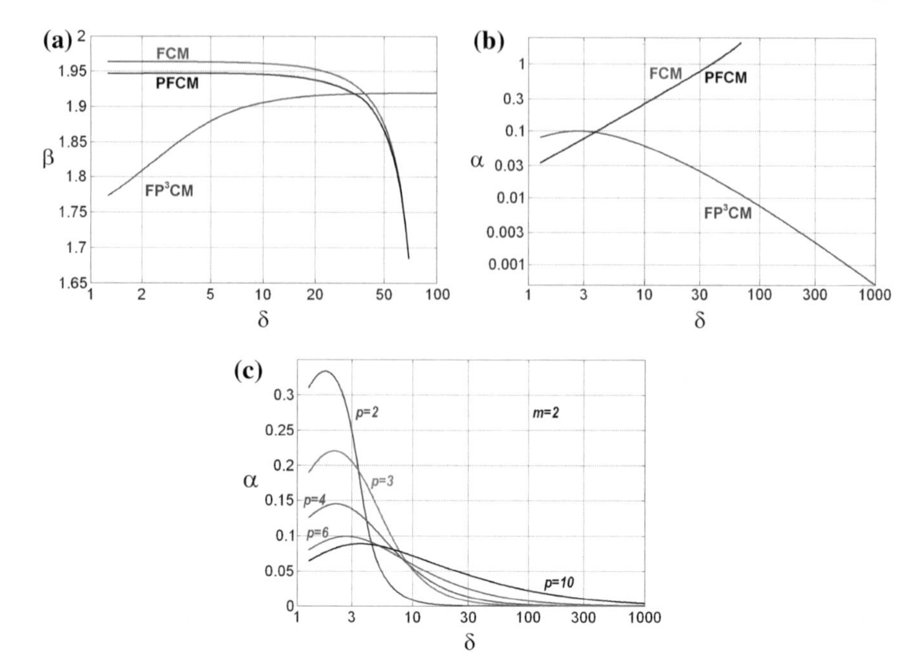

Fig. 2 **a**, **b** Position of the two symmetrical cluster prototypes at $m = 2$ and $p = 6$: **a** the β coordinate plotted against the position of the outlier δ, **b** logarithmic plot of α coordinate against the distance of the outlier. Some of these graphs end at the threshold value of δ where the algorithms fail. In case of FP^{3}CM, the further the outlier wanders, the less influence it has upon cluster prototypes; **c** The α coordinate produced by the proposed algorithm FP^{3}CM, at $m = 2$ and various values of p. The algorithm manages to suppress the effect of extreme outliers

Table 1 The limit distance δ, in case of various algorithms and circumstances, where the tested algorithm fails to produce nine accurate clusters

Algorithm	Circumstances					Limit distance	Algorithm	Circumstances					Limit distance
	m	p	$\sqrt{\eta_i}$	a	b			m	p	$\sqrt{\eta_i}$	a	b	
FCM	2					361	PFCM	2	3	1.0	1	5	437
FPCM	2	5				361	PFCM	2	3	1.5	1	5	521
FPCM	2	2				367	PFCM	2	3	2.0	1	5	593
FPCM	2	1.2				401	PFCM	2	3	2.5	1	5	546
PFCM	2	2	1.0	2	3	410	PFCM	2	2	1.0	1	5	459
PFCM	2	2	1.5	2	3	479	PFCM	2	2	1.5	1	5	602
PFCM	2	2	2.0	2	3	563	PFCM	2	2	2.0	1	5	789
PFCM	2	2	2.5	2	3	649	PFCM	2	2	2.5	1	5	1001
PFCM	2	5	1.0	1	5	394	PFCM	2	2	3.0	1	5	1220
PFCM	2	5	1.5	1	5	421	PFCM	2	2	4.0	1	5	1354
PFCM	2	5	2.0	1	5	428	PFCM	2	2	5.0	1	5	1089
PFCM	2	5	2.5	1	5	370	FP^{3}CM	wide range					$+\infty$

Table 2 Partition accuracies and confusion matrices in various scenarios

Circumstances	IRIS type	FCM			PFCM			FP^{3}CM			Correct decisions
		v_1	v_2	v_3	v_1	v_2	v_3	v_1	v_2	v_3	
no	Setosa	50	0	0	50	0	0	50	0	0	FCM → 136
outlier	Versicolor	0	47	3	0	47	3	0	48	2	PFCM → 136
added	Virginica	0	11	39	0	11	39	0	7	43	FP^{3}CM → 141
outlier	Setosa	50	0	0	50	0	0	50	0	0	FCM → 134
added	Versicolor	0	50	0	0	50	0	0	47	3	PFCM → 135
at20	Virginica	0	16	34	0	15	35	0	7	43	FP^{3}CM → 140
outlier	Setosa	50	0	0	50	0	0	50	0	0	FCM → 128
added	Versicolor	1	49	0	1	49	0	0	47	3	PFCM → 131
at30	Virginica	0	21	29	0	18	32	0	7	43	FP^{3}CM → 140
outlier	Setosa	50	0	0	50	0	0	50	0	0	FCM crashes
added at	Versicolor	3	47	0	3	47	0	0	47	3	PFCM crashes
50or10^6	Virginica	0	50	0	0	50	0	0	7	43	FP^{3}CM → 140

seems to efficiently suppress the influence of the outlier vector, leading to accurate partitions for any high value of δ.

Numerical tests using IRIS data. In the following, we analyze the accuracy and robustness of the investigated clustering models using the IRIS data set [25], which consists of 150 labeled feature vectors of four dimensions (sepal length and width, petal length and width), organized in three clusters ("setosa","versicolor", and "virginica") of fifty vectors each. It is a reported fact, that conventional clustering models like FCM produce 133-134 correct decisions when classifying IRIS data. PFCM produced the best reported accuracy with 140 correct decisions using $a = b = 1, m = p = 3$, and initializing $\boldsymbol{v}_i$ with terminal FCM prototypes [10]. Under less advantageous circumstances, PFCM reportedly produced 136-137 correct decisions.

We have tested the FP^{3}CM clustering model in a wide range of the fuzzy and the possibilistic exponents. The best partition achieved by FP^{3}CM has 141 correct decisions, which is above any reported result. We also need to remark, that almost any parameter setting leads to good partition quality. To make sure FP^{3}CM clusters accurately, the possibilistic term should not be too strong, it is recommendable to keep the possibilistic exponent at $p \geq 2$.

A series of numerical tests using the IRIS data targeted the clustering robustness. We artificially inserted an outlier vector into the input data set, with coordinates $\boldsymbol{x}_{151} = (\delta, \delta, \delta, \delta)^T$, and proceeded all vectors to clustering into $c = 3$ groups. Table 2 gives us an overview upon accuracy, confusion matrices, and sensibility to the outlier's position. As we can see in the table, most existing clustering models failed somewhere between $\delta = 30$ and $\delta = 50$, while the FP^{3}CM algorithm led to high quality partition even at $\delta = 10^6$, being less affected by distant outliers. All these tests were performed at $m = 2.0$, $p = 3.5$, $\sqrt{\eta_i} = 0.7\ \forall i = 1 \ldots c$, $a = 1$, and $b = 5$. Further details can be found in [11].

Application in blind speaker grouping. The FP^{3}CM clustering algorithm was employed in a human speech processing framework, to distinguish the voice signals of various speakers in a small group. Also in this application, the FPPP partition based c-means clustering outperformed the FCM and PFCM algorithms in terms of accuracy. Details are reported in [26].

6.2 Evaluation of the c-Spherical Shells Clustering Algorithms

In the following, we will perform some numerical tests to evaluate the robustness and accuracy of the c-spherical shell algorithms. We will compare the performance of FP^{3}CSS with counter candidates like the FCSS [13], and the unpublished possibilistic-fuzzy c-spherical shell (PFCSS) clustering we derived from the ultimate robust and accurate PFCM [10]. It is necessary to remark that most shell clustering algorithms were published without testing them in noisy environment. This study investigates the accuracy of algorithms in noisy environment, in order to emphasize the advantages of the FPPP partition.

Four circles and one outlier. Let us consider four sets of ν data points each, uniformly distributed in the proximity of four circles of equal radius, as shown in Fig. 3. The input data set also includes an outlier, situated at $\boldsymbol{x}_{4\nu+1} = (0, \delta)^T$, where δ is a positive real valued parameter. We will attempt to find $c = 4$ two-dimensional shell clusters among these $n = 4\nu + 1$ vectors. These input vectors were fed to three algorithms: FCSS, PFCSS, and the proposed FP^{3}CSS, setting the initial clusters in the proximity of the ideal solution using the FCM-based technique mentioned in Sect. 4. The question is, how these algorithms will identify the four circles within the data set.

Figure 3 exhibits the obtained circles for various values of the δ parameter. The circles identified by FCSS are drawn with dash-and-dotted line, the ones of PFCSS with dashed lines, while the circles of the FP^{3}CSS algorithm are represented with continuous lines. Although FCSS gives the outlier approximately $1/c = 0.25$ probability to belong to each of the classes, this does not visibly affect the identified circles when outlier stays at short distance ($\delta \approx 10$). However, as the outlier goes further, the identified circles drift away from the actual position of the input vectors, and at a certain level of δ, one of the identified circles will jump to cross the outlier. At this level, FCSS can be considered failed. PFCSS can keep the identified circles in the proximity of the ideal solution somewhat further than FCSS, but when it loses the control (at $\delta \approx 14$) it starts acting similarly to FCSS. Tests have revealed that the product partition based algorithm identifies all circles with high accuracy even at high values of δ in order of 10^3–10^6. These results were produced at the following parameter settings:

- FCSS used the most popular value of the fuzzy exponent, $m = 2$;
- PFCSS was performed using a stronger possibilistic factor caused by fuzzy exponent set to $m = 6$ and possibilistic exponent to $p = 2$. Trade-off parameters were

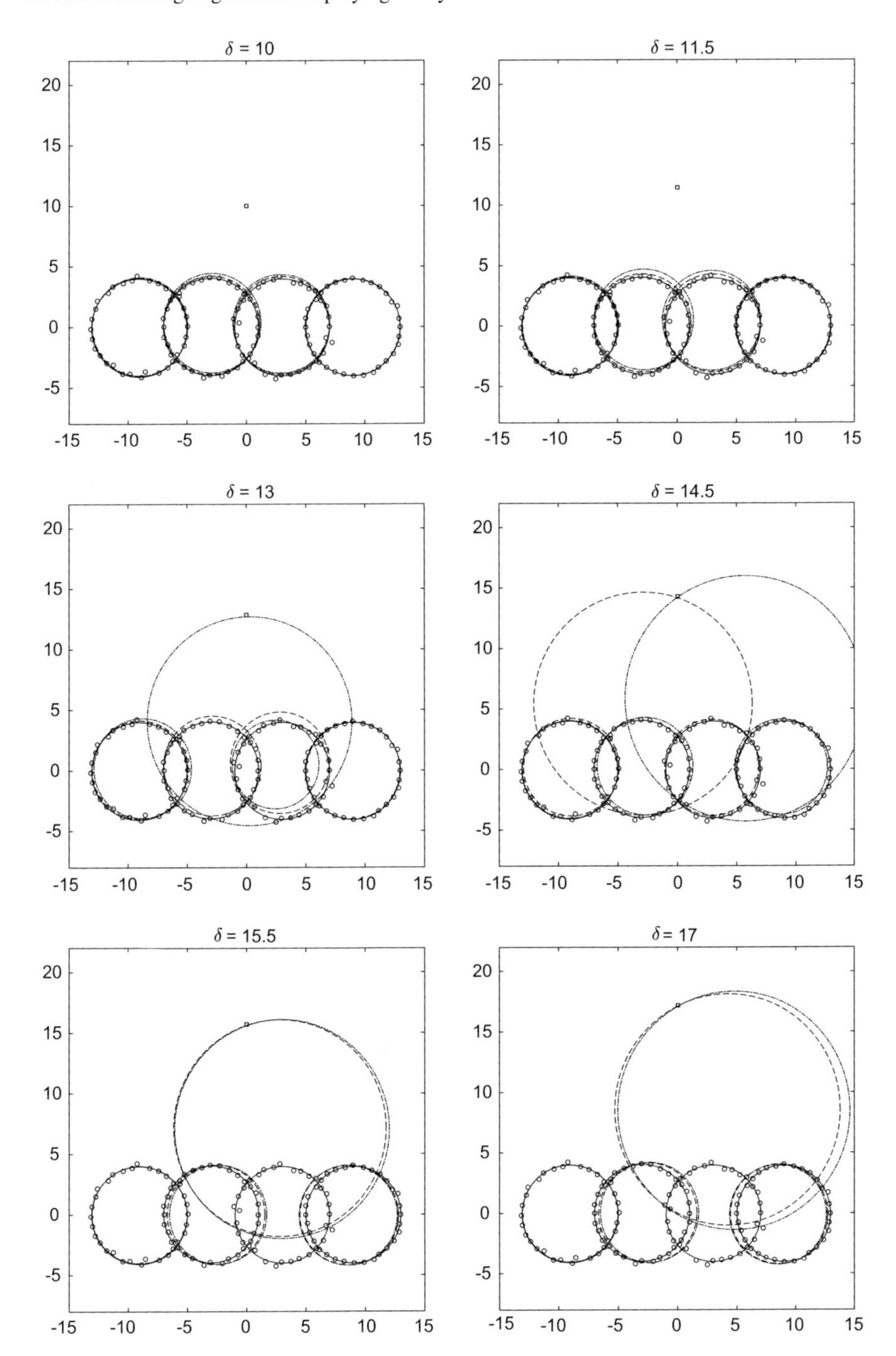

Fig. 3 Four identified circles in various circumstances. Circles of FCSS are drawn with *dash-and-dotted lines*, the result of PFCSS with *dashed lines*, while the identified circles of the FP^3CSS are drawn with *continuous lines*

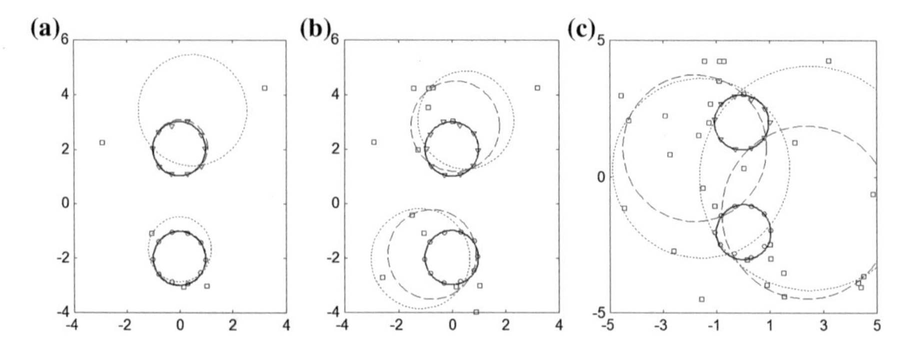

Fig. 4 Identified circles in case of two circle clusters of $\nu = 10$ vectors each, and an increasing number of outliers: **a** 5 outliers; **b** 15 outliers; **c** 30 outliers. Cluster prototypes were initially set in the proximity of the ideal solution. Outliers vectors are indicated by *squares* (□). Identified *circles* of FCSS are drawn with *dotted lines*, the result of PFCSS with *dashed lines*, while the identified circles of the proposed FP^{3}CSS are represented with *continuous lines*

also set to favor the possibilistic term: $a = 1$ and $b = 5$; the possibilistic penalty terms η_i were established by the rules set in [6].

- FP^{3}CSS was using the same settings as PFCSS for m, p, and η_i.

Two circles and several outliers. In a new example, we use two circle shaped clusters defined by $\nu = 12$ data points each. This time we add several outliers at random positions within a larger rectangular box that incorporates both circles. The number of added outliers gradually grows from 5 to 30, and the clustering is performed using initialization of the cluster prototypes in the proximity of the ideal solutions.

Figure 4 summarizes the outcome of the tested three algorithms in various circumstances. The chance of producing errors visibly grows with the number of added outliers. FCSS fails earlier than PFCSS, but practically there is no qualitative difference in their outcome. These algorithms cannot keep the cluster prototypes even if they are initialized with the ideal positions. On the other hand, the product partition based algorithm can handle the case even if the number of added outliers exceeds the count of correct input vectors. It is necessary to remark that all algorithms fail if initialized randomly. Consequently, choosing good estimates at the beginning is indeed a key issue. Further details and test cases can be found in [23], including a three-dimensional problem with two spheres and several outliers.

6.3 *Evaluation of the c-Elliptical Shells Clustering Algorithms*

Let us define a problem with $c = 12$ ellipses to be identified, of random radii and orientation, each represented by $\nu = 18$ data points situated along the boundary, as shown in the middle panel of Fig. 5. The ellipse centers are situated on a circle

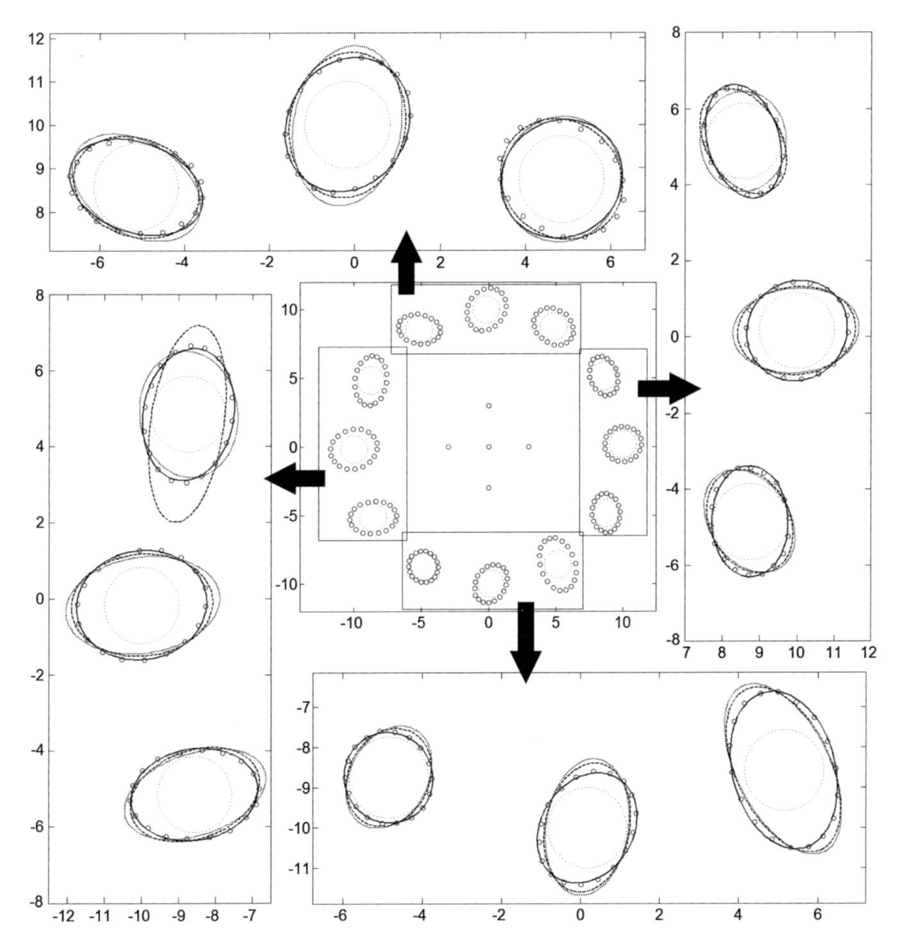

Fig. 5 Magnified view of identified *ellipses* in case of $12\nu + 5 = 221$ input vectors and $c = 12$ elliptic shell clusters. Cluster prototypes were initialized as unit *radius circles* centered in ideal position. *Ellipses* identified by FCES are drawn with *dotted lines*, the result of PFCES with *dashed lines*, while the identified ellipses of the proposed FP^{3}CES are represented with *continuous lines*

of 10-unit radius, thus the center of ellipse number i ($i = 1 \dots c$) is given by $\overline{\boldsymbol{v}}_i = (10\cos(2\pi i/c),\, 10\sin(2\pi i/c))^T$. Five further data points (not too distant outliers) are added to the input data set, as shown in Fig. 5. The resulting data set is clustered to $c = 12$ elliptic shell clusters via FCES, PFCES and FP^{3}CES algorithms. Initial cluster prototypes are set as unit radius circles placed at $\overline{\boldsymbol{v}}_i$ ($i = 1 \dots c$), drawn in Fig. 5 with light and narrow dotted lines. The four outer panels of Fig. 5 exhibits the outcome of all three algorithms.

Again, we can easily notice the superiority of the FP^{3}CES algorithm as it identifies all ellipses with hardly visible mistakes, while the outliers bring considerable obstacles for the counter candidate algorithms. In general, FCES produces stronger distortions than PFCES, but none of them can be called accurate in the presence of outliers. These results were produced using the following parameter settings:

- FCES was using the most popular value of the fuzzy exponent, $m = 2$;
- PFCES was performed using a stronger possibilistic factor (as recommended by its authors [10]) caused by fuzzy exponent set to $m = 2.5$ and possibilistic exponent to $p = 1.5$. Trade-off parameters were set equal: $a = b = 1$; the possibilistic penalty terms η_i were established by the rules set in [6].
- FP^3CES was using the same settings as PFCES for m, p, and η_i.

The initialization of cluster prototypes is a key issue that strongly influences the outcome of the clusters. As the cost function may have several local minima, it is important to start the algorithm in the neighborhood of the global optimum. Without proper initialization, none of the tested elliptic shell clustering algorithms can produce accurate results.

7 Conclusion

This chapter presented several variants of fuzzy-possibilistic product partition based clustering algorithms, and showed their improved behavior compared to previous clustering models, emphasizing the efficient outlier rejection. Besides the better accuracy and robustness, another advantage of the fuzzy-possibilistic product partition compared to the possibilistic-fuzzy partition [10], is the lack of trade-off parameters between probabilistic and possibilistic terms, making the proposed algorithm easier to tune. Consequently, algorithms derived from FP^3CM are good candidates for various data mining problems in noisy environment.

References

1. Bezdek, J. C.: Pattern recognition with fuzzy objective function algorithms. Plenum, New York (1981)
2. Davé, R. N.: Characterization and detection of noise in clustering. Patt. Recogn. Lett. 12, 657–664 (1991)
3. Menard, M., Damko, C., Loonis, P.: The fuzzy $c + 2$ means: solving the ambiguity rejection in clustering. Patt. Recogn. 33, 1219–1237 (2000)
4. Chintalapudi, K. K., Kam, M.: A noise-resistant fuzzy c-means algorithm for clustering. IEEE World Congr. Comput. Intell. 1458–1463 (1998)
5. Alanzado, A. C., Miyamoto, S.: Fuzzy c-means clustering in the presence of noise cluster for time series analysis. Proc. Modeling Decisions in Artificial Intelligence (MDAI), Lect. Notes Comp. Sci. 3558, 156–163 (2005)
6. Krishnapuram, R., Keller, J. M.: A possibilistic approach to clustering. IEEE Trans. Fuzzy Syst. 1, 98–110 (1993)
7. Barni, M., Capellini, V., Mecocci, A.: Comments on a possibilistic approach to clustering. IEEE Trans. Fuzzy Syst. 4, 393–396 (1996)

8. Timm, H., Borgelt, C., Döring, C., Kruse, R.: An extension to possibilistic fuzzy cluster analysis. Fuzzy Sets and Systems 147, 3–16 (2004)
9. Pal, N. R., Pal, K., Bezdek, J. C.: A mixed c-means clustering model. Proc. IEEE Int'l Conf. Fuzzy Systems (FUZZ-IEEE), pp. 11–21 (1997)
10. Pal, N. R., Pal, K., Keller, J. M., Bezdek, J. C.: A possibilistic fuzzy c-means clustering algorithm. IEEE Trans. Fuzzy Syst. 13, 517–530 (2005)
11. Szilágyi, L.: Fuzzy-Possibilistic Product Partition: a novel robust approach to c-means clustering. Proc. Modeling Decisions in Artificial Intelligence (MDAI), Lect. Notes Comp. Sci. 6820, 150–161 (2011)
12. Gunderson, R.: An adaptive FCV clustering algorithm. Int. J. Man-Mach. Stud. 19, 97–104 (1983)
13. Krishnapuram, R., Nasraoui, O., Frigui, H.: A fuzzy c spherical shells algorithm: a new approach. IEEE Trans. Neur. Netw. 3, 663–671 (1992)
14. Davé, R. N.: Generalized fuzzy c-shells clustering and detection of circular and elliptical boundaries. Pattern Recogn. 25(7), 713–721 (1992)
15. Frigui, H., Krishnapuram, R.: A comparison of fuzzy shell clustering methods for the detection of ellipses. IEEE Trans. Fuzzy Syst. 4, 193–199 (1996)
16. Krishnapuram, R., Frigui, H., Nasraoui, O.: New fuzzy shell clustering algorithms for boundary detection and pattern recognition. SPIE Proc. Robot. Comp. Vis. 1607, 1460–1465 (1991)
17. Davé, R. N., Bhaswan, K.: Adaptive fuzzy c-shells clustering and detection of ellipses. IEEE Trans. Neural Netw. 3(5), 643–662 (1992)
18. Krishnapuram, R., Frigui, H., Nasraoui, O.: Fuzzy and possibilistic shell clustering algorithms and their application to boundary detection and surface approximation - Part I. IEEE Trans. Fuzzy Syst. 3, 29–43 (1995)
19. Bezdek, J.C., Hathaway, R.J., Pal, N.R.: Norm induced shell prototype (NISP) clustering. Neur. Parall. Sci. Comput. 3, 431–450 (1995)
20. Höppner, F.: Fuzzy shell clustering algorithms in image processing: fuzzy c-rectangular and 2-rectangular shells. IEEE Trans. Fuzzy Syst. 5, 599–613 (1997)
21. Steinhaus, H.: Sur la division des corp materiels en parties. Bull. Acad. Pol. Sci. C1 III. (IV) 801–804 (1956)
22. Szilágyi, L., Szilágyi, S. M., Benyó, B., Benyó, Z.: Intensity inhomogeneity compensation and segmentation of MR brain images using hybrid c-means clustering models. Biomed. Sign. Proc. Contr. 6, 3–12 (2011)
23. Szilágyi, L.: Robust spherical shell clustering using fuzzy-possibilistic product partition. Int. J. Intell. Syst. 28, 524–539 (2013)
24. Szilágyi, L., Varga, Zs. R., Szilágyi, S. M.: Application of the fuzzy-possibilistic product partition in elliptic shell clustering. Proc. Modeling Decisions in Artificial Intelligence (MDAI), Lect. Notes Comp. Sci. 8825, 158–169 (2014)
25. Anderson, E.: The IRISes of the Gaspe peninsula. Bull. Amer. IRIS Soc. 59, 2–5 (1935)
26. Gosztolya, G., Szilágyi, L.: Application of fuzzy and possibilistic c-means clustering models in blind speaker clustering. Acta Polytech. Hung. 12(7), 41–56 (2015)

Consensus-Based Agglomerative Hierarchical Clustering

José Luis García-Lapresta and David Pérez-Román

Abstract In this contribution, we consider that a set of agents assess a set of alternatives through numbers in the unit interval. In this setting, we introduce a measure that assigns a degree of consensus to each subset of agents with respect to every subset of alternatives. This consensus measure is defined as 1 minus the outcome generated by a symmetric aggregation function to the distances between the corresponding individual assessments. We establish some properties of the consensus measure, some of them depending on the used aggregation function. We also introduce an agglomerative hierarchical clustering procedure that is generated by similarity functions based on the previous consensus measures.

Keywords Consensus · Clustering · Aggregation functions · OWA operators

1 Introduction

When a group on agents show their opinions about a set of alternatives, an important issue is to know the homogeneity of these opinions. In this chapter we consider that agents evaluate each alternative by means of a number in the unit interval. For measuring the consensus in a group of agents over a subset of alternatives, we propose to aggregate the distances between the corresponding individual assessments through an appropriate symmetric aggregation function. This outcome measures the dispersion of individual opinions in a similar way to the Gini index [19] measures the inequality of individual incomes (see Yitzhaki [32]). The consensus measure we propose is just 1 minus the mentioned dispersion measure.

J.L. García-Lapresta (✉)
PRESAD Research Group, BORDA Research Unit, IMUVA,
Departamento de Economía Aplicada, Universidad de Valladolid, Valladolid, Spain
e-mail: lapresta@eco.uva.es

D. Pérez-Román
PRESAD Research Group, BORDA Research Unit, Departamento de Organización de Empresas y C.I.M., Universidad de Valladolid, Valladolid, Spain
e-mail: david@emp.uva.es

V. Torra et al. (eds.), *Fuzzy Sets, Rough Sets, Multisets and Clustering*,
Studies in Computational Intelligence 671, DOI 10.1007/978-3-319-47557-8_8

The most important is not to know the degree of consensus in a specific group of agents, but comparing the consensus of different group of agents with respect to an alternative or a subset of alternatives. This is the starting point of the agglomerative hierarchical clustering procedure we propose. We consider as linkage clustering criterion one generated by a consensus-based similarity function that merges clusters or individuals by maximizing the consensus.

The rest of the chapter is organized as follows. Section 2 contains some notation and basic notions. In Sect. 3, we introduce and analyze the proposed consensus measures. Section 4 contains our proposal of consensus-based agglomerative hierarchical clustering. In Sect. 5, we illustrate the introduced procedures with an example. Finally, in Sect. 6, we conclude with some remarks.

2 Preliminaries

Given $\boldsymbol{y}, \boldsymbol{z} \in [0,1]^k$, by $\boldsymbol{y} \geq \boldsymbol{z}$ we mean $y_i \geq z_i$ for every $i \in \{1, \ldots, k\}$. Given $\boldsymbol{y} \in [0,1]^k$, the decreasing reordering of the coordinates of $\boldsymbol{y}$ is indicated as $y_{[1]} \geq \cdots \geq y_{[k]}$. In particular, $y_{[1]} = \max\{y_1, \ldots, y_k\}$ and $y_{[k]} = \min\{y_1, \ldots, y_k\}$.

Given a real number y, by $\lfloor y \rfloor$ we denote the *integer part* of y, i.e., the greatest integer number smaller than or equal to y.

With $\#I$ we denote the cardinality of I. With $\mathscr{P}_2(A) = \{I \subseteq A \mid \#I \geq 2\}$ we denote the family of subsets of at least two elements.

We begin by defining standard properties of real functions on $[0,1]^k$. For further details the interested reader is referred to Fodor and Roubens [12], Calvo et al. [6], Beliakov et al. [4], Torra and Narukawa [27], Grabisch et al. [20] and Beliakov et al. [3].

Definition 1 Let $F : [0,1]^k \longrightarrow [0,1]$ be a function.

(1) F is *idempotent* if for every $y \in [0,1]$ it holds $F(y \cdot \mathbf{1}) = y$.
(2) F is *symmetric* if for every permutation π on $\{1, \ldots, k\}$ and every $\boldsymbol{y} \in [0,1]^k$ it holds $F(y_{\pi(1)}, \ldots, y_{\pi(k)}) = F(\boldsymbol{y})$.
(3) F is *monotonic* if for all $\boldsymbol{y}, \boldsymbol{z} \in [0,1]^k$ it holds $\boldsymbol{y} \geq \boldsymbol{z} \Rightarrow F(\boldsymbol{y}) \geq F(\boldsymbol{z})$.
(4) F is *compensative* if for every $\boldsymbol{y} \in [0,1]^k$ it holds $y_{[k]} \leq F(\boldsymbol{y}) \leq y_{[1]}$.
(5) F is *self-dual* if for every $\boldsymbol{y} \in [0,1]^k$ it holds $F(\mathbf{1} - \boldsymbol{y}) = 1 - F(\boldsymbol{y})$.
(6) F is *stable for translations* if for all $\boldsymbol{y} \in [0,1]^k$ and $t \in [0,1]$ such that $\boldsymbol{y} + t \cdot \mathbf{1} \in [0,1]^k$ it holds $F(\boldsymbol{y} + t \cdot \mathbf{1}) = F(\boldsymbol{y}) + t$.

Definition 2

(1) Given $k \in \mathbb{N}$, a function $F^{(k)} : [0,1]^k \longrightarrow [0,1]$ is called an k-ary aggregation function if it is monotonic and satisfies the boundary conditions $F^{(k)}(\mathbf{0}) = 0$ and $F^{(k)}(\mathbf{1}) = 1$. In the extreme case of $k = 1$, the convention $F^{(1)}(y) = y$ for every $y \in [0,1]$ is considered.
(2) An *aggregation function* is a sequence $F = \left(F^{(k)}\right)_{k \in \mathbb{N}}$ of k-ary aggregation functions.

(3) An aggregation function $F = \left(F^{(k)}\right)_{k \in \mathbb{N}}$ satisfies a property (in particular, those appearing in Definition 1) whenever $F^{(k)}$ satisfies the same property for every $k \in \mathbb{N}$.

It is easy to see that for every k-ary aggregation function, idempotency and compensativeness are equivalent.

For the sake of simplicity, the k-arity is omitted whenever it is clear from the context.

An interesting class of aggregation functions is the family of OWA operators, introduced by Yager [29].

A *weighting vector* of dimension k is a vector $\boldsymbol{w} = (w_1, \ldots, w_k) \in [0, 1]^k$ such that $\sum_{i=1}^{k} w_i = 1$.

Definition 3 Given a weighting vector $\boldsymbol{w}$ of dimension k, the *OWA operator associated with* $\boldsymbol{w}$ is the aggregation function $F_{\boldsymbol{w}} : [0, 1]^k \longrightarrow [0, 1]$ defined as

$$F_{\boldsymbol{w}}(y_1, \ldots, y_k) = \sum_{i=1}^{k} w_i \cdot y_{[i]}.$$

Some well-known aggregation functions are specific cases of OWA operators. With appropriate weighting vectors $\boldsymbol{w} = (w_1, \ldots, w_k)$ we obtain

(1) The *maximum*, for $\boldsymbol{w} = (1, 0, \ldots, 0)$.
(2) The *minimum*, for $\boldsymbol{w} = (0, \ldots, 0, 1)$.
(3) The *arithmetic mean*, for $\boldsymbol{w} = \left(\frac{1}{k}, \ldots, \frac{1}{k}\right)$.
(4) The *t-trimmed means*:

- If $t = 1$, for $\boldsymbol{w} = \left(0, \frac{1}{k-2}, \ldots, \frac{1}{k-2}, 0\right)$.
- If $t = 2$, for $\boldsymbol{w} = \left(0, 0, \frac{1}{k-4}, \ldots, \frac{1}{k-4}, 0, 0\right)$.
- ....

(5) The *median*:

(a) If k is odd, for $w_i = \begin{cases} 1, \text{ if } i = \frac{k+1}{2}, \\ 0, \text{ otherwise.} \end{cases}$

(b) If k is even, for $w_i = \begin{cases} 0.5, \text{ if } i \in \left\{\frac{k}{2}, \frac{k}{2} + 1\right\}, \\ 0, \quad \text{otherwise.} \end{cases}$

(6) The *mid-range*, for $\boldsymbol{w} = (0.5, 0, \ldots, 0, 0.5)$.

OWA operators are continuous, idempotent (hence, compensative), symmetric, and stable for translations. They have been characterized by Fodor et al. [11].

Centered OWA operators have been introduced by Yager [31] in order to give "the most weight to the central scores in the argument tuples and less weighting to the extreme values". We now introduce a more general notion than that provided by Yager. It was introduced by García-Lapresta and Martínez-Panero [14].

Definition 4 Given a weighting vector $\boldsymbol{w}$ of dimension k, the OWA operator associated with $\boldsymbol{w}$ is *centered* if the following two conditions are satisfied:

(1) $w_{k+1-i} = w_i$ for every $i \in \{1, \ldots, k\}$.
(2) $w_i \le w_j$ whenever $i < j \le \lfloor \frac{k+1}{2} \rfloor$ or $i > j \ge \lfloor \frac{k+1}{2} \rfloor$.

The first condition is equivalent to the property of self-duality (see García-Lapresta and Llamazares [13, Proposition 5]). The second condition is weaker than the original of Yager [31], called *strongly decaying*, that requires strict inequalities $w_i < w_j$.

Yager [31] requires a third condition in the definition of centered OWA operators, *inclusiveness*: $w_i > 0$ for every $i \in \{1, \ldots, k\}$. That condition is very restrictive for our purposes, since it eliminates some interesting OWA operators as median and trimmed means, among others.

Definition 5 An *extended OWA (EOWA) operator* is a sequence of OWA operators $(F_{\boldsymbol{w}^k})_{k \in \mathbb{N}}$ with associated weighting vectors $\boldsymbol{w}^k = (w_1^k, \ldots, w_k^k)$, one for each dimension $k \in \mathbb{N}$.

Following Mayor and Calvo [24], Calvo and Mayor [7], Beliakov et al. [4, pp. 54–56] and Beliakov et al. [3, pp. 73–76]), we can show graphically an EOWA operator as a weighting triangle where the entries in each row add up to one:

$$
\begin{array}{ccccccccc}
 & & & & w_1^1 & & & & \\
 & & & w_1^2 & & w_2^2 & & & \\
 & & w_1^3 & & w_2^3 & & w_3^3 & & \\
 & w_1^4 & & w_2^4 & & w_3^4 & & w_4^4 & \\
w_1^5 & & w_2^5 & & w_3^5 & & w_4^5 & & w_5^5 \\
\cdots & \cdots & \cdots & \cdots & \cdots & \cdots & \cdots & \cdots & \cdots
\end{array}
$$

A very useful approach for obtaining the EOWA weights is the functional method introduced by Yager [30, 31]. Given a *BUM function*, i.e., a monotonic function $f : [0, 1] \longrightarrow [0, 1]$ such that $f(0) = 0$ and $f(1) = 1$, the associated EOWA weights are defined as

$$
w_i^k = f\left(\frac{i}{k}\right) - f\left(\frac{i-1}{k}\right), \quad i = 1, \ldots, k. \tag{1}
$$

Yager [31] proposes to generate BUM functions by means of centering functions.

A *centering function* is a function $g : [0, 1] \longrightarrow \mathbb{R}$ satisfying the following conditions:

1 $g(x) > 0$ for every $x \in [0, 1]$.
2 $g(0.5 + x) = g(0.5 - x)$ for every $x \in [0, 0.5]$.
3 $g(x) < g(y)$ for $x < y \le 0.5$ and $g(x) < g(y)$ for $x > y \ge 0.5$.

Then, the function $f : [0, 1] \longrightarrow [0, 1]$ defined as

$$f(x) = \frac{\displaystyle\int_0^x g(y)\,dy}{\displaystyle\int_0^1 g(y)\,dy} \tag{2}$$

is a BUM function.

3 Consensus

For measuring the degree of consensus among a group of agents that provide their opinions on a set of alternatives, different proposals can be found in the literature (see Martínez-Panero [23] for an overview of different notions of consensus).

In the social choice framework, the notion of *consensus measure* was introduced by Bosch [5] in the context of linear orders. Additionally, Bosch [5] and Alcalde-Unzu and Vorsatz [1] provided axiomatic characterizations of several consensus measures in the context of linear orders. García-Lapresta and Pérez-Román [15] extended that notion to the context of weak orders and they analyzed a class of consensus measures generated by distances. Alcantud et al. [2] provided axiomatic characterizations of some consensus measures in the setting of approval voting. In turn, Erdamar et al. [8] extended the notion of consensus measure to the preference-approval setting through different kinds of distances, and García-Lapresta et al. [18] introduced another extension to the framework of hesitant linguistic assessments.

Let $A = \{1, \ldots, m\}$, with $m \geq 2$, be a set of agents and let $X = \{x_1, \ldots, x_n\}$, with $n \geq 2$, be the set of alternatives which have to be evaluated in the unit interval.

A *profile* is a matrix

$$V = \begin{pmatrix} v_1^1 & \cdots & v_i^1 & \cdots & v_n^1 \\ \cdots & \cdots & \cdots & \cdots & \cdots \\ v_1^a & \cdots & v_i^a & \cdots & v_n^a \\ \cdots & \cdots & \cdots & \cdots & \cdots \\ v_1^m & \cdots & v_i^m & \cdots & v_n^m \end{pmatrix} = \left(v_i^a\right)$$

consisting of m rows and n columns of numbers in $[0, 1]$, where the element v_i^a represents the assessment given by the agent $a \in A$ to the alternative $x_i \in X$.

Let $V = \left(v_i^a\right)$ be a profile, π a permutation on A, σ a permutation on $\{1, \ldots, n\}$, $I \in \mathscr{P}_2(A)$ and $\emptyset \neq Y \subseteq X$. The profiles V^π, V_σ and V^{-1}, and the subsets I^π and Y_σ are defnined as follows:

(1) $V^\pi = \left(u_i^a\right)$ where $u_i^a = v_i^{\pi(a)}$.
(2) $V_\sigma = \left(u_i^a\right)$ where $u_i^a = v_{\sigma(i)}^a$.
(3) $V^{-1} = \left(u_i^a\right)$ where $u_i^a = 1 - v_i^a$.
(4) $I^\pi = \left\{\pi^{-1}(a) \mid a \in A\right\}$, i.e., $a \in I^\pi \;\Leftrightarrow\; \pi(a) \in I$.
(5) $Y_\sigma = \{x_{\sigma^{-1}(i)} \mid x_i \in Y\}$, i.e., $x_i \in Y_\sigma \;\Leftrightarrow\; x_{\sigma(i)} \in Y$.

We now introduce a consensus measure associated with a symmetric aggregation function. Given a profile, it assigns a degree of consensus in each subset of at least two agents with respect to a subset of alternatives.

Definition 6 Let $F = \left(F^{(k)}\right)_{k\in\mathbb{N}}$ be a symmetric aggregation function. Given a profile $V = (v_i^a)$, the *degree of consensus* in a subset of agents $I \in \mathscr{P}_2(A)$ over a subset of alternatives $\emptyset \neq Y \subseteq X$ is defnined as

$$C_F(V, I, Y) = 1 - F\left(\left|v_i^a - v_i^b\right|_{\substack{a,b\in I\,,\,a<b\\ x_i\in Y}}\right).$$

In Proposition 1 we establish some properties of the consensus notion introduced in Definition 6. Normalization means that the degree of consensus is always in the unit interval. Anonymity means that all agents are treated in the same way. Unanimity establishes necessary and sufficient conditions for reaching maximum consensus. Maximum dissension establishes necessary and sufficient conditions for reaching minimum consensus in two agents. Positiveness establishes that with more than two agents the degree of consensus is never minimum. Neutrality means that all alternatives are treated in the same way. And reciprocity means that if all the agents reverse their assessments, then the degree of consensus does not change.

Proposition 1 *Let $F = \left(F^{(k)}\right)_{k\in\mathbb{N}}$ be a symmetric aggregation function. The following properties are satisfied:*

(1) Normalization: $C_F(V, I, Y) \in [0, 1]$.
(2) Anonymity: $C_F(V^\pi, I^\pi, Y) = C_F(V, I, Y)$ for every permutation π on A.
(3) Unanimity: If for every $x_i \in Y$ there exists $t_i \in [0, 1]$ such that $v_i^a = t_i$ for every $a \in I$, then $C_F(V, I, Y) = 1$.
Additionally, if $F^{(k)}(\mathbf{y}) = 0 \Leftrightarrow \mathbf{y} = \mathbf{0}$, for all $k \in \mathbb{N}$ and $\mathbf{y} \in [0, 1]^k$, and $C_F(V, I, Y) = 1$, then for every $x_i \in Y$ there exists $t_i \in [0, 1]$ such that $v_i^a = t_i$ for every $a \in I$.
(4) Maximum dissension: If $\left(\left(v_i^a = 0 \text{ and } v_i^b = 1\right) \text{ or } \left(v_i^a = 1 \text{ and } v_i^b = 0\right)\right)$ for all $x_i \in Y$, then $C_F(V, \{a, b\}, Y) = 0$.
Additionally, if $F^{(k)}(\mathbf{y}) = 1 \Leftrightarrow \mathbf{y} = \mathbf{1}$, for all $k \in \mathbb{N}$ and $\mathbf{y} \in [0, 1]^k$, and $C_F(V, \{a, b\}, Y) = 0$, then $\left(\left(v_i^a = 0 \text{ and } v_i^b = 1\right) \text{ or } \left(v_i^a = 1 \text{ and } v_i^b = 0\right)\right)$ for all $x_i \in Y$.
(5) Positiveness: If $F^{(k)}(\mathbf{y}) = 1 \Leftrightarrow \mathbf{y} = \mathbf{1}$, for all $k \in \mathbb{N}$ and $\mathbf{y} \in [0, 1]^k$, and $\#I > 2$, then $C_F(V, I, Y) > 0$.
(6) Neutrality: $C_F(V_\sigma, I, Y_\sigma) = C_F(V, I, Y)$ for every permutation σ on $\{1, \ldots, n\}$.
(7) Reciprocity: $C_F(V^{-1}, I, Y) = C_F(V, I, Y)$.

Proof It is straightforward.

Remark 1 Let $(F_{\boldsymbol{w}^k})_{k\in\mathbb{N}}$ an EOWA operator with associated weighting vectors $\boldsymbol{w}^k = (w_1^k, \ldots, w_k^k)$, $k \in \mathbb{N}$. It is easy to check that $F_{\boldsymbol{w}^k}(\boldsymbol{y}) = 0 \Leftrightarrow \boldsymbol{y} = \boldsymbol{0}$, for every $\boldsymbol{y} \in [0, 1]^k$, if and only if $w_1^k > 0$; and $F_{\boldsymbol{w}^k}(\boldsymbol{y}) = 1 \Leftrightarrow \boldsymbol{y} = \boldsymbol{1}$, for every $\boldsymbol{y} \in [0, 1]^k$, if and only if $w_k^k > 0$.

Consequently, any EOWA operator satisfying $w_1^k > 0$ and $w_k^k > 0$ for every $k \in \mathbb{N}$ verifies all the properties included in Proposition 1. Therefore, when considering the EOWA operators generated by the maximum, the minimum, the trimmed means and the median, the corresponding consensus measures do not satisfy the strong versions of unanimity and maximum dissension.

For our purposes, an interesting class of EOWA operators is the one generated by centered OWA operators (in the sense of Definition 4) satisfying $w_1^k = w_k^k > 0$ for every $k \in \mathbb{N}$.

4 Clustering

There are many clustering algorithms (see Ward [28], Jain et al. [21] and Everitt et al. [9], among others). Most methods of hierarchical clustering use an appropriate metric (for measuring the distance between pairs of observations), and a linkage criterion which specifies the similarity/dissimilarity of sets as a function of the pairwise distances of observations in the corresponding sets.

Ward [28] proposed an agglomerative hierarchical clustering procedure, where the criterion for choosing the pair of clusters to merge at each step is based on the optimization of an objective function.

Usually, clusters are merged by minimizing a distance between clusters. The complete, single and average linkage clustering take into account the maximum, minimum and mean distance between elements of each cluster, respectively. In turn, centroid linkage clustering is based on the distances between the clusters centroids.

In all the mentioned linkage clustering criteria there is a loss of information. In our proposal, clusters are merged when maximizing the consensus and, consequently, all the information is used for merging clusters.

Definition 7 Let $F = \left(F^{(k)}\right)_{k\in\mathbb{N}}$ be a symmetric aggregation function. Given a profile $V = (v_i^a)$, the *similarity function* relative to a subset of alternatives $\emptyset \neq Y \subseteq X$

$$S_F^Y : \left(\mathscr{P}(A) \setminus \{\emptyset\}\right)^2 \longrightarrow [0, 1]$$

is defined as

$$S_F^Y(I, J) = \begin{cases} C_F(V, I \cup J, Y), & if\ \#(I \cup J) \geq 2, \\ 1, & if\ \#(I \cup J) = 1. \end{cases}$$

Remark 2 In the extreme case of two agents and a single alternative, the similarity between these agents on that alternative is just 1 minus the distance between their assessments. More formally, given an alternative $x_i \in X$ and two different agents $a, b \in A$, we have

$$S_F^{\{x_i\}}(\{a\},\{b\}) = C_F(V,\{a,b\},\{x_i\}) = 1 - \left|v_i^a - v_i^b\right|.$$

The agglomerative hierarchical clustering procedure we propose has some similarities to the ones provided by García-Lapresta and Pérez-Román [16, 17], in different settings. Given an aggregation function $F = \left(F^{(k)}\right)_{k\in\mathbb{N}}$ and a profile $V = (v_i^a)$, our proposal consists of a sequential process addressed by the following stages:

(1) The initial clustering is $\mathscr{A}_0^Y = \{\{1\},\ldots,\{m\}\}$.
(2) Calculate the similarities between all the pairs of agents, $S_F^Y(\{a\},\{b\})$ for all $a, b \in A$.
(3) Select the two agents $a, b \in A$ that maximize S_F^Y and construct the first cluster $A_1^Y = \{a, b\}$.
(4) The new clustering is $\mathscr{A}_1^Y = \left(\mathscr{A}_0^Y \setminus \{\{a\},\{b\}\}\right) \cup \left\{A_1^Y\right\}$.
(5) Calculate the similarities $S_F^Y(A_1^Y, \{c\})$ and take into account the previously computed similarities $S_Y(\{c\},\{d\})$, for all $\{c\},\{d\} \in \mathscr{A}_1^Y$.
(6) Select the two elements of $\mathscr{A}_1^Y$ that maximize S_F^Y and construct the second cluster A_2^i.
(7) Proceed as in previous items until obtaining the next clustering $\mathscr{A}_2^i$.

The process continues in the same way until obtaining the last cluster, $\mathscr{A}_{m-1}^Y = \{A\}$.

In the case of several pairs of agents or clusters are in a tie, then proceed in a lexicographic manner in $1,\ldots,m$.

5 An Illustrative Example

In order to illustrate the agglomerative hierarchical clustering procedure introduced in Sect. 4, consider a set of eight experts $A = \{1, 2, 3, 4, 5, 6, 7, 8\}$ assessing a set of six alternatives $X = \{x_1, x_2, x_3, x_4, x_5, x_6\}$ through the following profile

$$V = \begin{pmatrix} 1.0 & 0.9 & 0.7 & 0.5 & 0.5 & 0.0 & 0.5 & 1.0 \\ 0.8 & 0.0 & 0.6 & 0.4 & 0.3 & 0.8 & 1.0 & 0.8 \\ 0.6 & 0.6 & 0.6 & 1.0 & 0.2 & 0.6 & 0.7 & 1.0 \\ 0.4 & 0.4 & 0.4 & 0.3 & 0.9 & 0.7 & 0.3 & 0.7 \\ 0.3 & 0.3 & 0.7 & 1.0 & 0.0 & 0.9 & 0.2 & 1.0 \\ 0.0 & 1.0 & 0.5 & 1.0 & 0.5 & 0.7 & 1.0 & 0.7 \end{pmatrix}.$$

In order to show the importance of the aggregation function for defnining the consensus measure that generates the cluster formation, we have considered four centered EOWA operators (in the sense of Definition 4): the arithmetic mean, the 1-trimmed mean (or olympic EOWA operator) and two specific cases generated by the functional method introduced by Yager [30, 31].

The clustering processes have been carried out for the case of all the alternatives ($Y = X$). The outcomes are summarized in the corresponding dendrograms.

The dendrograms generated by the arithmetic mean and the 1-trimmed mean are shown in Figs. 1 and 2, respectively.

Figure 3 shows the dendrogram that corresponds to consider the EOWA operator whose weights are given by applying Eq. (1) to the BUM function generated by Eq. (2) with the piecewise linear centering function g_1 defnined as

$$g_1(x) = \begin{cases} 2x, & \text{if } 0 \leq x \leq 0.5, \\ 2 - 2x, & \text{if } 0.5 \leq x \leq 1. \end{cases}$$

In this case, the weights are $w_i^k = \dfrac{2(2i-1)}{k^2}$ for $i \leq \dfrac{k+1}{2}$, and $w_i^k = w_{k+1-i}$ if $i \geq \dfrac{k+1}{2}$.

Similarly, Fig. 4 shows the dendrogram that corresponds to consider the EOWA operator whose weights are given by applying Eq. (1) to the BUM function generated by Eq. (2) with the parabolic centering function $g_2(x) = 4(x - x^2)$. Now the

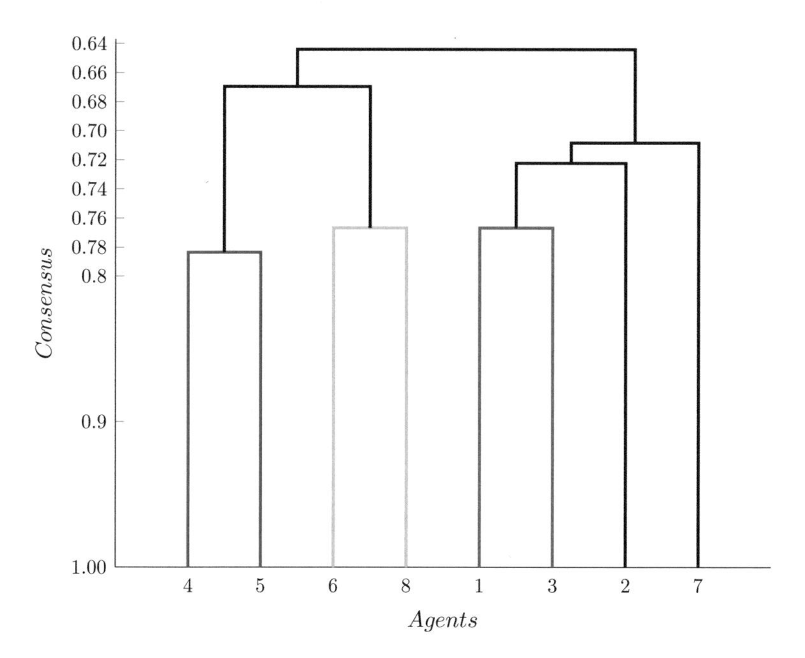

Fig. 1 Dendrogram obtained with the arithmetic mean

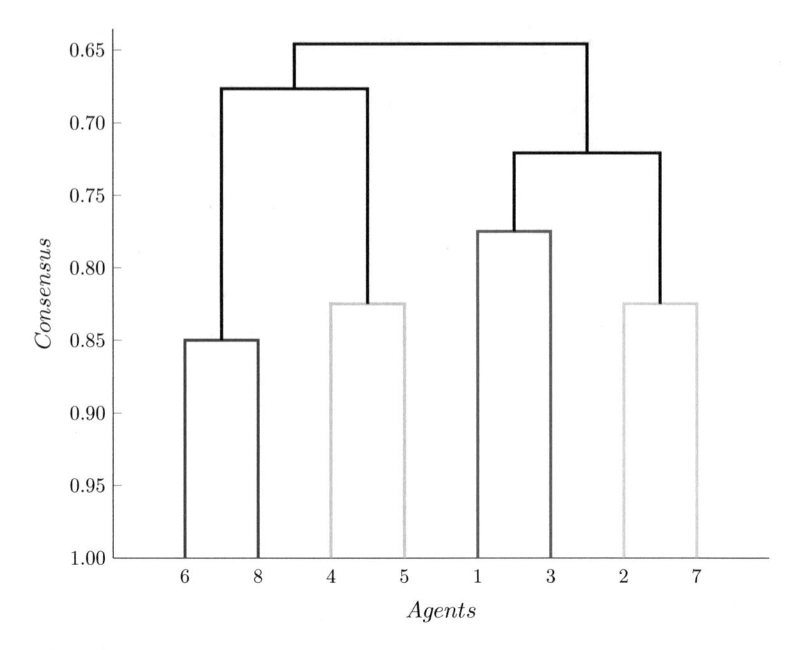

Fig. 2 Dendrogram obtained with the 1-trimmed mean

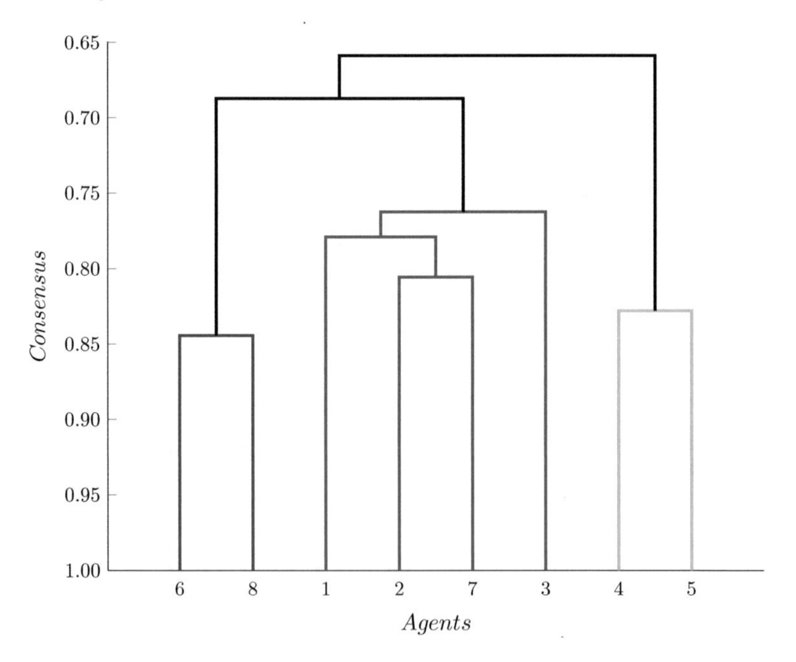

Fig. 3 Dendrogram obtained with the EOWA operator generated by g_1

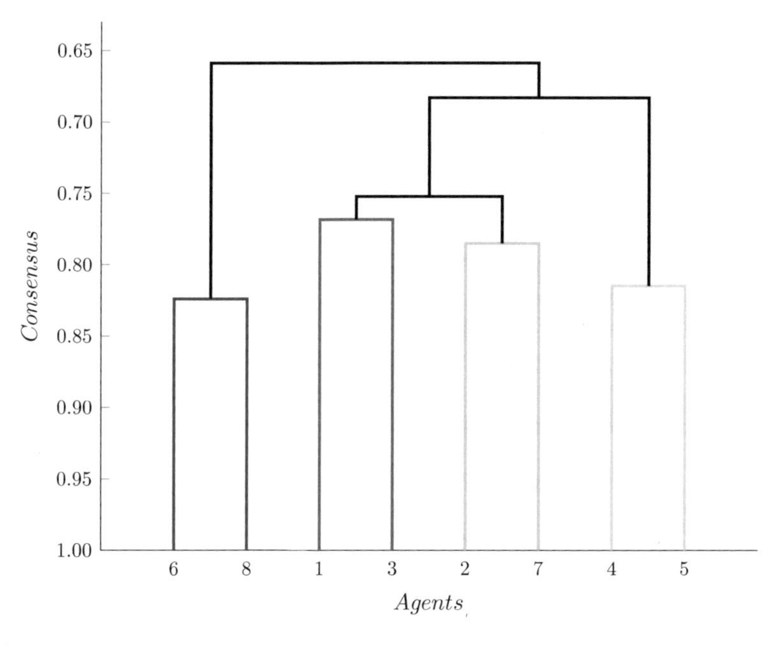

Fig. 4 Dendrogram obtained with the EOWA operator generated by g_2

weights are $w_i^k = \dfrac{3k(2i-1) - 6i(i-1) - 2}{k^3}$ for $i \leq \dfrac{k+1}{2}$, and $w_i^k = w_{k+1-i}$ if $i \geq \dfrac{k+1}{2}$.

6 Concluding Remarks

When a group of agents show their opinions about a set of alternatives, an interesting problem is to know what is the consensus in the whole group or in a subset of agents with respect to one or several alternatives. The most important is not to know the corresponding degrees of consensus, but to compare the consensus in different subsets of agents and alternatives. With our proposal, this information can be easily achieved. Even more, the proposed consensus-based clustering and the corresponding dendrograms provide a rich and visual picture of the homogeneity in the individual opinions.

The mentioned consensus and clustering procedures are static. However, in consensus reaching processes the degree of consensus in a specific situation is only the starting point of a dynamic and iterative process that pursues to increase the agreement among agents. A consensus reaching process consists of several rounds where a human or virtual moderator may invite some agents to modify their opinions in order to increase the collective agreement (see Fedrizzi et al. [10], Saint and Lawson [26], Martínez and Montero [22] and Palomares et al. [25], among others).

These consensus reaching processes and the corresponding clustering analyses can be carried out in the setting of this contribution. The fact that the proposed consensus measure is associated with an aggregation function provides flexibility to the process. Once are determined the alternatives x_i where the degree of consensus in the whole group of agents, $C_F(V, A, \{x_i\})$, is smaller than the overall degree of consensus $C_F(V, A, X)$, the moderator may invite those agents whose opinions over x_i are quite different to the median assessment to properly modify their assessments. If these agents move their assessments on the selected alternatives towards the corresponding median assessments, then the degree of consensus increases. Due to the monotonicity of the aggregation function, the overall degree of consensus increases as well. All these changes can be visualized through the corresponding dendrograms.

Acknowledgements The authors gratefully acknowledge the funding support of the Spanish *Ministerio de Economía y Competitividad* (project ECO2012-32178) and *Consejería de Educación de la Junta de Castilla y León* (project VA066U13).

References

1. Alcalde-Unzu, J., Vorsatz, M. (2013). Measuring the cohesiveness of preferences: an axiomatic analysis. *Social Choice and Welfare* 41, pp. 965–988.
2. Alcantud, J.C.R., de Andrés, R., Cascón, J.M. (2013). On measures of cohesiveness under dichotomous opinions: some characterizations of Approval Consensus Measures. *Information Sciences* 240, pp. 45–55.
3. Beliakov, G., Bustince Sola, H., Calvo Sánchez, T. (2016). *A Practical Guide to Averaging Functions*. Springer, Heidelberg.
4. Beliakov, G., Pradera, A., Calvo, T. (2007). *Aggregation Functions: A Guide for Practitioners*. Springer, Heidelberg.
5. Bosch, R. (2005). *Characterizations of Voting Rules and Consensus Measures*. Ph. D. Dissertation, Tilburg University.
6. Calvo, T., Kolesárova, A., Komorníková, M., Mesiar, R. (2002). Aggregation operators: Properties, classes and construction methods. In: T. Calvo, G. Mayor, R. Mesiar (eds) *Aggregation Operators: New Trends and Applications*, pp. 3–104. Physica-Verlag, Heidelberg.
7. Calvo, T., Mayor, G. (1999). remks on two types of extended aggregation functions. *Tatra Mountains Mathematical Publications* 16, pp. 235–253.
8. Erdamar, B., García-Lapresta, J.L., Pérez-Román, D., Sanver, M.R. (2014). Measuring consensus in a preference-approval context. *Information Fusion* 17, pp. 14–21.
9. Everitt, B.S., Landau, S., Leese, M. (2001). *Cluster Analysis*, 4th Edition. Oxford University Press, New York.
10. Fedrizzi, M., Kacprzyk, J., Owsińnski, J.W., Zadrożny, S. (1994). Consensus reaching via a GDSS with fuzzy majority and clustering of preference profiles. *Annals of Operations Research* 51, pp. 127–139.
11. Fodor, J., Marichal, J.L., Roubens, M. (1995). Characterization of the ordered weighted averaging operators. *IEEE Transtactions on Fuzzy Systems* 3, pp. 236–240.
12. Fodor, J., Roubens, M. (1994). *Fuzzy Preference Modelling and Multicriteria Decision Support*. Kluwer Academic Publishers, Dordrecht.
13. García-Lapresta, J.L., Llamazares, B. (2001). Majority decisions based on difference of votes. *Journal of Mathematical Economics* 45, pp. 463–481.

14. García-Lapresta, J.L., Martínez-Panero, M. (2009). Linguistic-based voting through centered OWA operators. *Fuzzy Optimization and Decision Making* 8, pp. 381–393.
15. García-Lapresta, J.L., Pérez-Román, D. (2011). Measuring consensus in weak orders. In: E. Herrera-Viedma, J.L. García-Lapresta, J. Kacprzyk, H. Nurmi, M. Fedrizzi, S. Zadrożny (eds.), *Consensual Processes*, STUDFUZZ, vol. 267. Springer-Verlag, Berlin, pp. 213–234.
16. García-Lapresta, J.L., Pérez-Román, D. (2015). Ordinal proximity measures in the context of unbalanced qualitative scales and some applications to consensus and clustering. *Applied Soft Computing* 35, pp. 864–872.
17. García-Lapresta, J.L., Pérez-Román, D. (2016). Consensus-based clustering under hesitant qualitative assessments. *Fuzzy Sets and Systems* 292, pp. 261–273.
18. García-Lapresta, J.L., Pérez-Román, D., Falcó, E. (2014). Consensus reaching processes under hesitant linguistic assessments. In: P. Angelov et al. (eds.), *Intelligent Systems'2014. Advances in Intelligent Systems and Computing* 322, pp. 257–268.
19. Gini, C. (1912). *Variabilità e Mutabilità*. Tipografia di Paolo Cuppini, Bologna.
20. Grabisch, M., Marichal, J.L., Mesiar, R., Pap, E. (2009). *Aggregation Functions*. Cambridge University Press, Cambridge.
21. Jain, A.K., Murty, M.N., Flynn, P.J. (1999). Data clustering: A review. *ACM Computing Surveys* 31 (3), pp. 264–323.
22. Martínez, L., Montero, J. (2007). Challenges for improving consensus reaching process in collective decisions. *New Mathematics and Natural Computation* 3, pp. 203–217.
23. Martínez-Panero, M. (2011). Consensus perspectives: Glimpses into theoretical advances and applications. In: E. Herrera-Viedma, J.L. García-Lapresta, J. Kacprzyk, H. Nurmi, M. Fedrizzi, S. Zadrożny (eds.), *Consensual Processes*, STUDFUZZ, vol. 267, Springer-Verlag, Berlin, pp. 179–193.
24. Mayor, G., Calvo, T. (1997). On extended aggregation functions. *Proceedings of IFSA 97*, vol. I, Prague, pp. 281–285.
25. Palomares, I., Estrella, F.J., Martínez, L., Herrera, F. (2014). Consensus under a fuzzy context: taxonomy, analysis framework AFRYCA and experimental case of study. *Information Fusion* 20, pp. 252–271.
26. Saint, S., Lawson, J.R. (1994). *Rules for Reaching Consensus. A Modern Approach to Decision Making*. Jossey-Bass, San Francisco.
27. Torra, V., Narukawa, Y. (2007). *Modeling Decisions: Information Fusion and Aggregation Operators*. Springer, Berlin.
28. Ward Jr., J.H. (1963). Hierarchical grouping to optimize an objective function. *Journal of the American Statistical Association* 58, pp. 236–244.
29. Yager, R.R. (1988). On ordered weighted averaging operators in multicriteria decision making. *IEEE Transactions on Systems Man and Cybernetics* 8, pp. 183–190.
30. Yager, R.R. (1996). Quantifier guided aggregation using OWA operators. *International Journal of Intelligent Systems* 11, pp. 49–73.
31. Yager, R.R. (2007). Centered OWA operators. *Soft Computing* 11, pp. 631–639.
32. Yitzhaki, S. (1998). More than a dozen alternative ways of spelling Gini. *Research on Economic Inequality* 8, pp. 13–30.

Using a Reverse Engineering Type Paradigm in Clustering. An Evolutionary Programming Based Approach

Jan W. Owsiński, Janusz Kacprzyk, Karol Opara, Jarosław Stańczak and Sławomir Zadrożny

To Saddaki, Professor Sadaaki Miyamoto, with a deep appreciation for his long time research and scholarly excellence and inspiration, and so many years of friendship

Abstract The aim of this work is to propose a novel view on the well-known clustering approach that is here dealt with from a different perspective. We consider a kind of a reverse engineering related approach, which basically consists in discovering the broadly meant values of the parameters of the clustering algorithm, including the choice of the algorithm itself, or even – more generally – its class, and some other parameters, that have possibly led to a given partition of data, known a priori. We discuss the motivation and possible interpretations related to such a novel reversed process. In fact the main motivation is gaining insight into the structure of the given data set or even a family of data sets. The use of the evolutionary strategies is proposed to computationally implement such a reverse analysis. The idea and feasibility of the proposed computational approach is illustrated on two benchmark type data sets. The preliminary results obtained are promising in terms of a balance between analytic and computational effectiveness and efficiency, quality of results obtained and their comprehensiveness and intuitive appeal, a high application potential, as well as possibilities for further extensions.

Keywords Clustering · Parameterizing clustering · Evolutionary programming · Distance functions · Variable choice · Reverse engineering

---

J.W. Owsiński · J. Kacprzyk · K. Opara · J. Stańczak · S. Zadrożny (✉)
Systems Research Institute, Polish Academy of Sciences, Newelska 6,
01-447 Warszawa, Poland
e-mail: zadrozny@ibspan.waw.pl

V. Torra et al. (eds.), *Fuzzy Sets, Rough Sets, Multisets and Clustering*,
Studies in Computational Intelligence 671, DOI 10.1007/978-3-319-47557-8_9

1 Introduction

Cluster analysis is one of the most rudimentary data analysis techniques and, at the same time, one of the most popular and powerful; cf., e.g., Kaufman and Rousseeuw [16], Miyamoto et al. [21]. It is most often used when no extra information is available and, thus, there is no clue as to the structure of the data set at hand. In order to execute a cluster analysis exercise one has to make a number of choices. First of all, the data set must be characterized in terms of attributes and their values for particular data points[1] which may be task dependent. Then, a clustering technique, some measures of distance/similarity, and other parameters related to a chosen clustering technique have to be assumed. The choice of parameters may be guided by the experience of a user, availability of data (values of the attributes to be used), some available metadata or may be based on results of some preliminary data analysis; cf., e.g., Torra et al. [28]. As soon as the choice is done, one can run a selected clustering algorithm and obtain some groups of data points, i.e., a partition of the data set under consideration. Usually, such an exercise is repeated several times for different configurations of the above mentioned parameters in order to find an "optimal" one. The whole process may be seen as a kind of a transformation which turns a data set of individual data points into groups of such data points.

In this paper we consider the problem which may be interpreted to be a kind of a reverse engineering related one that is applied to the results of the previously described source clustering process. Namely, *we assume that a partition of the data set is given and we want to discover parameters of the process (transformation) that have resulted in the given partition*. It is very common to consider the data sets that are divided into subsets (clusters, classes, groups, ...) in a certain manner, which is more or less "certain" or "justified", and attempt to reconstruct the given division using some "other" data than just the respective labels. This is most often done in order to validate or check the quality of the classification, clustering, machine learning, etc. schemes. More advanced purposes may involve model building and checking, as well as – quite to the contrary – verification of the original, prior division of the data set. There may also be other objectives, some of them quite classical, like the detection of outliers, and also very specific ones, like the assessment of adequacy of the classification data to the labels or forming some descriptions of known groups of data in terms of the values of some of the attributes characterizing them.

It is easy to notice that the results of the above reverse engineering type clustering process can provide the analyst and user with a very useful information. However, to be usable in practice, that is, also for novice users, domain specialists, very often with a limited command of data analysis, clustering, etc., who are now presumably the largest target group of users in virtually all real world applications, they must be presented in some aggregated form that involves a broadly perceived granulation of

---

[1] It is possible to start with a data similarity/distance matrix, if available, without an explicit characterization of the data in terms of some attributes values, and such a setting also seems to provide a reasonable context for the paradigm proposed in this paper, but we will leave this case for a possible further study.

data and information, and some summarization, both of a numerical and linguistic character.

The problem of comprehensiveness of data analysis, data mining, machine learning, etc. results (patterns) had been known for some time, and it had been presumably Michalski who already in the early 1980s devised the so called *postulate of comprehensibility* whose essence can be summarized as (cf. [19]): "… The results of computer induction should be symbolic descriptions of given entities, semantically and structurally similar to those a human expert might produce observing the same entities. Components of these descriptions should be comprehensible as single 'chunks' of information, directly interpretable in natural language, and should relate quantitative and qualitative concepts in an integrated fashion…". Michalski's vision has had a great impact on machine learning, data mining, data analysis, etc. research, and has been further developed by many authors, cf. Craven and Shavlik [5], Zhou [29], Pryke and Beale [23], Fish, Gruber and Sick [10], to name just a few. A recent study on the comprehensiveness of linguistic summaries by Kacprzyk and Zadrożny [15] combines many ideas from those works, and recasts them in the context of a very natural representation of results via natural language.

Most of the above mentioned works on the comprehensiveness of data analysis/mining results emphasize as the main reasons for the importance of comprehensibility, cf. Kacprzyk and Zadrożny [15], to name a few: (1) To be confident in the performance and usefulness of the algorithms, and to be willing to use them, the users have to understand how the result is obtained and what it means, (2) The results obtained should be novel and unexpected, in one sense or another, and these results can only be accessible to the human if they are understandable, (3) Usually, the results obtained may imply some action to be taken, and hence their comprehensiveness is clearly crucial, (4) The results obtained may provide much insight into a potential better feature representation which, to be meaningful, should be comprehensible, (5) The results obtained can be employed for refining knowledge about a domain or field in question, and the more comprehensible, the better.

As we will see, the reverse engineering type approach to clustering can be viewed to be following at the conceptual level the above mentioned philosophy of attaining comprehensiveness. First, the clustering process itself implies an increase of comprehensiveness of data as it produces *per se* representations of results that are closer to human perception. Then, which is the essence of the approach proposed in this paper, we go further and try to find those parameters of the clustering algorithm employed that have led to the results obtained. That is, an extremely important additional knowledge is derived about the algorithms, parameters, types of distance/similarity functions, etc. This all is useful and greatly helps a potential user to understand (comprehend) intrinsic relations between many aspects of the data analysis process.

Clearly, this paper is just a first step into a deeper comprehensiveness related clustering analysis, for instance, following ideas proposed in Kacprzyk and Zadrożny [15]. The problem needs a further study which will be done in next papers.

We performed the experiments, meant to show the feasibility of the proposed idea and the nature of results, on two data sets. One of them was the classic Fisher's Iris data set and the other one – an empirical data set measuring the numbers of vehicles passing by definite points on motorways, originating from Germany. In both these cases it could be assumed that the prior partition was available, but possibly of a different standing, what is discussed in Sect. 2.2.

The structure of the paper is the following. In the next section we formally define the proposed concept of the reverse engineering type clustering and give some related motivations. In Sect. 3 we briefly point out some related work. Section 4 presents and discusses the results of the computational experiments and, finally, we conclude by pointing out possible directions of the further research.

2 The Concept of the Reverse Engineering in Clustering

2.1 The Concept

This paper presents a study, in which we try to develop a reverse engineering type procedure of reconstructing a certain partition[2] of a data set, X, $X = \{x_i\}_i, i = 1, \ldots, n$, into p subsets (clusters), $A_q, q = 1, \ldots, p$. We assume that each object, indexed i, is characterized by m variables, so that $x_i = (x_{i1}, \ldots, x_{ik}, \ldots, x_{im})$. Having the partition $P_A = \{A_q\}_q$ given in some definite manner, we now try to figure out details of the clustering procedure which, when applied to X, could have produced the partition P_A or its possibly accurate approximation. That is, we search in the space of *configurations*, spanned by

(i) the choice of the clustering algorithm, and characteristic parameters of the respective algorithm(s);
(ii) the selection or other operations on the set of variables (e.g. subsetting, aggregation), and
(iii) the definition of a similarity/distance measure between objects, used in the algorithm.

The partition, resulting from applying the clustering procedure with a candidate configuration of the above parameters is denoted P_B and is composed of clusters $B_{q'}, q' = 1, \ldots, p'$, $P_B = \{B_{q'}\}_{q'}$. The search is performed by minimizing a certain criterion, denoted $Q(P_A, P_B)$. Thus, if we denote the set of parameters that is being optimized in the search by Z (notwithstanding the potential differences in the actual content of Z), and space of values of these parameters by Ω, then we are looking in Ω for a Z^* that minimizes $Q(P_A, P_B)$. Formally, we can treat Z as a transformation of the data set X (a cluster operator) and thus denote the optimization problem for a given data set X and its known partition P_A as follows:

$$Z : X \to \mathscr{P}(X), \quad Z(X) = P \tag{1}$$

[2]The concept of a Reverse Cluster Analysis has been introduced by Ríos and Velásquez [25] in case of the SOM based clustering but it is meant there in a rather different sense as associating original data points with the nodes in the trained network.

$$Z^* = \arg\ \min_{Z \in \Omega} Q(P_A, Z(X)) \tag{2}$$

where $\mathscr{P}(X)$ denotes the set of all possible partitions of the data set X.

Notice that this optimization problem is in line with the reverse engineering paradigm. Because of the irregularity of circumstances of this search (the nature of the search space and the values of the performance criterion), the solution of the optimization problem defined above is a challenging task. In our experiments, presented later on in the paper, this is performed with the use of evolutionary algorithms.

In order to bring the problem formulated here closer to some real life situation, let us consider the following example: assume a secondhand car dealer disposes of a set of data on (potential) customers, who visit the website of the dealer, call this set Y, and the set of data on those, who actually bought a car, the set X. Naturally, the set X is much smaller than Y (it may constitute, say, less than 1 % of Y). In this case, P_A might be based on the makes and/or types of cars purchased (data set X). The dealer might wish to identify the partition of X, disregarding the labeling by car makes/types, that approximates possibly well P_A. The obvious objective would be to identify the "classes" of (potential) customers, at whom it would be effective to address the promotional offers or just information, regarding definite makes/types of cars. Upon finding the Z^* that produces P_B that is the closest to P_A, one might hope that by applying Z^* to Y it would be possible to define the classes of the (potential) customers, at whom appropriate offers could be addressed during their search through the website. These classes would form the partition $Z^*(Y)$.

2.2 Some Interpretations

The idea of the proposed paradigm can be understood in a variety of manners. In a way, it reminds of identifying the best classifier, expressed in this case through Z^*. Namely, the current setting may be seen as typical for the supervised learning in that a known a priori grouping P_A is the starting point. However, we are not interested here in a further classification of incoming data points but in the understanding of the process through which the grouping P_A emerged. This is strongly related to the issue of comprehensibility. On the other hand, Z^* may be indeed interpreted as a part of an interesting, though usually utterly inefficient, classification scheme. Namely, in general, the classification would be carried out for the subsequent x_i, $i = n + 1, \ldots,$ in such a way that:

$$P_B = Z^*(X \cup \{x_i\}) \tag{3}$$

would be computed which would place x_i in one of the clusters of P_B. An unorthodox character of such a classification scheme consists in that the partition P_B may be, in general, different from P_A. This will be especially true when we consider the batch classification, i.e., when $\{x_i\}$ in (3) is replaced with a whole set of data points to be classified simultaneously, like in the example, described at the end of the preceding section.

The classification scheme sketched above may be formally written down as:

$$\mathscr{P}(X) \rightarrow \Omega \rightarrow \mathscr{P}(X'), \quad X \subseteq X' \tag{4}$$

and would be usually prohibitively expensive from the computational point of view. However, in some cases an excessive cost may be avoided. As pointed out by Miyamoto [20], in case of some clustering algorithms, referred to as *inductive clustering* techniques, appropriate classification rules R may be derived based on the partition those algorithms yield, so that a direct classification of a new data point x_i using the rule R yields exactly the same result as by starting over the whole clustering procedure for the data set X enlarged with x_i, i.e.,:

$$Z^*(X \cup \{x_i\})[x_i] = R(P, x_i) \tag{5}$$

where $P[x_i]$ denotes a cluster to which x_i belongs in the partition P (here $P = Z^*(X \cup \{x_i\})$, i.e., a partition resulting from the application of the "classifier" (clustering configuration) Z^* to the enlarged data set X), and

$$R : \Omega \times X \rightarrow 2^X \tag{6}$$

denotes a rule which for a given partition P (here Ω is equated with the set of all partitions of the set X) and a data point x_i yields a cluster of P to which x_i should be assigned.

Thus, if Z^* obtained in our reverse engineering process (2) for P_A refers to an inductive clustering technique (e.g., belonging to the family of k-means algorithms), then new data points can be directly classified to one of the clusters of P_A (in this case using the 1-nn classification technique with respect to the centroids of the clusters of P_A) with the same effect which would result from going through the classification scheme (4).

A different view of the applicability of our approach is the following one: based on a given partition P_A of a set X we want to derive a clustering scheme which then may be justified to be applied to a whole family of sets $\{X_j\}$ of which X is just an exemplary member. It may be understood:

- more narrowly, i.e. as if a given Z^* (rather than P_B), obtained for the assumed Ω, or
- more broadly, if i.e. as the entire procedure, leading from some P_A through choices, related to Ω and Z, down to $P_B = Z(X)$,

were a specific kind of classification, namely when we intend to group, or partition, much bigger sets of data than X, underlying P_A and P_B. Under such circumstances:

(a) we do not expect an absolute or ideal accuracy of results over large data sets but we wish to preserve the essential character of the prior partition (or "distribution") P_A, and
(b) we would like to check (and preserve) the validity of the Z^* for various, even apparently similar P_A's.

Another important aspect of the approach considered consists in the status of the prior partition P_A. Two issues are essential here:

(I) where does this partition come from (what is its relation to the data set X)?
(II) what is the degree of certainty of this prior partition (degree of our belief in its validity)?

Depending upon the answers to these questions, we deal with entirely different cases, or tasks, even if the methodologies applied may remain the same. An extreme, "absolute" case takes place when

A. with respect to (I) above: the partitioning P_A has been imposed on the data set X irrespectively of the values of the particular attributes,[3] , i.e., P_A has (at least apparently) nothing to do with the attributes of the data set X (i.e. either it is simply given, and we do not know anything about the relation between P_A and the attributes $k = 1, \ldots, m$, or we know that the division has been performed on the basis of the attribute(s) not accounted for in X), and, at the same time
B. with respect to (II) above: the partition is fully certain and we are therefore fully convinced of its validity – it is the only correct partition of X in a given context.

This extreme case should be softened to account for the partitions, which are produced by experts, so it may be assumed that:

C. with respect to (I) above: they take into account, even if implicitly, the actual data from X, and
D. with respect to (II) above: their opinions can be put to doubt, or at least under discussion. Thereby, we come to the situation in which P_B may give rise to a feedback, enriching our knowledge that has led to P_A.[4]

Thus, in the scenario when A and D are true, i.e., a given partition P_A is somehow transcendent with respect to the set of attributes characterizing X and, at the same time, we are not fully convinced in its validity, we would be interested if it is possible to recover partition P_A using some Z^* but we should expect that the best P_B obtained may be quite different from P_A. Moreover, we can think of P_B as a legitimate replacement for P_A, which may be therefore treated just as the starting point for getting to a "real" partition of X.

In another scenario, when B and C are true, i.e., when we treat the partition P_A as a valid one and at the same time we know that it has been established with the reference to the actual values of the attributes characterizing the data set X, we will be more concerned with recovering exactly P_A and the benefits from carrying out the reverse engineering procedure would be primarily related to getting a better insight into the very meaning of the particular attributes and their role in justifying the real/valid

[3]Possibly except for an identifier attribute, which makes it possible to distinguish particular elements of X.

[4]Actually, even in the "absolute" case, doubts may arise, if the situation resembles the one of multiple overlapping distributions, i.e. although P_A is well established and "certain", it is hardly reflected in the data, represented by X, so that many objects x_i might be equally well assigned to different clusters.

partition P_A; this all has a clear relation to the issue of comprehensibility mentioned above.

The above can well be illustrated in terms of the example, described at the end of Sect. 2.1. Thus, although we know that labeling by car makes/types is "certain", this might not be true of the actually imposed partition P_A, which, in addition, is definitely associated with the customer data in X. Hence, the iterative procedure, involving consecutively obtained partitions P_B, and corresponding configurations Z^*, might be treated as a learning procedure of finding the best match between the customer data and the cars purchased.

3 Related Work

The idea of a reverse engineering type clustering as presented in Sect. 2 is obviously novel. However, the procedure, formally represented by (2), may be seen as referring to many facets which are individually addressed in the literature by multiple authors in various contexts and settings. Let us briefly remind the main aspects which are relevant for our considerations.

The choice of the attributes has been thoroughly studied, in particular, in the context of the classification [13], but also in the area of more broadly meant data analysis. Many different approaches have been proposed, which are applicable for our purposes. Some of them take into account the information on classes, to which elements of X are assigned, some not. In our case, both modes are possible as we start with a partition which may be interpreted as the classification. The choice of an appropriate family of techniques may be based on the aspects discussed at the end of Sect. 2.2. Namely, if the partition P_A is to be seen as the valid one, then taking into account the information on class assignments is more justified than in the other cases. In our experiments, reported in the next section, we associate weights with the attributes and the values of the weights are optimized during the evolutionary procedure. This may effectively lead to completely ignoring some of the attributes characterizing the data set X. Notice that the choice of attributes has also been discussed in the literature on the comprehensibility of data analysis, data mining, machine learning, etc. results, and Craven and Shavlik [5] may be here a good source of information.

Another important decision concerns the choice of the distances/similarities from among the plethora of those proposed in the literature [6]. This choice has, of course, to take into account the scale with which a given attribute is endowed, i.e., nominal, ordinal, interval or ratio. For the latter type of attributes it may be convenient to assume a parametrized family of distances, e.g., Minkowski distance, what makes simpler the representation for the purposes of an evolutionary optimization. One can even go further, using a fuller survey of similarity/dissimilarity measures presented in Choi, Cha and Tappert [4], in which those measures are classified into classes, and a similar reverse type analysis is performed. This will not be, however, considered in this paper and be left for a further study.

The essence of the problem of the reverse engineering clustering formulation as meant in this paper is the formulation and solution of the optimization problem defined in (2). Its important component is the performance criterion denoted as Q which usually will be identified with a measure of the fit between two partitions. In particular, we will often interpret Q in such a way that it should measure how well the partition P_B, produced using Z^*, matches the originally given partition P_A. Such measures belong to a broader family of the cluster validity measures, which are meant to evaluate the quality of the partition produced by a clustering algorithm; cf., e.g., Halkidi et al. [12], Arbelaitz et al. [1]. According to Brun et al. [2] three broad classes of such measures may be distinguished which not necessarily refer to a golden standard partition, in our case denoted as P_A. The first class comprises internal measures which are based on the properties of the clusters produced. The second class of the relative measures "is based on comparisons of partitions generated by the same algorithm with different parameters or different subsets of the data" (cf. [2]). And finally, the third class comprises the external measures referring to the comparison of partitions produced by the algorithm and a partition known a priori to be a valid one. As our primary goal is the reconstruction of a cluster operator which could have produced a given partition P_A for a given data set X, then we are first of all interested in the usage of the external validity measures (cf. the next section and the use of the Rand validity measure therein). However, it should be stressed that in different scenarios, discussed in Sect. 2, also other types of measures may be of use. In particular, if our belief in the validity of a given partition P_A is not full, then we can define the quality criterion as a combination of an external and internal one, for instance, looking for P_B which provides a balance between the matching of P_A and having a high quality (e.g., internal consistency) in terms of one or more internal measures.

Another parameter of the clustering procedure whose choice attracted a lot of attention in the literature is the fixed number of clusters assumed, e.g., for the k-means family of clustering algorithms. The choice of the value for this parameter has evidently a far reaching influence on the obtained partition while it may seem rather arbitrary. Thus, a number of approaches has been proposed which usually base on the earlier mentioned validity measures. Namely, the optimal number of clusters is recognized as the number for which a given validity measure attains its extremum or satisfies some specific formula (Charrad et al. [3]). In our general approach, the actual number of clusters present in the partition P_A is a natural candidate for the value of this parameter. However, such a choice may be questioned when taking into account the assumption (A) mentioned earlier or considering the reverse engineering of P_A to obtain Z^* as a first step towards partitioning other, possibly much larger, datasets using Z^*.

An example of a software package which combines all the above mentioned main aspects and, thus, is very relevant for our idea of the reverse engineering type clustering, is the NbClust package [3] available on the R platform. It is primarily oriented at supporting the choice of the number of clusters. However, this package actually implements a number of:

- clustering algorithms,
- cluster validity indexes (measures), and
- distance measures (dissimilarity measures)

and makes it possible to use them in various configurations, together with a varying number of clusters, where appropriate, to search for the "best" one for a given data set. The configuration is pointed out as the best when the majority of the validity measures involved confirm its superiority. Thus, the NbClust package may be seen as a valuable and extremely relevant tool to carry out the endeavor laid out in this paper. However, our proposal provides a broader framework for the emerging type of data analysis and makes it possible to envision some interesting directions for the further research. Moreover, it adds to the analysis an important aspect of comprehensibility of results obtained.

4 The Experiments

4.1 The General Outline

On the algorithmic side, the experiments we have performed and report here made use of the classical clustering techniques, including:

1. k-means,
2. the DBSCAN algorithm, and
3. the progressive merger procedures.

These were parameterized along the following lines:

1. k-means by the number of clusters assumed,
2. DBSCAN by the proximity and neighbor number parameters, and
3. the progressive merger procedures – by the coefficients of the Lance–Williams formula [17].

Furthermore, concerning the variables, either a possibility of choice among them was introduced (actually, weighting schemes were used with the aim of potentially eliminating some variables), or some aggregations were defined. The distances were parameterized by the exponent in the Minkowski distance definition.

We also employed two different setups for the evolutionary algorithm, serving to search through Ω to find Z^*, one entirely developed by one of the authors, another one originating from the R library. The fitting function used for guiding the search was either based on:

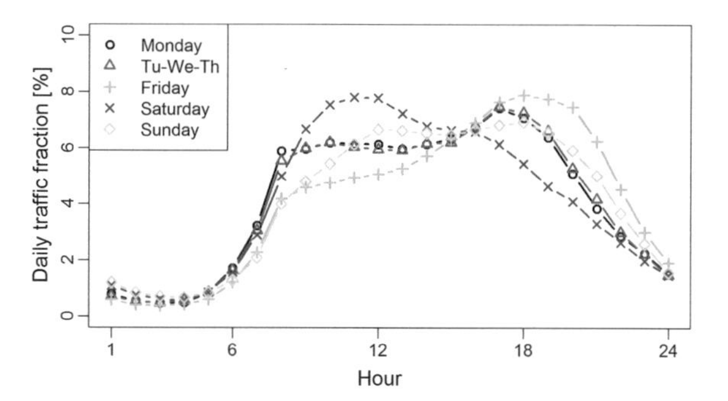

Fig. 1 Average hourly profiles of traffic for the classes of weekdays

- the number of incorrectly classified objects, or
- the basic Rand index [14], relying on the counting of pairs of objects being in the same and in different clusters in the two partitions compared:

$$Q_R = 2(r + s)/(n(n - 1)),$$

where r is the *number of pairs of objects of X which are in the same cluster* in both partitions, and s is the *number of pairs of objects of X which are in different clusters* in both partitions compared. Since the expected value of this criterion for the randomly generated partitions is not zero, an adjusted form, taking into account the bias of expectation, provided by Hubert and Arabie [14], is also used. In actual computations, the Rand-based indices were complemented with the weighted sum of the parameters involved in Z, so as to balance the importance of the two aspects.

As said, the experiments that we report here involve two data sets (Iris and vehicle traffic). In the case of the Iris data, the status of P_A appears to be relatively close to the "absolute" one, mentioned before, although definite doubts may also arise in this case. In the case of the traffic data, which represent the numbers of vehicles passing in consecutive hours of the 24-h cycle, the partition P_A is based on the classification of the days of the week, holidays etc., and reflects the expert's opinion on how these days ought to be treated. This case is illustrated in Fig. 1. In this case a single observation or object is, therefore, a vector of the numbers of vehicles passing a given point on the road during a given day, at subsequent hours (24 numbers), plus the label of the day of the week.

We have performed two series of experiments, using various choices in these two series. The subsequent two sections summarize the results from these two series of experiments.

4.2 Experiment Series 1

In this series the following assumptions have been made: the algorithms used have been from the k-means and hierarchical agglomeration families. Their implementation in the R package "cluster" by Maechler et al. [18], based on Kaufman and Rousseeuw [16], has been employed.

The "pam" (partitioning around medoids) method is a variant of the k-means with the number of clusters, "k" (p in our notation) as the sole parameter. The algorithm "agnes" (agglomerative nesting) is a hierarchical clustering method, parameterized by the number of clusters, p, and a scalar a. The latter is used to obtain coefficients of the Lance–Williams [16] formula in a highly simplified manner, as $a = a_1 = a_2$, $b = 1 - 2a$, and $c = 0$, where the original formula is given as

$$d(\{C_i, C_j\}, C_l) = a_1 d(C_i, C_l) + a_2 d(C_j, C_l) + bd(C_i, C_j) + c|d(C_i, C_l) - d(C_j, C_l)|.$$

where $d(C_i, C_j)$ denotes the distance between two clusters C_i and C_j, and $\{C_i, C_j\}$ denotes a cluster resulting from the merging of two clusters C_i and C_j, respectively.

The distances between the data points $x_i, x_j \in X$ has been defined as weighted Minkowski distances, i.e.

$$d(x_i, x_j) = \left(\sum\nolimits_k w_k |x_{ik} - x_{jk}|^v\right)^{1/v},$$

where w_k are the weights assigned to the variables, which can also assume the value of 0. In this manner, both the vector $\{w_k\}_k$, and the exponent v could be treated as parameters, defining the space Ω, along with the coefficients of the clustering algorithms.

To solve the optimization problem we used the Differential Evolution (DE) [27] metaheuristic, which is an evolutionary global optimization algorithm, and its implementation in the "DEoptim" R package by Mullen et al. [22]. DE is a state-of-the-art real-parameter global optimizer whose various modifications and applications are surveyed in Das and Suganthan [7].

Chromosomes are represented as vectors of real numbers $Z = (\pi, a, v, w_1, \ldots, w_n) \in \Omega$. After rounding, the first parameter represents the number of clusters $p = \text{round}(\pi)$, the second is a parameter a of the Lance–Williams formula (which is not used in the k-means algorithm), the third is an exponent v of the Minkowski distance, and the next are weights $w_1, \ldots, w_n$ of the variables. The search space Ω was defined by constraining the possible values of individual elements to: $p \in [1, 10]$, $a \in [0, 1]$, $v \in [0.1, 4]$, and $w_i \in [0, 1]$.

The distinctive feature of DE is the differential mutation operator which boils down to the addition of a scaled difference between two randomly chosen individuals Z_2 and Z_3 to a third randomly picked vector Z_1

$$Z' = Z_1 + F \cdot (Z_2 - Z_3)$$

where F is a scalar parameter known as the scaling factor, typically $F \in [0.4, 1)$. By assuming that the creation of new individuals is based on differences between population members, we arrive at the adaptation of the search direction to the local shape of the objective function.

Table 1 Summary of results for the first series of experiments

Data	Algorithm	Optimized parameters (Z)	Adjusted Rand index	Rand index
Iris	pam	$p, v, w_1, \ldots, w_4$	**0.758**	**0.892**
		p	0.730	0.880
	agnes	$p, a, w_1, \ldots, w_4$	**0.922**	**0.966**
		p, a	0.759	0.892
Traffic	pam	$p, v, w_1, \ldots, w_{24}$	**0.600**	0.821
		p	0.581	**0.823**
	agnes	$p, a, w_1, \ldots, w_{24}$	**0.654**	**0.850**
		p, a	0.625	0.837

For each dataset and algorithm we have run the DE algorithm for 1000 iterations. In Table 1 we provide the values of the best Rand index and its adjusted variant obtained. In the reference cases only the optimal values of the basic parameters of algorithms (i.e. p and a) were sought for. In the test case we have also optimized the exponent v from the Minkowski measure and the vector of weights of the attributes.

In all cases the optimization of the whole configuration Z of the clustering procedure leads to better results. Sometimes the difference is significant. We have also observed that the hierarchical clustering allows for a more accurate reconstruction of the reference partitions in our datasets than the k-means (in the form of "pam").

The optimized value of the exponent v was varied over the range from 0.8 to 2.3 which confirms that the whole family of these measures is useful, despite the loss of formal properties for the values of v below 1. This is in agreement with the recent paper by De Amorim [8], in which feature rescaling with the use of L_p norm proves useful for the hierarchical clustering.

The regularization with the L_1 norm resulted in the attribute selection of weights in the traffic dataset which has implied some of them to be zero.

Table 2 provides highly interesting results for the traffic data dealt with using the full Z and choosing the "agnes" algorithm.

Thus, this result, to a large extent induced by the conditions of computation, rather than a "natural" tendency of the method, shows that perhaps the expert's opinion as to the original classes, ought to be verified (Monday profiles being classified along with those for Tuesday, Wednesday and Thursday). This result is shown in Fig. 2 where, indeed, the hourly traffic distribution on Mondays is not different from distributions for other weekdays except for Friday. This is exactly the instance of the potential feedback we mentioned before.

Cluster 2 is visualized in Fig. 2 as subplot (e). In this case, apart from the typical morning and afternoon peaks we observe a very high traffic intensity late in the night. The days for which such unusual phenomenon is observed are scattered throughout the year and represent different days of the week. This could be an effect of measurement errors and will undergo additional plausibility checks.

Table 2 Results for traffic data for the entire vector of parameters Z, with the use of hierarchical aggregation (values of Rand index = 0.850, of adjusted Rand = 0.654). The upper part of the table shows the coincidence of patterns in particular A_q, based on the days of the week, and obtained B_q

Days of the week	Clusters obtained					
	1	2	3	4	5	
Friday	1	2	42	0	3	
Monday	45	2	0	0	2	
Saturday	0	1	0	46	1	
Sunday	0	1	0	1	47	
Tu-We-Th	140	3	0	0	4	
Parameters						
p	5	w_1 through w_{24} – weights of variables, corresponding to the consecutive hours of the day				
a	0.78					
v	0.91					
w_1–w_6	0.47	0.45	0.62	0.17	0.48	0.84
w_7–w_{12}	1.00	0	0.90	0.58	0.83	0.33
w_{13}–w_{18}	0.07	0.09	0.48	0	0	0.96
w_{19}–w_{24}	0.30	0.43	0.79	0.53	0.25	0.90

In case of the traffic intensity data anomalies are typically detected through the assessment of specially trained operators. The ability to approximate their partitions by means of an appropriately tuned clustering algorithm makes it possible to automate this procedure. The anomaly detection differs from the classification because the training samples consist nearly entirely of typical observations. Moreover, a new observation can be anomalous in a great variety of ways. This case represents a one-class learning problem consisting of identifying whether a given observation is typical rather than distinguishing between different classes.

4.3 Experiment Series 2

In this series three kinds of clustering algorithms have been used: DBSCAN, the k-means, and the progressive merger. The evolutionary algorithm used has been developed by one of the present authors [26].

The use of specialized genetic operators requires the application of a selection method to execute them in all iterations of the algorithm. The traditional method with a small probability of mutation and a high probability of crossover is not applicable in this case because the number of operators is greater than 2 and their properties cannot be easily described as the exploration or exploitation. In the approach used here [26] it is assumed that an operator that generates good results should have a higher

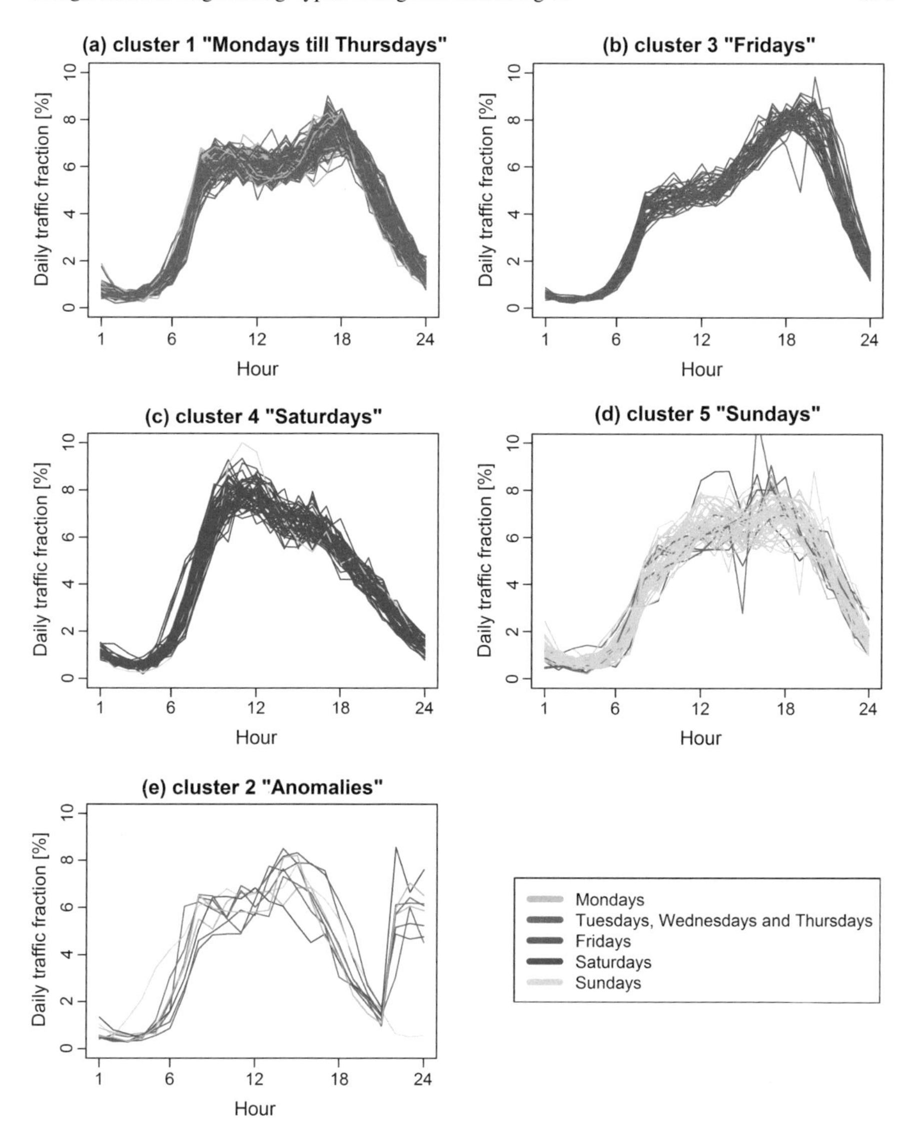

Fig. 2 Interpretation of clusters described in Table 2

probability of execution and more frequently affect the population. But it is very likely that the operator, proper for one individual, would give worse effects for another one, for instance because of its location in the domain of possible solutions. Thus, each individual may have its own preferences. So, each individual has a vector of floating point numbers in addition to the encoded solution. Each number corresponds to one genetic operation. It is a measure of quality of the genetic operator (a quality factor). The higher the factor, the higher the probability of using the operator. The ranking of qualities becomes a base to compute the probabilities of appearance and execution of the genetic operators. The simple normalization of the vector of quality coefficients turns it into a vector of operators' execution probabilities. This set of probabilities can be treated as a base of experience of each individual and according to it, an operator is chosen in each epoch of the algorithm. Due to the experience gathered one can maximize the chances of its offspring to survive.

The essential criterion used was simply the number of "misclassified" objects for a given data set, i.e. with respect to the prior P_A. Because of this difference, and also because of different evolutionary strategies, as well as parameter vectors Z, somewhat different results have been obtained (although, like in series 1, all variables have been explicitly weighted, and the distance exponent has varied as well). For clarity, we will characterize these result below in a more general terms.

The results obtained with DBSCAN, parameterized with the number of neighbors and the maximum distance, have been the poorest, namely not less than 12 misclassified objects in the Iris case, out of the total of 150, depending upon the settings of the evolutionary algorithm, and not less than 104 misclassified objects in the traffic case, out of the total of 341.

For the classical k-means, the results, with respect to the criterion assumed, but also quite intuitively, have been much better: three clusters and only 4 objects misclassified for the Iris data, and 60 misclassified objects for the traffic data – but with the "optimum" number of clusters equal 8! One can compare these results with 64 misclassified objects, shown in Table 2 for series 1, when the "proper" number of clusters, i.e. 5, was actually forced.

In the case of the hierarchical merger procedures the difference with series 1 has additionally boiled down to the complete parameterization of the procedure, according to the Lance–Williams formula (five coefficients). The results obtained have been comparable with those for the k-means in terms of their "quality": only 2 misclassified objects for the Iris data (three clusters), and 64 for the traffic data (with 5 clusters as the "optimum" number).

4.4 Computational Aspects

An effort oriented towards the rationalization of the computational aspect was explicitly made only in series 2 of experiments. This effort has consisted in trying out various comparison strategies, variants of representations, etc. All in all, the results, shortly characterised here, have been obtained using a modest workstation equipment,[5] without any parallelization, although in both series the search often has taken a long time (in days). In the case of series 2 most of the respective calculations have not been actually terminated in the strict sense, even though the changes in the last iterations performed have not shown any significant changes.

[5]Series 2: PC class 4-core processor station, 3.2 GHz, under 64 bit Linux Fedora 17; simulation software written in C and compiled with g++.

Thus, it appears obvious that in the case of larger prior data sets some specially devised schemes have to be applied, including also parallelization, and even some massively parallel computational paradigms. Yet, if the results of the nature exemplified here turned out sufficient for our purposes, even the relatively modest equipment similar to that used here, might be adequate.

5 Conclusions

The study performed demonstrated, first of all, the computational feasibility of the proposed reverse engineering type clustering approach. Then, the results themselves turned out to be both promising in terms of their accuracy, and in the possibility of drawing meaningful substantive conclusions, notably as shown, for instance, for the case of traffic data in series 1 of experiments.

There are several directions, in which our further work could be done. One of them is related to the exploration of the space Ω in terms of the dependence of results on its particular dimensions. Thus, for instance, Table 3 shows the values of the exponent v obtained for various experiments. It can be easily seen that these values vary a lot, even if some of them "cluster" within a definite segment. Thus, experiments have to be designed in order to explicitly check the sensitivity of partitions obtained and their quality Q with respect to changes in the particular variables, entering Z. At the same time, the results obtained with this respect seem to imply that taking a relatively broad interval of feasible values of v could be justified.

Another direction would consist in the explicit inclusion of the feedback loop, mentioned already here. In case the P_B indicates a definite manner of partitioning the set X that distinctly differs in some aspect(s) from P_A though the question might arise of how to reformulate P_A along appropriate lines, and to check whether an iterative procedure converges (quickly) to some terminal, but also meaningful, partition P^*.

A pragmatic direction of work would lead to the design of such schemes of a reverse engineering type clustering which are possibly most efficient numerically,

Table 3 The values of the exponent v determined in a sample of experiments

Experiment	Series 1, pam	Series 1, agnes	Series 2, DBSCAN1	Series 2, DBSCAN2	Series 2, DBSCAN3	Series 2, k-means
Iris data	1.02	2.98	1.88	3.94	2.11	3.17
Traffic data	1.35	0.91	1.72	0.70	1.03	2.08
Experiment	Series 2, aglom1	Series 2, aglom2				
Iris data	1.88	2.93				
Traffic data	1.58	1.58				

DBSCAN1, DBSCAN2, DBSCAN3 and aglom1, aglom2 denote, respectively, experiments differing by some aspects, especially related to the calculation of the quality criterion and can be used for bigger data sets to obtain Z^*, and then to effectively cluster yet bigger data sets with Z^*.

From the point of view of implementability and usefulness of our results to a larger group of users who are the target group in virtually all nontrivial real world applications, that is, novice users or domain experts with a limited command of formal and computational data analysis, data mining, clustering, etc. tools and techniques, a very important area of future research will certainly be related to an increase of comprehensibility. This would help those users better understand the very relations between sets of data, clustering methods and their parameters, distance/similarity functions employed, etc. and in such a way make the users more convinced in the operation of the tools and techniques and what they can provide, and then persuaded them to use them in practice.

Acknowledgments Partially supported by the National Science Centre under Grant UMO-2012/05/B/ST6/03068.

References

1. Arbelaitz, O., Gurrutxaga, I., Muguerza, J., Pérez, J.M., Perona, I. (2013) An extensive comparative study of cluster validity indices. *Pattern Recognition* 46 (1), 243–256.
2. Brun, M., Sima, Ch., Hua, J., Lowey, J., Carroll, B., Suh, E., Dougherty, E.R. (2007) Model-based evaluation of clustering validation measures. *Pattern Recognition* 40(3), 807–824,
3. Charrad, M., Ghazzali, N., Boiteau, V., Niknafs, A. (2014) NbClust: An R Package for Determining the Relevant Number of Clusters in a Data Set. *Journal of Statistical Software*, 61(6), 1–36.
4. Choi, S.-S., Cha, S.-K., Tappert, Ch.C. (2010) A survey of binary similarity and distance measures. *Journal of Systemics, Cybernetics and Informatics*, vol. 8, no 1, 43–48.
5. Craven, M.W., Shavlik, J.W. (1995). Extracting comprehensible concept representations from trained neural networks. In: *Working Notes of the IJCAI'95 Workshop on Comprehensibility in Machine Learning*, Montreal, Canada, 61–75.
6. Cross, V. Sudkamp, Th.A. (2002) *Similarity and compatibility in fuzzy set theory: assessment and applications*. Physica-Verlag, Heidelberg; New York.
7. Das, S., Suganthan, P.N. (2011) Differential evolution: a survey of the state-of-the-art. *Evolutionary Computation, IEEE Transactions on,* 15(1), 4–31.
8. De Amorim, R. (2015) Feature Relevance in Ward's Hierarchical Clustering Using the Lp Norm. *Journal of Classification* 32, 46–62.
9. Denœud, L., Guénoche, A. (2006) Comparison of distance indices between partitions. In *Data Science and Classification*, Springer, Berlin Heidelberg, 21–28.
10. Fisch, D., Gruber, T., Sick, B. (2011) SwiftRule: Mining Comprehensible Classification Rules for Time Series Analysis. *IEEE Transactions on Knowledge and Data Engineering*: 23 (5), 774–787.
11. Fisher, R.A. (1936) The use of multiple measurements in taxonomic problems. *Annals of Eugenics* 7 (2), 179–188.
12. Halkidi, M., Batistakis, Y., Vazirgiannis, M. (2001) On clustering validation techniques. *Journal of Intelligent Information Systems*, 17(2–3), pp. 107–145.
13. Hastie, T., Tibsihrani, R., Friedman, J. (2009) *The Elements of Statistical Learning Data Mining, Inference, and Prediction*. Second Edition. Springer-Verlag New York.
14. Hubert, L., Arabie, P. (1985) Comparing partitions. *Journal of Classification*, 2(1), 193–218.

15. Kacprzyk, J., Zadrożny, S. (2013) Comprehensiveness of Linguistic Data Summaries: A Crucial Role of Protoforms. In: Christian Moewes and Andreas Nürnberger (Eds.): *Computational Intelligence in Intelligent Data Analysis*. Springer-Verlag, Berlin, Heidelberg, 207–221.
16. Kaufman, L., Rousseeuw, P.J. (1990) *Finding Groups in Data: An Introduction to Cluster Analysis*. Wiley, New York.
17. Lance, G.N.,Williams. W.T. (1966) A General Theory of Classificatory Sorting Strategies. 1. Hierarchical systems. *Computer Journal*, 9, 373–380.
18. Maechler, Martin, et al. (2015) "*cluster: cluster analysis extended Rousseeuw et al.*" R package, version 2.0.3.
19. Michalski, R. (1983) A theory and methodology of inductive learning. *Artificial Intelligence:* 20(2), 111–161.
20. Miyamoto, S. (2014) Classification Rules in Methods of Clustering (featured article). *IEEE Intelligent Informatics Bulletin*, 15(1), 15–21.
21. Miyamoto, S., Ichihashi, H., Honda, K. (2008) *Algorithms for Fuzzy Clustering: Methods in c-Means Clustering with Applications*. Springer-Verlag, Berlin Heidelberg, Studies in Fuzziness and Soft Computing 229.
22. Mullen, K.M, Ardia, D., Gil, D., Windover, D., Cline, J. (2011) DEoptim: An R Package for Global Optimization by Differential Evolution. *Journal of Statistical Software*, 40(6), 1–26.
23. Pryke, A., Beale, R. (2004) Interactive Comprehensible Data Mining. In: Y. Cai (ed.): *Ambient Intelligence for Scientific Discovery*, Springer, LNCS 3345, 48–65.
24. R Core Team (2014) R: A Language and Environment for Statistical Computing. R Foundation for Statistical Computing, Vienna, Austria. URL http://www.R-project.org/.
25. Ríos, S.A., Velásquez J.D. (2011) Finding Representative Web Pages Based on a SOM and a Reverse Cluster Analysis. International Journal on Artificial Intelligence Tools 20(1) 93–118.
26. Stańczak, J. (2003) Biologically inspired methods for control of evolutionary algorithms. *Control and Cybernetics*, **32** (2), 411–433.
27. Storn, R., Price, K. (1997) Differential evolution–a simple and efficient heuristic for global optimization over continuous spaces. *Journal of Global Optimization*, 11(4), 341–359.
28. Torra, V., Endo, Y., Miyamoto, S. (2011) Computationally intensive parameter selection for clustering algorithms: The case of fuzzy *c*-means with tolerance. *International Journal of Intelligent Systems*, 26 (4), 313–322.
29. Zhou, Z.H. (2005) Comprehensibility of data mining algorithms. In: J. Wang (ed.): *Encyclopedia of Data Warehousing and Mining*, IGI Global: Hershey, 190–195.

On Hesitant Fuzzy Clustering and Clustering of Hesitant Fuzzy Data

Laya Aliahmadipour, Vicenç Torra and Esfandiar Eslami

Abstract Since the notion of hesitant fuzzy set was introduced, some clustering algorithms have been proposed to cluster hesitant fuzzy data. Beside of hesitation in data, there is some hesitation in the clustering (classification) of a crisp data set. This hesitation may be arise in the selection process of a suitable clustering (classification) algorithm and initial parametrization of a clustering (classification) algorithm. Hesitant fuzzy set theory is a suitable tool to deal with this kind of problems. In this study, we introduce two different points of view to apply hesitant fuzzy sets in the data mining tasks, specially in the clustering algorithms.

Keywords Hesitant fuzzy sets · Data mining · Clustering algorithm · Fuzzy clustering

1 Introduction

Clustering and classification are important tasks in data mining. Clustering methods divide a data set into different clusters such that elements of the same cluster are as similar as possible and elements of different clusters are as dissimilar as possible. Clustering algorithms classify the data into k partitions, which together satisfy the following requirements: (1) each partition must contain at least one object, and (2) each object must belong to exactly one partition. Notice that the second requirement can be ignored in some fuzzy clustering techniques.

L. Aliahmadipour (✉) · E. Eslami
Faculty of Mathematics and Computer, Department of Mathematics,
Shahid Bahonar University of Kerman, Kerman, Iran
e-mail: L.Aliahmadipour@math.uk.ac.ir

E. Eslami
e-mail: Esfandiar.Eslami@uk.ac.ir

V. Torra
School of Informatics, Skövde University, Skövde, Sweden
e-mail: Vicenc.Torra@his.se

V. Torra et al. (eds.), *Fuzzy Sets, Rough Sets, Multisets and Clustering*,
Studies in Computational Intelligence 671, DOI 10.1007/978-3-319-47557-8_10

In general, based on [1] the major clustering methods for crisp data can be classified into the following categories: Partitioning methods such as K-means [2] and K-medoids [3], Hierarchical methods such as AGNES (AGglomerative NESting) and DIANA (DIvisive ANAlysis) [3], Density-based methods, such as DBSCAN and OPTICS [4], Grid-based methods such as STING [5] and Model-based methods such as SOM and EM [6, 7]. The choice of a clustering algorithm depends both on the type of available data and on the particular purpose of the application.

Since fuzzy set theory was introduced [8], this issue has been much considered in data mining methods such as classification [9] and clustering [10]. Later, K.T. Atanassov introduced the notion of intuitionistic fuzzy set (A-IFS) [11] as a generalization of fuzzy set. A-IFS is characterized by a membership function and a non-membership function. Recently, in order to consider the importance of hesitation and uncertainty in the nature of some problems, a new extension of fuzzy set was introduced. It is hesitant fuzzy set (HFS) [12]. It was introduced to deal with hesitant situations which were not well managed by the previous tools. A HFS is defined in terms of a function that returns a set of membership values for each element in the domain.

Fuzzy sets, intuitionistic fuzzy sets and hesitant fuzzy sets can be applied in the clustering tasks from two points of view:

1. Consider uncertain data and, in particular, fuzzy data (FD) [13], intuitionistic fuzzy data (IFD) [14] and hesitant fuzzy data (HFD) [15].
2. Consider crisp data set but uncertain clusters and, in particular, fuzzy partition (FP) [16], intuitionistic fuzzy partition (IFP) [17, 18] and hesitant fuzzy partition (HFP) [19].

Figure 1 illustrates these two perspectives. In the following sections we focus on hesitant fuzzy clustering algorithms and review the algorithms proposed in the literature from both points of view. First, Sect. 2 presents some basic definitions and descriptions we need later. Section 3 reviews some clustering algorithms for hesitant fuzzy data. Section 4 introduces hesitant fuzzy clustering algorithms on crisp data sets. Finally, we summarize the results and give some ideas for future work.

2 Preliminaries

In this section, we present some basic concepts related to hesitant fuzzy clustering algorithms.

2.1 Hesitant Fuzzy Sets

The membership degree of a HFS is represented by several possible values in [0, 1]. The definition is as follows:

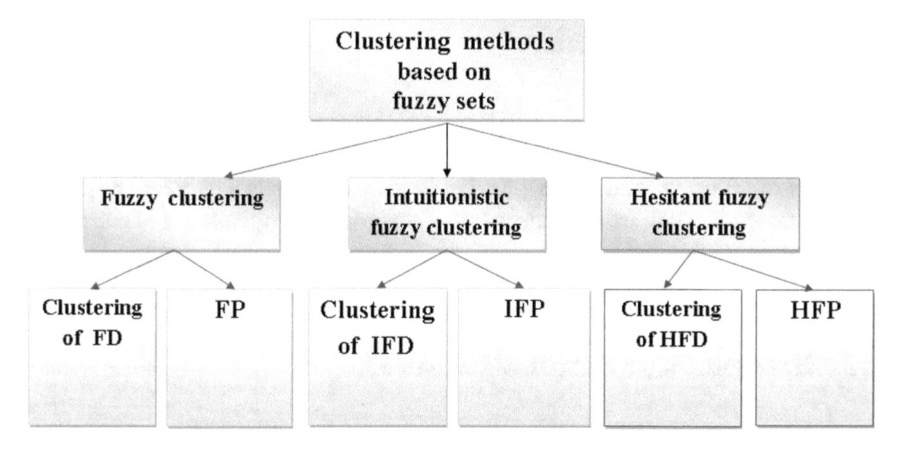

Fig. 1 The classification of fuzzy and clustering methods

Definition 1 ([12]) Let X be a fixed set and $\mathcal{P}(U)$ be the power set of U, then a hesitant fuzzy set (HFS) on X in terms of a function h is such that when applied to X returns a subset of [0, 1], i.e., $h : X \rightarrow \mathcal{P}([0, 1])$.

We use the term typical HFS [20] when the subsets $h(x)$ are finite. Furthermore, given a set of fuzzy sets, a HFS can be defined in accordance with the union of their memberships as follows:

Definition 2 ([12]) Let $M = \{\mu_1, \mu_2, \ldots, \mu_n\}$ be a set of n membership functions on a reference set X. The HFS associated with M, h_M is defined as:

$$h_M(x) = \bigcup_{\mu \in M} \{\mu(x)\} \qquad for\, all\, x \in X$$

Xia and Xu [21] called $h(x)$ a hesitant fuzzy element (HFE). A hesitant fuzzy element (HFE) is a set of values in [0, 1], and a HFS is a set of HFEs, for each $x \in X$.

Definition 3 ([12]) Given a HFE, h, we define the intuitionistic fuzzy value (IFV) $Aenv(h)$ as the envelope of h, where $Aenv(h)$ can be represented as $(h^-, 1 - h^+)$, with $h^- = \inf\{\gamma | \gamma \in h\}$ and $h^+ = \sup\{\gamma | \gamma \in h\}$.

Based on the relationship between the HFEs and IFVs, Xia and Xu in [21] defined some new operations on the HFEs. Let h, h_1 and h_2 be HFEs and λ be a real number then,

- $h^\lambda = \cup_{\gamma \in h}\{\gamma^\lambda\}$,
- $\lambda h = \cup_{\gamma \in h}\{1 - (1 - \gamma)^\lambda\}$,
- $h_1 \oplus h_2 = \cup_{\gamma_1 \in h_1, \gamma_2 \in h_2}\{\gamma_1 + \gamma_2 - \gamma_1\gamma_2\}$,

- $h_1 \otimes h_2 = \cup_{\gamma_1 \in h_1, \gamma_2 \in h_2} \{\gamma_1 \gamma_2\}$,
- $h^c = \cup_{\gamma \in h} \{1 - \gamma\}$,
- $h_1 \cup h_2 = \cup_{\gamma_1 \in h_1, \gamma_2 \in h_2} \max\{\gamma_1, \gamma_2\}$,
- $h_1 \cap h_2 = \cup_{\gamma_1 \in h_1, \gamma_2 \in h_2} \min\{\gamma_1, \gamma_2\}$.

Here, we recall some concepts involved in hesitant fuzzy sets which will be used in the present work. They are distance measures, aggregation operators and correlation coefficients.

2.2 *Hesitant Fuzzy Distance Measure*

Recall that for any two objects A_1, A_2 in a reference set X, a distance is a mapping, $d : X \times X \to \mathcal{R}$. $d(A_1, A_2)$ is the distance between A_1 and A_2 when,

1. $0 \leq d(A_1, A_2) \leq 1$,
2. $d(A_1, A_2) = 0$ if and only if $A_1 = A_2$,
3. $d(A_1, A_2) = d(A_2, A_1)$.

Let A_1 and A_2 be two HFSs on $X = \{x_1, \ldots, x_n\}$, some examples of distances used on HFSs in clustering problems follow.

1. Hesitant normalized Hamming distance [22]:

$$d_1(A_1, A_2) = \frac{1}{n} \sum_{i=1}^{n} \left[\frac{1}{l_{x_i}} \sum_{j=1}^{l_{x_i}} |h_{A_1}^{\sigma(j)}(x_i) - h_{A_2}^{\sigma(j)}(x_i)| \right] \tag{1}$$

2. Hesitant normalized Euclidean distance [22]:

$$d_2(A_1, A_2) = \left[\frac{1}{n} \sum_{i=1}^{n} \left(\frac{1}{l_{x_i}} \sum_{j=1}^{l_{x_i}} |h_{A_1}^{\sigma(j)}(x_i) - h_{A_2}^{\sigma(j)}(x_i)|^2 \right) \right]^{\frac{1}{2}} \tag{2}$$

where $h_{A_1}^{\sigma(j)}(x_i)$ and $h_{A_2}^{\sigma(j)}(x_i)$ are the jth largest values in $h_{A_1}(x_i)$ and $h_{A_2}(x_i)$, respectively, which will be used thereafter. In most cases, $l_{h_{A_1}}(x_i) \neq l_{h_{A_2}}(x_i)$, and for convenience, let $l_{x_i} = \max\{l_{h_{A_1}}(x_i), l_{h_{A_2}}(x_i)\}$ for each x_i in X being $l_{h_{A_1}}(x_i)$ and $l_{h_{A_2}}(x_i)$ the number of values $h_{A_1}(x_i)$ and $h_{A_2}(x_i)$. To operate correctly, we should extend the hesitant fuzzy set with the lesser number of values until both of them have the same length when we compare them. To extend the shorter one, we can extend it by adding any value in it, which mainly depends on the user preferences. We can add maximum value (as optimist), minimum value (as pessimist) or the average value of the existed memberships. All these are determined based on the appropriate score and accuracy function. Also Xu and Xia in [22] further extended the above distance measures and defined a generalized hesitant normalized distance as follows:

3. Generalized hesitant normalized distance:

$$d_3(A_1, A_2) = \left[\frac{1}{n} \sum_{i=1}^{n} \left(\frac{1}{l_{x_i}} \sum_{j=1}^{l_{x_i}} |h_{A_1}^{\sigma(j)}(x_i) - h_{A_2}^{\sigma(j)}(x_i)|^{\lambda} \right) \right]^{\frac{1}{\lambda}} \quad (3)$$

where $\lambda > 0$. It is noted that the parameter λ provides the decision makers more choices and can be assigned different values according to different decision makers. Several novel distance and similarity measures between hesitant fuzzy sets (HFSs) are developed, in which both the values and the numbers of values of HFE are taken into account.

2.3 *Hesitant Fuzzy Aggregation Operators*

Xia and Xu presented in [21] some aggregation operators, such as hesitant fuzzy weighted averaging and hesitant fuzzy weighted geometric which are defined as follows. Other aggregation operators for HFS can be found in [23].

Definition 4 ([21]) Let $h_1, h_2, \ldots, h_n$ be a collection of HFEs the hesitant fuzzy weighted averaging (HFWA) operator is defined by

$$HFWA(h_1, \ldots, h_n) = \oplus_{i=1}^{n}(w_i h_i), \quad (4)$$

where $w = (w_1, \ldots, w_n)^T$ is a weighting vector (i.e., $w_i \in [0, 1]$ and $\sum_{i=1}^{n} w_i = 1$). In case of $w = (1/n, \ldots, 1/n)^T$, then the (HFWA) operator reduces to the hesitant fuzzy averaging (HFA) operator:

$$\begin{aligned} HFA(h_1, \ldots, h_n) &= \oplus_{i=1}^{n}\left(\frac{1}{n} h_i\right) \\ &= \bigcup_{\gamma_1 \in h_1, \ldots, \gamma_n \in h_n} \{1 - \prod_{i=1}^{n}(1 - \gamma_i)^{\frac{1}{n}}\}. \end{aligned} \quad (5)$$

Definition 5 Let $h_i (i = 1, \ldots, n)$ be a collection of HFEs, $w = (w_1, \ldots, w_n)^T$ be a weighting vector of them. Then, the generalized hesitant fuzzy weighted geometric (GHFWG) operator is defined as

$$GHFWA_{\lambda}(h_1, \ldots, h_n) = \left(\oplus_{i=1}^{n} (w_i h_i)^{\lambda}\right)^{\frac{1}{\lambda}}. \quad (6)$$

2.4 Hesitant Fuzzy Correlation Coefficient

We recall that two variables are correlated if there is a (linear) relation between them. It is an important concept in data analysis. Because of this, different correlation coefficients have been defined to different types of information (see [23] for details on its application to HFSs). Chen et al. [24] defined the informational energy for HFSs and a related correlation between HFSs. We review them below.

Definition 6 ([24]) Let H be a typical HFS (i.e., with a finite number of membership degrees) on $X = \{x_1, \ldots, x_n\}$, the informational energy of the HFS H, is defined as follows,

$$E_{HFS}(H) = \sum_{i=1}^{n} \left(\frac{1}{l(h_M(x_i))} \sum_{j=1}^{l_{x_i}} h^2_{M_\sigma(j)}(x_i) \right), \tag{7}$$

where $h_M(x_i)$ is a HFE and $h_{M_\sigma(j)}(x_i)$ are the jth largest values of $h_M(x_i)$ and $l(h_M(x_i))$ is the number of values in $h_M(x_i)$.

Definition 7 ([24]) Let H_1 and H_2 be two typical HFSs on $X = \{x_1, \ldots, x_n\}$, the correlation between M and N is defined by,

$$C_{HFS}(H_1, H_2) = \sum_{i=1}^{n} \left(\frac{1}{l_{x_i}} \sum_{j=1}^{l_{x_i}} h^1_{M_\sigma(j)}(x_i) h^2_{M_\sigma(j)}(x_i) \right), \tag{8}$$

where $l_{x_i} = max\{l(h^1_M(x_i)), l(h^2_M(x_i))\}$ and $h^1_M(x_i)$, $h^2_M(x_i)$ are a HFE for H_1, H_2, respectively.

Let M, N be two HFSs, the correlation satisfies:

- $C_{HFS}(M, M) = E_{HFS}(M)$;
- $C_{HFS}(M, N) = C_{HFS}(N, M)$.

By using Definitions 6 and 7 the following correlation coefficient is obtained.

Definition 8 Let M, N be two typical HFSs on $X = \{x_1, \ldots, x_n\}$ the correlation coefficient between M and N is,

$$\rho_{HFS}(M, N) = \frac{C_{HFS}(M, N)}{[C_{HFS}(M, M)]^{1/2}[C_{HFS}(N, N)]^{1/2}}. \tag{9}$$

3 Clustering of Hesitant Fuzzy Data

In this section, we present the proposed clustering algorithms that deal with uncertain information. Thus, the given input information is a set of hesitant fuzzy sets (HFSs), on feature space $X = \{x_1, x_2, \ldots, x_n\}$ in the following form:

$$S = \{D_1, D_2, \ldots, D_m\}$$

$$D_j = \{(x_i, h_{D_j}(x_i)) | i = 1, 2, \ldots, n\}, j = 1, 2, \ldots, m,$$

$$h_{D_j}(x_i) = \{\mu_k | \mu_k \in [0, 1], k = 1, 2, \ldots, K\},$$

where S is a set of HFSs as HFD and $h_{D_j}(x_i)$ is a set of experts assessments with arbitrary cardinality K which states that D_j satisfies feature x_i. Until now, four clustering algorithms have been proposed to cluster HFD. Brief descriptions of them follow. See also [15] for a comparison.

Hesitant Fuzzy Data Clustering Based on Correlation Coefficients of HFSs. Chen et al. in [24] propose some correlation coefficient formulas for HFSs and apply them to clustering under hesitant fuzzy environments. They used the derived correlation coefficient formulas to calculate the degrees of correlation among HFSs aiming at clustering different data. They cluster the data based on hesitant fuzzy equivalence matrices and the transitive closure technique, which is computationally costly and thus clustering needs a significant amount of time.

Hesitant Fuzzy Agglomerative Hierarchical Clustering. Authors in [25] propose a novel hesitant fuzzy agglomerative hierarchical clustering algorithm for HFD. Their algorithm has four steps and it considers each of the given HFDs as a unique cluster. In the first step, a distance matrix is calculated by utilizing the weighted Hamming distance or the weighted Euclidean distance between HFSs, and distances of each pair of HFSs are compared. In the second step, the two clusters with the smallest distance are merged and a new center is calculated from them with the equation in Definition 4. In the third step, the distance matrix is updated. Steps 2 and 3 are repeated until the desirable number of clusters is achieved. Also, they extend the algorithm to cluster the interval-valued hesitant fuzzy sets. They illustrate the effectiveness of the clustering algorithm by experimental results.

Hierarchical Hesitant Fuzzy *K*-Means Clustering Algorithm. Chen et al. in [26] propose a new clustering algorithm based on a combination of hierarchical and K-means methods for hesitant fuzzy data. They considered that K-means algorithm is sensitive to initial environment, i.e., initial clusters. So, for obtaining a better performance they used the results of hierarchical clustering as initial clusters for K-means algorithm. They show that such utilization can greatly reduce the computational cost arising from selecting the initial seeds randomly as original K-means algorithm. The hybrid method can significantly accelerate the clustering process. As a summary proposed algorithm has two procedures: (i) hierarchical procedure similar to hesitant fuzzy agglomerative hierarchical clustering and (ii) K-means procedure for hesitant fuzzy data using the hesitant fuzzy distance measure and the hesitant fuzzy weighted averaging (HFWA) operator. They have shown that this approach is a valuable tool to do clustering for HFDs.

Clustering of Generalized Hesitant Fuzzy Data. Authors in [27] propose a new hierarchical algorithm for clustering of generalized hesitant fuzzy (GHF) information. Qian et al. [28] introduced generalized hesitant fuzzy sets (GHFSs), on feature space $X = \{x_1, x_2, ..., x_n\}$ in the following form:

$$S = \{D_1, D_2, \ldots, D_m\}$$

$$D_j = \{(x_i, h_{D_j}(x_i)) | i = 1, 2, \ldots, n\}, \; j = 1, 2, \ldots, m,$$

$$h_{D_j}(x_i) = \{\alpha_i^k | \alpha_i^k =< \mu_i^k, \nu_i^k >, 0 \leq \mu_i^k, \nu_i^k \leq 1, 0 \leq \mu_i^k + \nu_i^k \leq 1, k \in K\},$$

where S is a data set of $D_j^{'}s$ and $h_{D_j}(x_i)$ is a set of experts assessments which states that D_j satisfies feature x_i. So, they developed a novel generalized hesitant fuzzy hierarchical clustering (GHFHC) algorithm [27] for GHFSs, which is based on the agglomerative hierarchical clustering algorithms and the intuitionistic fuzzy basic operators. The main difference between the former hesitant fuzzy clustering algorithms [24–26] and the GHFHC algorithm is that the latter deals with GHF information and needs less computational efforts. One of the advantages of applying hesitant fuzzy set is that clustering hesitant and vague information permits us to find patterns among hesitant fuzzy data.

4 Hesitant Fuzzy Clustering

In this section, we apply hesitant fuzzy sets to a clustering context, and use them to cluster crisp data sets. As a matter of fact, there are several kind of hesitations. In this chapter we just consider the following two cases:

Case 1: Hesitation arises from uncertain initial parameterizations.
Case 2: Hesitation arises from the choice of a suitable clustering algorithms.

Hesitant fuzzy partitions (HFPs) [19] are introduced to consider case 1. Authors in [29] proposed a hesitant fuzzy decision making method to choose a suitable clustering algorithm for a given data set to consider case 2.

4.1 H-Fuzzy Partition

Authors in [19] define a method to construct H-fuzzy partitions from a set of FPs obtained from several executions of fuzzy clustering algorithms with various initialization of their parameters. Their purpose is to consider some local optimal solutions to find a global optimal solution also letting the user to consider various reliable membership values and cluster centers to evaluate her/his problem using different cluster validity indices. To introduce HFP, we present definitions of FP and IFP in the following:

Definition 9 ([18]) Let X be a reference set. Then, a set of membership functions $M = \{\mu_1, \ldots, \mu_n\}$ on X is a fuzzy partition of X if for all $x \in X$, $\sum_{i=1}^{n} \mu_i(x) = 1$ holds.

Definition 10 Let X be a reference set. Then, a set of AIFSs $A = \{A_1, \ldots, A_m\}$ where $A_i = \langle \mu_i, \pi_i \rangle$ is an I-fuzzy partition if

1. $\sum_{i=1}^{m} \mu_i(x) = 1$ for all $x \in X$,
2. for all $x \in X$, there is at most one i such that $\nu_i(x) = 0$ (there is at most one IFS such that $\mu_A(x) + \pi_A(x) = 1$ for all x).

Proposition 1 ([18]) *I-fuzzy partitions generalize fuzzy partitions.*

The definition of HFP is for typical hesitant fuzzy sets [20] as follows:

Definition 11 ([19]) Let $X = \{x_1, \ldots, x_n\}$ be a reference set. Let H^* be a HFS on X $H^* = \{\langle x, \hat{h}_j \rangle \mid j = 1, 2, \ldots, m\}$, where m is the number of clusters and $\hat{h}_j = \{\mu_j^k \mid k = 1, 2, \ldots, \kappa\}$ are hesitant fuzzy elements. That is $\hat{h}_j$ is a finite set such that $\hat{h}_j \subseteq [0, 1]$ and κ is the number of membership degrees in $\hat{h}_j$ (i.e., the cardinality of $\hat{h}_j$ is κ. We can use κ_j for $j = 1, 2, \ldots, m$. However for the sake of simplicity they use $\kappa_j = \kappa$ for all j). Then H^* is a hesitant fuzzy partition (H-fuzzy partition) if

$$\frac{\sum_{j=1}^{m} \sum_{k=1}^{\kappa} \mu_j^k(x)}{\kappa m} \leq 1 \qquad \forall x \in X, \quad 0 \leq \mu_j^k(x) \leq 1. \tag{10}$$

Also they consider a more general case in which the set $\hat{h}_j$ is infinite. This is a generalization of the former definition and they use it later to prove that this definition generalizes the one for I-fuzzy partitions discussed above (see Definition 10).

Definition 12 ([19]) Let $X = \{x_1, \ldots, x_n\}$ be a reference set. Then a set of HFE $H = \{\hat{h}_1, \ldots, \hat{h}_m\}$, where $\hat{h}_j$ is an infinite set, is a hesitant fuzzy partition if the following holds (note that the following inequality is a reformulation of Eq. (10)):

$$\sum_{j=1}^{m} \frac{\int_0^1 y \mu_j(x)(y) dy}{m \int_0^1 \mu_j(x)(y) dy} \leq 1 \qquad \forall x \in X, \tag{11}$$

where $\mu_j(x)$ is the characteristic function of the set $\hat{h}_j(x)$.

Authors underline that this definition is not only valid for hesitant fuzzy set but also for type-2 fuzzy sets as $\mu_j(x)$ can be a fuzzy set. Note that if $\hat{h}_j(x)$ is a single value then we have type-1 fuzzy sets, if $\hat{h}_j(x)$ is a finite set we have typical hesitant fuzzy sets, and when $\hat{h}_j(x)$ is a fuzzy set we have type-2 fuzzy sets. In this case the membership degree of x to cluster j is given by the membership function $\mu_j(x)(y)$ for $y \in [0, 1]$ instead of having $\mu_j(x)$ a single value.

Proposition 2 ([19]) *H-fuzzy partitions generalize I-fuzzy partitions.*

In [19] a method is described for construction of H-Fuzzy partitions.

4.2 Data Clustering Based on Hesitant Fuzzy Decision Making

Authors in [29] deal with hesitation to choose a clustering algorithm for a data set without any prior knowledge about it. So, they model this hesitation by a hesitant fuzzy decision making problem. That is, the problem of choosing a proper clustering algorithm. To this end two categories of clustering algorithms are considered. They are (i) partitioning methods and (ii) hierarchical clustering methods. Each of them is suitable for a particular type of data distribution. So, they consider FCM [16] and agglomerative hierarchical (linkage family) [3] as representative of partitioning methods and hierarchical clustering methods, respectively.

To determine whether the FCM family or the hierarchical clustering algorithms is the most suitable for the data, authors introduce a procedure based on a new hesitant fuzzy decision making method for the data clustering. It is a hesitant fuzzy decision making problem which deals with the clustering task. So, the two clustering algorithms play the role of experts in a hesitant fuzzy decision making problem. Each of them assigns various membership degrees for each data to the each cluster. Then, a procedure is proposed to choose an appropriate clustering algorithm. This procedure is based on Neutrosophic FCM (NFCM) clustering algorithm [30]. The proposed procedure for a given data set S has two phases [29]:

Phase 1. Determine suitable cluster number.

Phase 2. Determine appropriate clustering algorithm through hesitant fuzzy decision making method and then apply the clustering algorithm to the data points.

5 Conclusion

In this study, we have considered the issue of fuzzy set membership hesitation in the data mining tasks and, in particular, in clustering methods. We review existing hesitant fuzzy clustering methods from two points of view. Some of them deal with the hesitation in the nature of data and others consider the hesitation to cluster the crisp data. The former methods can be useful when we collect the data in an uncertain environment and final clusters are analysed by experts. The latter methods can be more useful than the former in real world problem.

The aim of this chapter was to show that hesitant fuzzy sets can be applied in different contexts within data clustering.

Acknowledgements The first author gratefully acknowledges the support of the Ministry of Science, Research and Technology of the Islamic Republic of Iran and Shahid Bahonar University of Kerman.

References

1. Han, J., Kamber, M.: Data Mining: Concepts and Techniques. Morgan Kaufmann Publishers, pp. 383–464 (2009).
2. Jain, A.K.: Data clustering: 50 Years Beyond K-means. Pattern Recognition Letters. 31, 651–666 (2010)
3. Kaufman, L., Rousseeuw, P.J.: Finding Groups In Data: An Introduction To Cluster Analysis. New York: John Wiley & Sons (1990)
4. Ester, M., Kriegel, H.P., Sander, J., Xu, X.: Density-Connected Sets And Their Application For Trend Detection In Spatial Databases. Conf. Knowledge Discovery and Data Mining (KDD 97), pp 10 –15 (1997)
5. Pilevar A.H., Sukumar, M.: GCHL: A Grid-Clustering Algorithm For High-Dimensional Very Large Spatial Data Bases. Pattern Recognition Letters. 6 999–1010 (2005)
6. Yang, M.S., Yo Lai, C., Lin, C.Y.: A Robust EM Clustering Algorithm for Gaussian Mixture Models. Pattern Recognition. 45 3950–3961 (2012)
7. Mingoti, S.A., Lima, J.O.: Comparing SOM Neural Network With Fuzzy C-Means, K-Means And Traditional Hierarchical Clustering Algorithms. Euro. J. of Operational Research. 174 1742–1759 (2006)
8. Zadeh, L.A.: Fuzzy sets. Information and Control. 8 338–353 (1965)
9. Scherer, R.: Multiple Fuzzy Classification Systems. Studies in Fuzziness and Soft Computing, Springer, 288 (2012)
10. Valente de Oliveira, J., Pedrycz, W.: Advances in Fuzzy Clustering and Its Applications. Wiley (2007)
11. Atanassov, K.T.: Intuitionistic Fuzzy Sets. Fuzzy Sets and Systems. 20 87–96 (1986)
12. Torra, V.: Hesitant Fuzzy Sets. Int J of Intelligent Syst. 25 529–539 (2010)
13. Sato, M., Sato, Y.: Fuzzy Clustering Model For Fuzzy Data. Fuzzy Systems , International Joint Conference of the Fourth, pp 2123–2128 (1995)
14. Xu, Z.S.: Intuitionistic Fuzzy Aggregation and Clustering. Studies in Fuzziness and Soft Computing. Springer 279 1–284 (2012)
15. Aliahmadipour, L., Taghavi, A., Eslami, E.: An Introduction To Hesitant Fuzzy Data Clustering. 4th Iranian Joint Congress on Fuzzy and Intelligent Systems (CFIS), Zahedan, Iran, pp.1–4 (2015)
16. Bezdek, J.C.: Pattern Recognition with Fuzzy Objective Function Algorithms. Plenum Press, New York pp.191–203 (1981)
17. Chaira, T.: A Novel Intuitionistic Fuzzy C Means Clustering Algorithm And Its Application To Medical Images. Applied Soft Computing. 11 1711–1717 (2011)
18. Torra, V., Miyamoto, S.: A definition for I–fuzzy partition. Soft Computing. 15 363–369 (2011)
19. Aliahmadipour, L., Torra, V., Eslami, E., Eftekhari, M.: A Definition for Hesitant fuzzy Partitions. Int.J.Copmpu. Intel.Syst, 9 497–505 (2016)
20. Bedregal, B., Reiser, R., Bustince, H., López-Molin, C., Torra, V.: Aggregation Functions For Typical Hesitant Fuzzy Elements and The Action Of Automorphisms. Information Sciences. 255 82–99 (2014).
21. Xia, M.M, Xu, Z.S.: Hesitant Fuzzy Information Aggregation In Decision Making. Int. J. Approximate Reasoning 52 395–407 (2011).
22. Xu, Z.S, Xia, M.M.: Distance And Similarity Measures For Hesitant Fuzzy Sets. Information Sciences. 181 2128–2138 (2011)

23. Rodríguez, R.M., Martínez, L., Torra, V., Xu, Z.S., Herrera, F.: Hesitant Fuzzy Sets: State of The Art and Future Directions. International J. of Intelligent Systems. 29 495–524 (2014)
24. Chen, N., Xu, Z.S, Xia, M.M.: Correlation Coefficients of Hesitant Fuzzy Sets And Their Applications To Clustering Analysis. Applied Mathematical Modelling. 37 2197–2211(2013)
25. Zhang, X., Xu, Z.S.: Hesitant Fuzzy Agglomerative Hierarchical Clustering Algorithms. International Journal of Systems Science. 1–16 (2013)
26. Chen, N., Xu, Z.S., Xia, M.M.: Hierarchical Hesitant Fuzzy K-Means Clustering Algorithm. Applied Mathematics-A Journal of Chinese Universities. 29 1–17 (2014)
27. Aliahmadipour, L., Eslami, E.: GHFHC: Generalized Hesitant Fuzzy Hierarchical Clustering Algorithm. International Journal of Intelligent Systems. In press (2016)
28. Qian, G., Wang, H., Feng, X.: Generalized Hesitant Fuzzy Sets And Their Application In Decision Support System. Knowledge-Based Syst. 37 357–365 (2012)
29. Aliahmadipour, L., Eslami, E., Eftekhari, M., Torra, V.: Data clustering Based on Hesitant Fuzzy Decision Making Method. (submitted)
30. Guo, Y., Sengur, A.: NECM: Neutrosophic Evidential C-Means Clustering Algorithm. Neural Computing & Application. 26 561–571 (2014)

Experiences Using Decision Trees for Knowledge Discovery

Eva Armengol, Àngel García-Cerdaña and Pilar Dellunde

Abstract Knowledge discovery is the process of identifying useful patterns from large data sets. There are two families of approaches to be used for knowledge discovery: clustering, when the classes of domain objects are not known; and inductive learning algorithms, when the classes are known and the goal is to construct a domain model useful to identify new unseen objects. Clustering algorithms have also been proposed to analyze the data when the classes are known. However, to our knowledge, inductive learning methods are not used to analyze the available data but only for prediction. What we propose here is a methodology, namely FTree, that uses a decision tree to analyze both the available data identifying patterns and some important aspects of the domain (at least from the domain's part represented by the data at hand) such as similarity between classes, separability, characterization of classes and even some possible errors on data.

1 Introduction

Knowledge Discovery (KD) is defined in [12] as the process of identifying valid, novel, useful and understandable patterns from large data sets. Its goal is to develop models useful for data analysis and prediction. However, both data analysis and prediction are two different tasks and depending in which of them we want to focus, the way to construct the model will be different. In some cases, the data available

E. Armengol (✉) · À. García-Cerdaña · P. Dellunde
IIIA, Artificial Intelligence Research Institute CSIC, Spanish National Research Council, Campus UAB, 08193 Bellaterra, Catalonia, Spain
e-mail: eva@iiia.csic.es

À. García-Cerdaña
Departament de Tecnologies de la Informació i les Comunicacions, Universitat Pompeu Fabra, Tànger, 122-140, 08018 Barcelona, Catalonia, Spain
e-mail: angel@iiia.csic.es

P. Dellunde
Philosophy Department, Universitat Autònoma de Barcelona, Campus UAB, 08193 Bellaterra, Catalonia, Spain
e-mail: pilar@iiia.csic.es

V. Torra et al. (eds.), *Fuzzy Sets, Rough Sets, Multisets and Clustering*, Studies in Computational Intelligence 671, DOI 10.1007/978-3-319-47557-8_11

are descriptions of domain objects and the goal is to find some regularities among groups of objects. In such situations, clustering algorithms are the most useful tools. Clustering algorithms [9] are unsupervised learning algorithms that group domain objects by similarity. Possibly, there is no prior information about neither how many classes can be defined in the domain nor about which are the important attributes that could characterize them. Commonly, the user has to give as input parameter the number of clusters, and the algorithm works with it. Once the clusters are formed, they can be described in terms of *prototypes*, i.e., a vector of characteristics representing the centroid of the cluster (see for instance [8]).

In many domains, even when facing large databases, experts know how many classes exist and which are they. In such situation, clustering algorithms can also be used although the most common choice are supervised learning algorithms such as the inductive ones. The goal of inductive learning algorithms is the construction of a domain theory from the known data, i.e., to characterize each one of the classes of the domain by means of discriminant descriptions. Commonly this domain theory is further used to predict the classification of unseen objects. Given a solution class C_i, the *discrimination task* for inductive learning methods is defined as follows:

- Given: a set E containing positive E^+ and negative E^- examples of a class C_i.
- Find: a description D_i such that it is satisfied by elements in E^+ and it is not satisfied by any of the elements in E^-.

A class C_i can be described by more than one description D_i. To build a model of a domain we have to perform the discrimination task over each one of the classes. As we have already mentioned before, unlike than clustering methods, the goal of inductive learning methods is to build a predictive model, i.e., a model capable to classify unseen domain objects correctly. One of the widely used inductive learning methods are the *decision trees*. We have two forms of decision trees:

- *classification trees*, when the predicted outcome is the class to which the data belong. The most used algorithm is ID3 [16], called C4.5 in its commercial version [17].
- *regression trees*, when the predicted outcome can be considered a real number. The most used algorithm is CART [5].

In this paper we will focus on classification problems and on classification trees. However, now we are not interested on prediction but in analysis of the available data. In particular, we want to focus on a problem that Pazzani [15] mentions: "most of literature is about validity and process, and very little is about utility, novelty and understability". We are also interested on dealing with huge databases but our primary goal is to discover knowledge that experts can understand and use. For this reason we do not speak here about the scalability of the approach we propose. We want to remark than the more examples we have, the more accurate inductive learning methods are. This means that they cannot be confident enough when we are facing small databases.

Decision trees have proved to be easy to understand by domain experts. Nevertheless, as Pazzani remarks [15], this point has not been objectively proved. However, our experience in working with physicians of different medical specialities, has shown us that experts quickly understand the knowledge coded in form of attribute-value and also are capable to assess the validity of the knowledge represented by either a decision tree or a rule set. In fact, most of our previous work focused on building predictive models [2, 3] but, during the interaction with the experts, we observed that sometimes the expert was more interested on the attributes taken into account during the construction of the tree than in the predictivity of the final model.

In [1] we pointed out that, given a decision tree, the path from the root to a leaf can be interpreted as an explanation of the classification since it contains the pairs attribute-value relevant for the classification. In the current paper we go beyond that idea and propose the use of decision trees for knowledge discovery. Our point is that the nodes of a tree (except leaves) contain sets of examples of different classes. In addition, the path from the root to that node gives an idea of the similarity between two or more classes.

In this paper we propose FTree (Filtered Tree), a methodology for analyzing the information available of a domain using decision trees. The idea is to use some known algorithm for decision tree growing, such as ID3, and analyze the tree shape. With FTree we can also take into account some background knowledge given by the expert and focus only on a subset of the available domain objects satisfying some pattern (i.e., a subset of pairs attribute-value). To get an accurate domain model, inductive learning methods and particularly decision trees need as many domain objects as possible. However, since FTree is an analysis tool, it can be used without being aware of the data base size. Moreover, with the FTree analysis we can identify lack of knowledge in the database and, therefore, to acquire appropriate examples.

The paper is organized as follows. Section 2 presents the basics on decision trees: what are they, which is their utility, how they are grown, measures used to grow them, etc. In Sect. 3 we introduce the FTree methodology useful to analyze a data set in accordance with the decision tree structure. Then we present two case studies of application of FTree on the domain. In Sect. 4 we analyze the domain of *malignant melanoma*. In Sect. 5 we analyze the domain of assessing the quality of life of people with intellectual disabilities. Section 6 discusses some aspects of the application of FTree. Final sections are devoted to related work and conclusions.

2 Classification Decision Trees

A *Decision Tree* (DT) is a directed acyclic graph in the form of a tree. The root of the tree has not incoming edges and the remaining ones have exactly one incoming edge. Nodes without outgoing edges are called *leaf* nodes and the others are *internal* nodes. A DT is a classifier expressed as a recursive partition of the set of known examples of a domain [12]. The goal is to create a domain model predictive enough to classify future unseen domain objects.

Each node of a tree has associated a set of examples that are those satisfying the path from the root to that node. For instance, the node size of the tree shown in Fig. 1

texture = *smooth* : EATABLE
texture = *spots*
| form = *planar* : POISONOUS
| form = *round*
| | size = *small* : EATABLE
| | size = big : POISONOUS

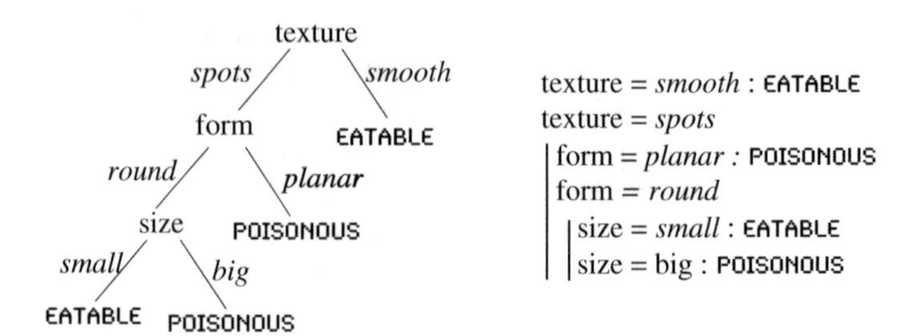

Fig. 1 Example of a DT for mushroom classification. Both representations are equivalent and we will use in the current paper the one in the *right side*

```
ID3 (examples, attributes)
   create a node
   if all examples belong to the same class  return class as the label for the node
   otherwise
        A ← best attribute
        for each possible value v_i of A
                  add a new tree branch below node
                  examples_vi ← subset of examples such that A = v_i
                  ID3(examples_vi, attributes - {A})
   return node
```

Fig. 2 ID3 algorithm for growing a decision tree

has associated all the examples having texture = *spots* and form = *round*. Notice also that the leaves of a tree determine a partition of the original set of examples, since each domain object only can be classified following one of the paths of the tree.

The construction of a decision tree is performed by splitting the source set of examples into subsets based on an attribute-value test. This process is repeated on each derived subset in a recursive manner called recursive partitioning. Figure 2 shows the ID3 algorithm [16, 18] commonly used to construct decision trees. To construct a decision tree, domain objects have to be represented by means of a set of pairs attribute-value. These values may be continuous-valued or categorical. Each tree node represents an attribute a_i selected by some criteria and each arch is followed according to the value v_{i1} of a_i. For instance, Fig. 1 shows an example classifying mushrooms as *eatable* or *poisonous*. Attributes describing a mushroom are texture, form and, size. The most relevant attribute for classifying a mushroom is texture because if it is *smooth* the mushroom can be classified as eatable. Otherwise the node has to be expanded. The next relevant attribute is form with two possible values: *planar* corresponding only to poisonous mushrooms; and *round* that is a characteristic shared by both classes of mushrooms. Finally, the attribute size allows a perfect classification of all the known mushrooms.

Notice that from a decision tree we can extract rules giving descriptions of classes. For instance, some eatable mushrooms are described by means of the rule:

$$\textsf{texture} = \textit{spots and } \textsf{form} = \textit{round and } \textsf{size} = \textit{small} \Rightarrow \textit{eatable}.$$

That is to say, each path from the root to a leaf form a description of a class. When all the examples of a leaf belong to the same class such description is *discriminant*. Notice also that intermediate nodes represent characteristics shared by several classes. For instance some eatable and poisonous mushrooms share a texture with spots and a round form.

A key issue of the construction of decision trees is the selection of the most relevant attribute to split a node. Each measure uses a different criteria, therefore the selected attribute could be different depending on it and, thus the whole tree could also be different. The most common measures are based on the *degree of impurity* of a node. They compare the impurity of a node, say t, with the impurity of the children nodes $t_1 \ldots t_k$ generated by an attribute a_i. This comparison is done for each one of the attributes used to represent the domain objects. The general expression to calculate the *gain* Δ associated to an attribute a_i is the following:

$$\Delta(a_i) = I(t) - \sum_{j=1}^{k} \frac{N(t_j)}{N} \cdot I(t_j)$$

where $I(\cdot)$ is an impurity measure, N is the total number of examples associated to the parent node t_j, k is the number of different values taken by a_i and $N(t_j)$ is the number of examples associated with the child node t_j. To build *regression trees*, the most common used impurity measure is the *Gini's index* that measures the divergences between the probability distributions of the attribute values according to the following expression:

$$Gini(t) = 1 - \sum_{i=1}^{c} p_i^2$$

where c are the class labels and p_i is the number of examples of the current node that belong to the class i. Concerning *classification trees* the most common impurity measure is the entropy H which is defined by:

$$H(t) = -\sum_{i=1}^{c} p_i \log_2 p_i$$

When the entropy is used as the impurity measure I, the difference in entropy is known as *information gain* [18] or also as *Quinlan's gain*.

A different kind of criteria to select the more relevant attribute is to split a node is the one proposed in [11]. This method consist in the comparison of the partition induced by an attribute, say a_i, with the *correct partition*, i.e., the partition that classifies correctly all the known examples. This comparison is done by using the López de Mántaras distance (LM). The best attribute is the one inducing the partition which is closest to the correct partition of the subset of training examples corresponding to this node. The distance LM is an entropy-based normalized metric defined in the set of partitions of a finite set. The entropy of a partition can be described as the information conveyed by the uncertainty that a randomly selected object belongs to a certain class. Given a finite set X and a partition $\mathcal{P} = \{P_1, \ldots, P_n\}$ of X in n sets, the entropy of $\mathcal{P}$ is defined as ($|\cdot|$ is the cardinality function):

$$H(\mathcal{P}) = -\sum_{i=1}^{n} p_i \cdot \log_2 p_i, \text{ where } p_i = \frac{|P_i|}{|X|}$$

and where the function $x \cdot \log_2 x$ is defined to be 0 when $x = 0$. The *López de Mántaras*' distance between two partitions $\mathcal{P} = \{P_1, \ldots, P_n\}$ and $\mathcal{Q} = \{Q_1, \ldots, Q_m\}$ is defined as:

$$\text{LM}(\mathcal{P}, \mathcal{Q}) = \frac{H(\mathcal{P}|\mathcal{Q}) + H(\mathcal{Q}|\mathcal{P})}{H(\mathcal{P} \cap \mathcal{Q})}, \tag{1}$$

where

$$H(\mathcal{P}|\mathcal{Q}) = -\sum_{i=1}^{n}\sum_{j=1}^{m} r_{ij} \cdot \log_2 \frac{r_{ij}}{q_j}, \quad H(\mathcal{Q}|\mathcal{P}) = -\sum_{j=1}^{m}\sum_{i=1}^{n} r_{ij} \cdot \log_2 \frac{r_{ij}}{p_i},$$

$$H(\mathcal{P} \cap \mathcal{Q}) = -\sum_{i=1}^{n}\sum_{j=1}^{m} r_{ij} \cdot \log_2 r_{ij},$$

$$\text{with } q_j = \frac{|Q_j|}{|X|}, \text{ and } r_{ij} = \frac{|P_i \cap Q_j|}{|X|}.$$

There are other different ways to select attributes as the one proposed by [13] that selects an attribute taking into account a cost criteria. For a summary on the subject we refer to [12].

There are at least three issues that have to be taken into account when constructing decision trees: the use of background knowledge, the presence of attributes with unknown values, and the scalability. Let us to briefly analyze them.

Background Knowledge. For many domains, there is some corpus of knowledge that can be used to extract some kind of new knowledge. However, decision trees in its primary form do not use background knowledge, that is to say, they cannot take benefit from already well known knowledge. For this reason there are many works focusing on how decision trees can exploit background knowledge. For instance, Ortega and Fisher [14] take into account expert's knowledge to rank the attributes that are candidates to expand a node, and the above mentioned work by Núñez [13] has a domain ontology relating domain concepts.

Management of unknown values. There are several options to deal with unknown values. One is to consider the most frequent value for that attribute among the objects belonging to a class. Another possibility is to calculate the probability for a class C_j than the value of a_i is v_i by means of the formula:

$$prob(a_i = v_i | class = C_j) = \frac{prob(a_i = v_i \& class = C_j)}{prob(class = C_j)}$$

Quinlan in [16] points out that these methods can produce bad results, for this reason he proposes the option of considering *unknown* as another value. Commonly, the management of unknown values (except in the last option) is considered as a data pre-processing independently of the tree construction.

Scalability. Algorithms for constructing decision trees assume that the whole database is in the main memory. However this assumption is not true in many applications and for this reason there are some approaches focused on using decision trees on huge databases. SPRINT [20] is a scalable version of CART [5] one of the most common algorithms used to construct regression trees. In order to avoid having all the information of attributes in the memory, SPRINT maintains attribute lists that are vertical partitions of the dataset. These lists are created in a preprocessing phase. RainForest [6] is a framework into which all the simple algorithms for constructing trees can be used. In order to reduce the necessary memory, the idea in RainForest is to define both the AVC-sets that are tables relating each attribute-value pair with a class label, and the AVC-groups that are groups of AVC-sets associated to each tree node.

3 FTree: Decision Trees for Knowledge Discovery

In this section we introduce FTree, a methodology for analyzing decision trees. This approach is based on our experience on interacting with domain experts. We see that decision trees are actually easy to understand for them since the attributes are the same used by experts. What we propose is to build a decision tree with some standard algorithm (in particular we use ID3 in our implementation) and use FTree to give some guidelines to analyze the tree together with a domain expert. In fact, this idea was also behind the assertion from [15] in the sense that, in addition to be consistent with the data, the induced model has also to be consistent with the expert's prior knowledge. FTree allows to introduce some filter with the idea of constructing a tree with only a subset of known examples. This modification allows the expert to focus on some characteristics that he considers important. For instance, one of the characteristics considered as relevant to diagnose a mole as a malignant melanoma is the presence of roundish pagetoid cells, therefore the experts want that this attribute appears in the tree and thus, only lesions having this kind of cells are considered in growing the tree.

Fig. 3 Example of decision tree for the *Iris* dataset

petalwidth $\leq$ 0.6: *Iris-setosa* (50.0)
petalwidth > 0.6
| petalwidth $\leq$ 1.7
| | petallength $\leq$ 4.9: *Iris-versicolor* (48.0/1.0)
| | petallength > 4.9
| | | petalwidth $\leq$ 1.5: *Iris-virginica* (3.0)
| | | petalwidth > 1.5: *Iris-versicolor* (3.0/1.0)
| petalwidth > 1.7: *Iris-virginica* (46.0/1.0)

A decision tree groups the set of known examples according to the values of some relevant attributes (those in the paths from the root to a leaf). ID3 produces a decision tree where each path from the root to a leaf gives the pairs attribute-value that are important to classify an example as belonging to a class. Each path can be interpreted as a general description d_i of a class C_i since d_i contains the features assessed as relevant for the classification. As we pointed out in [1] the tree paths can be interpreted as the explanation of the classification but they can also be interpreted as the similarity among a subset of examples of a class. For instance, from the tree shown in Fig. 3 we can interpret that all *iris setosa* are similar in that they have petalwidth ≤ 0.6. Also, we see that both *iris versicolor* and *iris virginica* are similar because they have $0.6 <$ petalwidth ≤ 1.7.

Our purpose goes beyond this interpretation since the FTree methodology uses decision trees to analyze the known examples of a domain. We will illustrate FTree with examples of three datasets from the Machine Learning repository of the Irvine's University [4]: *Iris, Splice* and *Soybean*. The *Iris* dataset has 150 domain objects distributed in three classes each one with 50 objects: *iris setosa, iris versicolor* and *iris virginica*. The *Splice* dataset has 3190 domain objects distributed in three classes: EI (767), IE (768) and N (1655). The *Soybean* dataset contains around 300 domain objects distributed on 18 solution classes, most of them having 10 objects, and four of the classes having 40 objects each.

The FTree methodology proposes the analysis of a decision tree at two levels: *global*, focusing on the shape of the tree; and *local*, focusing on the length of the paths. Thus, from the shape of the tree we can extract global information about the data set at hand, specifically referred to the separability and characterization of the classes. We can extract the following situations:

- *Width of the tree*. Relation between the number of classes and the number of branches of the tree. Cases:

1. $\sharp$*classes* $\simeq$ $\sharp$*branches*. If a tree has a number of branches similar to the number of classes then we can assume that the classes are clearly separable.
2. $\sharp$*classes* $\ll$ $\sharp$*branches*. There is a lot of variability among the elements of a class, therefore it has been necessary to construct many branches in order to characterize all the elements, i.e., there is overfitting. A possibility to be discussed with the domain expert could be the division of one or more classes.

3. $\sharp classes > \sharp branches$. The elements cannot be separated, therefore this means that several classes are very similar. A possibility is that, in fact, these classes are the same.

- *High depth of the tree*. If the tree depth is similar to the number of attributes used to represent the domain objects, we can assume that the classes are difficult to separate. For instance, objects of the *iris* dataset are described with a set of four attributes. The tree in Fig. 3 has depth 4, therefore we can interpret that some of the classes are very similar because it has been necessary to use all the attributes to separate them. Particularly, some *iris versicolor* and *iris virginica* are only different in the attribute petalwidth.
- *Low depth of the tree*. If the tree depth is much more smaller than the number of attributes used to represent the objects, the classes are easy to separate. For instance, Fig. 4 shows an sketch of the tree grown from the dataset *Splice*. The depth of this tree is 4, whereas domain objects are described using 60 attributes, therefore the classes are easy to separate.

We want to remark that the situations mentioned above show a general landscape of the domain since they may not be applicable to some particular cases. For instance, leaf 4 of the tree in Fig. 4 is satisfied by examples of the classes EI and N, that is to say, the path is short but it is not enough to separate both classes.

Commonly, a tree completely expanded (i.e., without trying to avoid overfitting) lies in intermediate cases that allow to analyze subparts of the domain. Let us to see now how the FTree methodology proposes a local analysis of the decision tree. As we already mentioned, this local analysis is mainly centered on the tree paths focusing on both their length and the number of examples associated to the nodes.

- *Overfitting*. Many paths with leaves containing only one element mean that all the classes of the domain are very similar and that probably the set of attributes used to represent the domain objects is not the most appropriate one. In such situation we need to perform a further domain analysis. This could be the case of leaves 6 and 7 in Fig. 5.
- *Pure populated leaves*. Paths with many elements on the leaf shows classes that are easy to characterize and to distinguish from the others, specially when the paths are short in comparison with the number of attributes that describe the domain objects. For instance, the class *iris setosa* is univocally characterized only using the attribute petalwidth having value lower than 0.6 (Fig. 3). In the sketch of the *Soybean* tree shown in Fig. 5 we see that all the 10 examples of the class *downy-mildew* are classified using only 3 attributes (i.e., leafspots-size, fruit-spots and mold-growth).
- *Pure non-populated leaves*. Paths with few elements mean that there are several classes with similar aspect, i.e., difficult to characterize and distinguish. For instance, the sketch shown in Fig. 6 serves to classify only 4 domain objects belonging to three different solution classes which are only different in the value of one of the attributes.

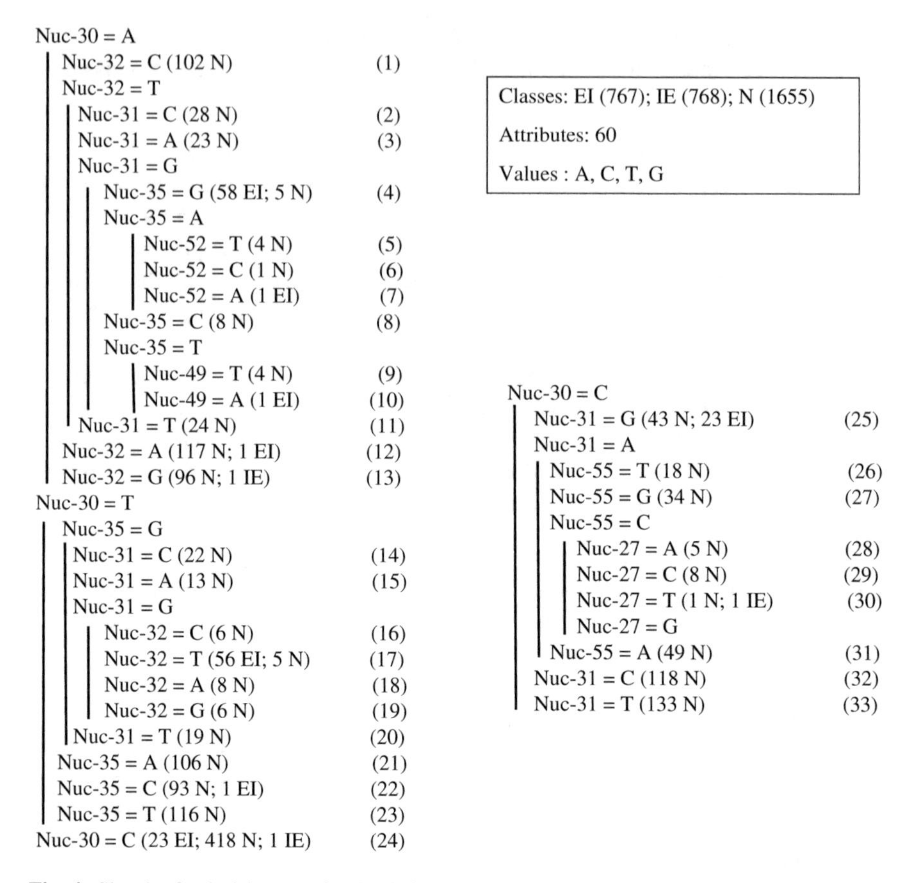

Fig. 4 Sketch of a decision tree for the *Splice* dataset. The *right-side* tree is the expansion of the leaf 24

- *Impure populated leaves*. Paths with leaves containing examples of several classes but with a majority of examples of one of the classes show us two possible situations: (1) there is some error on the input data or (2) the examples of the minority classes are special cases that need to be analyzed in detail. For instance, the leaves 12, 13, and 22 in Fig. 4 contain many examples of class N (117, 96, and 93 respectively) and only one of the class EI. Therefore the experts should focus on these examples to detect possible input errors.
- *Mixed leaves*. Paths with leaves containing examples of several classes but none of them has a clear majority. This means that with the attributes used to describe the domain objects, the classes cannot be separated, i.e., they are very similar. Therefore, the expert needs to analyze these classes. This could be, for instance, the case of the leaf 25 in Fig. 4.

It is important to remark that the attributes forming the tree depend on the measure chosen to determine which of them are relevant. Consequently, the picture proposed

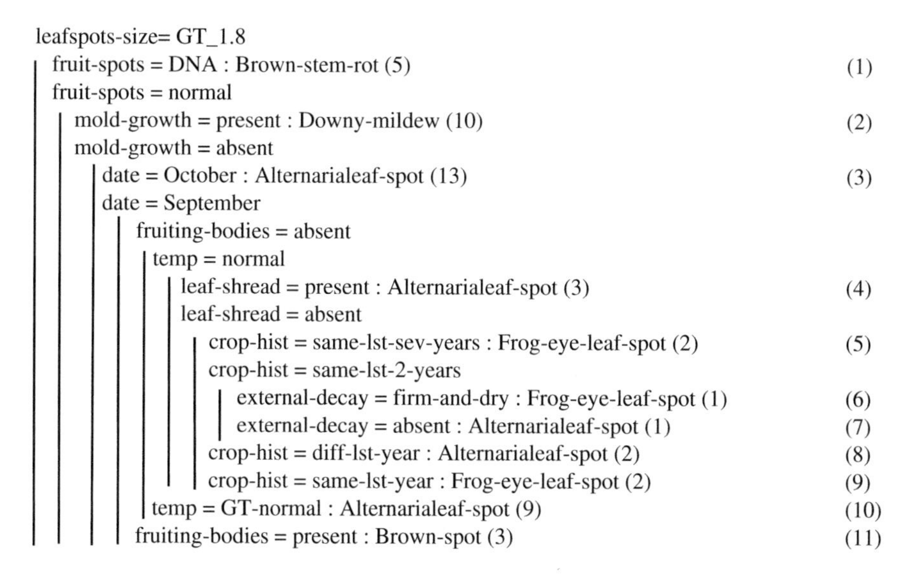

Fig. 5 Sketch of a decision tree for the *Soybean* dataset

```
leafspots-size= GT_1.8
| fruit-spots = normal
| | mold-growth = absent
| | | date = July
| | | | severity = pot
| | | | | area-damaged = scattered
| | | | | | crop-hist = same-lst-sev-years : Frog-eye-leaf-spot (1)
| | | | | | crop-hist = diff-lst-year : Alternarialeaf-spot (2)
| | | | | area-damaged = whole-field : Brown-spot (1)
```

Fig. 6 Sketch of a decision tree for the *Soybean* dataset

by the tree will be different depending on that measure. Therefore, a first issue to take into account is that the expert has to agree with the majority of the attributes used in the tree, otherwise a different measure should be used. It is also possible that the expert has in mind some patterns (i.e., background knowledge) that should be used, for instance he could be interested on forcing the presence of some attribute with some particular value. An example will be shown in the next section were we will use decision trees to analyze a database containing descriptions of skin lesions.

Let us suppose now that the domain expert has some domain knowledge that considers important and, therefore, it has to be used during the process of construction of the tree. If the resulting tree has high overfitting degree, this means that the knowledge introduced has to be reconsidered since it is shared by many different classes. Conversely, if the resulting tree separates well the classes, the knowledge can be considered as valid.

In the next section we show our experience in using decision trees to support dermatologists in building a domain theory capable to differenciate malignant melanoma from benignant skin lesions.

4 Case Study I: Identification of *Malignant Melanoma*

We use a decision tree to analyze a data set containing descriptions of 144 melanocytic lesions, 43 out of them are *malignant melanoma* (MM) and the remaining 101 are benignant. The goal of the experts is to identify MM in early stages since in these situations the characteristics of MM are not clearly developed. We want to remark that all these analysis of the current database, the results and the comments have been performed together with an expert dermatologist.

Figure 7 shows the decision tree over the whole data base. This tree (as all others in the experiments) has been grown using the LM distance as measure to select the most relevant attribute. The numbers between parenthesis show how many lesions of each class are identified by the path from the root to a node. First of all we want to remark that the expert agrees the attributes assessed as the most relevant. As we will see later, most of them are the same used by experts to identify whether or not a skin lesion is malignant.

An analysis of the subtree of root typical-basal = *mild* reveals that this is an important characteristic to identify benignant lesions since only two out of 27 are MM. Because it is very important to avoid false negatives, i.e., to identify all possible MM, the tree path shows that only lesions such that, in addition to typical-basal = *mild*, have dermal-nests = *no* and PG-global = *none* are MM. Notice, however, that this path also has one false positive, i.e., a benignant lesion with the

Fig. 7 A decision tree from a data set containing descriptions of 144 melanocytic lesions, 144 MM and 101 benignant

```
melanocytic = yes (43 MM; 101B)
  typical-basal = mild (2MM; 25B)
    dermal-nests= yes (14B)
    dermal-nests = no (2MM; 11B)
        PG-global = pleomorphic (4B)
        PG-global = monomorphic (6B)
        PG-global = none (2MM; 1B)   <------
  typical-basal = marked (19MM; 6B)
    dermal-cells = nucleated-typical and non-nucleated (1MM)
    dermal-cells = nucleated-atypical and plump (11MM; 1B)
        asymmetry= symmetric (1B)
        asymmetry = one-axis (2MM)
        asymmetry = two-axes (9MM)
    dermal-cells = nucleated-atypical (5MM)
    dermal-cells = nucleated-typical and plump (1MM; 2B) <------
    dermal-cells = no (1MM; 3B)
  typical-basal = typical (7MM; 14B)
    PG-global = pleomorphic (1B)
    PG-global = monomorphic (7MM; 13B)   <------
```

same important characteristics than MM. The consulted experts agree with this result and they will have to try to find other additional characteristics to avoid the false positive.

The subtree of root typical-basal = *marked* reveals that there are 25 lesions satisfying such characteristic 6 out of them are benignant, therefore it is a good characteristic to identify MM. The same subtree shows that dermal-cells = *nucleated-typical and plump* is satisfied by two benignant lesions and one MM; and when dermal-cells = *no* is satisfied by one MM and 3 benignant lesions. Because the goal is to avoid false negatives, the expert prefers to diagnose a melanocytic lesion with marked atypia as a MM although this can produce false positives (at least 4 in the data base at hand).

Finally, the subtree of root typical-basal = *typical* shows an important issue for experts' discussion. This subtree describes 20 lesions that have, in addition PG-global = *monomorphic*. We see that 7 of them are MM and 13 of them are benignant. Therefore, this subtree supports the experts on identifying a subset of lesions very susceptible from errors, at least using the current set of attributes for describing a lesion.

The experts have some background knowledge about the important characteristics identifying a MM. They use the algorithm in Fig. 8 based on the addition and substraction of points according to the characteristics of a melanocytic lesion. In particular, the algorithm adds 1 when the lesion has roundish pagetoid cells; it also adds 1 when dermal cells are atypical nucleated; rests 1 when the dermal papilla is edged; and rests 1 when the basal cells are typical. The sum of these assessments result in a number S such that:

- if $S = -2$ the lesion probably is a nevus (benignant);
- if $S = -1$ the lesion probably is a nevus but it may also be a MM;
- otherwise the lesion is probably a MM.

The specificity of such algorithm is 95 % and the sensitivity is 86 % (on the data base at hand it produces 5 false negatives).

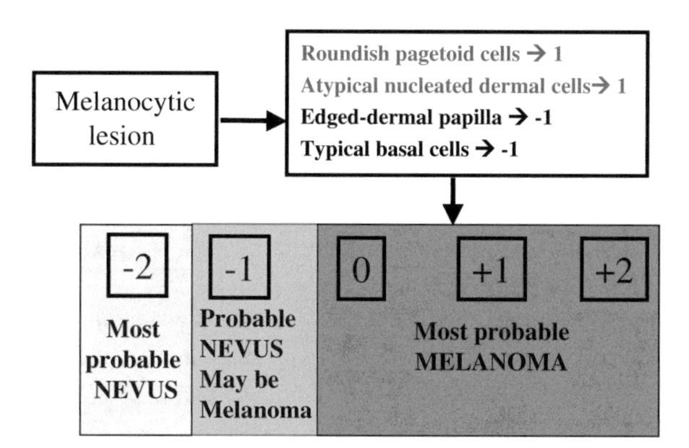

Fig. 8 Algorithm followed by dermatologists on the database at hand, to differentiate MM from benignant skin lesions

```
Typical-basal = typical and DP_papilla = edged
  | PG_global= pleomorphic (1 B)
  | PG_global = monomorphic (4 MM; 4 B)
```

```
DP_papilla = edged
  | pagetoid-infiltration= no
  |   | globules = yes (25 B)
  |   | globules = no
  |   |   | corneal = 1 (1 MM)
  |   |   | corneal = 0 (6 B)
  | pagetoid-infiltration = yes
  |   | junctional-nests-features = thickenings (1 MM)
  |   | junctional-nests-features = clusters (4 MM)
  |   | junctional-nests-features = no (1 MM; 9 B)
```

Fig. 9 Two decision trees grow from subsets of examples according to the algorithm used by dermatologists. The *upper part* shows the DT from examples with typical-basal = *typical*, dermal-papilla = *edged*. The *lower part* shows the DT from examples with dermal-papilla = *edged*

The next experiment consisted on take as basis the algorithm followed by the experts to diagnose malignant melanoma (Fig. 8). If we give as initial pattern typical-basal = *typical* and dermal-papilla = *edged* (i.e., the attribute-value pairs that according to the expert's knowledge are signals of benignancy), then FTree constructs the tree shown in the upper part of Fig. 9 that is satisfied by 4 MM and 28 benignant lesions. Therefore, these characteristics seem to be the appropriate to distinguish benignant lesions from malignant ones. However, we seen that the tree is not able to classify all the objects satisfying the pattern because 19 of them have the value of the attribute PG-global *unknown*.

Taking as initial pattern only the characteristics considered as important to assess malignancy (i.e., dermal-cells = *atypical and nucleated* and pagetoid-cells = *roundish*) we see that the database contains only 3 lesions (2 MM and 1 benign). Therefore, no conclusion can be obtained from this little sample.

Let us analyze now the lesions with $S = -1$, i.e., those that *a priori* are benignant but that could be MM. A possibility for $S = -1$ is that the lesion has only one characteristic of malignity, so we have analyzed the subset of lesions having dermal-papilla = *edged*. There are 48 lesions satisfying this pattern, 7 of them are MM and the remaining ones are benignant. When expanding the tree (see lower part of Fig. 9) we have found that when, in addition, they have not *pagetoid-infiltration* 32 of them are benignant and 1 is MM. The remaining 15 lesions have pagetoid infiltration and 6 of them are MM and 9 of them are benignant. Therefore, we have found here another subset of lesions on which the expert have to focus in order to distinguish between MM and benignant. The database does not contain enough information to analyze other subsets of lesions having $S = -1$.

Finally, concerning lesions having 0 as result of the dermatologist's algorithm, none of the possible combinations of attribute-values has enough elements to perform a relevant analysis. In fact, this means that the hypothesis of diagnosing a skin lesion as MM when the result of the algorithm is zero has not enough evidence on the database at hand.

From this preliminary work, we can conclude that the database has not enough information to extract predictive conclusions but allows to focus on some aspects as, for instance, the attributes considered relevant for the experts. Our analysis has allowed to determine areas of the domain knowledge on which more information is needed. For instance, hypothesis about which attribute-values are signals of malignancy (those included in the expert's algorithm in Fig. 8) cannot be confirmed because the database has not enough information. Therefore, experts need to obtain additional skin descriptions to cover this area. A similar situation occurs for lesions with $S = 0$. Our analysis also shows that even in the case that the attribute-values used by the experts as characteristics of benignancy are correct, they are not enough since the attribute PG-global is unknown in most of the benignant lesions. Thus, on the one hand it is necessary to analyze what happens with lesions having typical-basal = *typical*, dermal-papilla = *edged* and PG-global = *monomorphic*; and on the other hand it is necessary to include the value of the attribute PG-global for all descriptions.

5 Case Study II: Quality of Life Assessment for People with Intellectual Disabilities

In this section we analyze a database with information about quality of life of people with intellectual disabilities. The concept of quality of life (QoL) was introduced into the fields of education, health care, and social services in the early 1980s. The notion of QoL includes both objective and subjective factors but far from seeing this subjective information a serious drawback, we regard it as an opportunity, both from a theoretical and from a practical point of view. The data we deal with is based on the Shalock and Verdugo [19] model to assess the QoL of a person. This model considers that there are 8 dimensions that have to be taken into account: emotional well-being (EW), interpersonal relations (IR), material well-being (MW), personal development (PD), physical well-being (PW), self-determination (SD), social inclusion (SI) and rights (RI). Experts are interested in knowing which kind of strategies could be done in order to improve the QoL. To achieve this goal it is important to know the interrelations among the different dimensions and we propose the use of FTrees to perform this analysis.

The database we use is formed by 5158 records corresponding to answers to the GENCAT questionnaire [7]. The questionnaire has 69 questions divided in 8 blocks, one for each dimension of the QoL model proposed by Shalock and Verdugo [8]. Each question could be answered by an integer from 1 to 4, where 4 means the maximum agreement and 1 the minimum agreement. Each dimension has associated between 8 and 10 questions. Because the database has the punctuation for each question separately, we performed a pre-processing phase where the questions corresponding to each one of the dimensions has been grouped. So, for instance, there are 8 questions related to the emotional well-being (EW); let us suppose that a person has given the punctuations 2, 3, 2, 3, 4, 2, 3, 2 to each one of the EW questions. After the pre-processing we considered that this person has 21 points in EW. The aggregation of the punctuation of the 8 dimensions gives the global index of quality of life (IQV) that has also been discretized. The records of the data base are distributed in three classes in the following way: 391 examples with IQV low, 3888 with IQV medium, and 879 with IQV high. We discretized the data based on a preliminary statistical analysis.

We grow a decision tree with the whole data base using the ID3 algorithm and the LM distance to select the relevant attributes at each level. For reasons of space we do not show the tree completely expanded. According to this methodology, the most relevant feature turns out to be SI. The study also shows that the tree has high overfitting although there are several impure populated leaves, mainly classifying many examples with medium IQV and few examples with low or high IQV. As an illustration of the result, Fig. 10 shows a part of the subtree of root SI = *medium*. When SD = *medium* the majority of the examples have *high* or *medium* IQV and only 4 examples have *low* IQV. From the tree we can extract symbolic rules relating dimensions, such as: if SI = *medium*, SD = *medium* and IR = *low* the IQV is always *medium*.

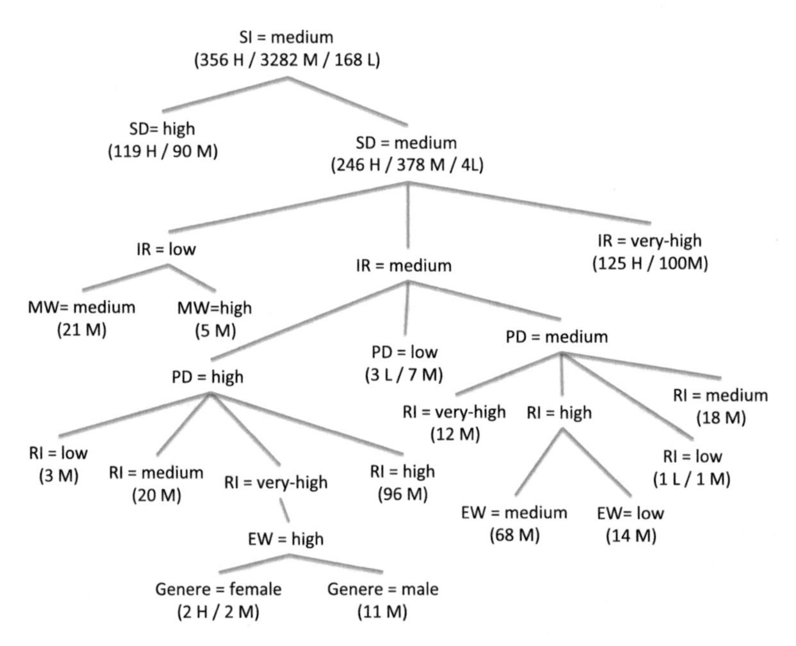

Fig. 10 Part of the decision tree grown from the whole database. In parenthesis the number of examples satisfying each node. H stands for *high* IQV, M stands for *medium*, and L stands for *low*

Figure 11 shows the development of the subtree classifying examples with SI = *medium*, SD = *medium* and IR = *high*. First of all we see that the subtree has high depth, that is, the depth of the tree (8) is almost equal to the number of attributes describing the domain objects. We show that the majority of leaves have few examples, this means that it is difficult to separate the classes. The main conclusion here is that a good classification can be obtained using only the three upper levels of the subtree (those upper the dashed line). Also we see that high values of IQV are related with high values of RI and EW since the 74 examples with IQV high are classified in this part of the tree.

We also used the FTree methodology to analyze parts of the database and trying to enlight some relations between the dimensions. The idea is to select a subgroup of examples satisfying some conditions and to see which are the features that allow the distinction among the groups having different IQV. In the global analysis we see that SI, SD and RI seem to be relevant dimensions, so now we give some trees constructed based on them.

The feature SD has been discretized in four intervals: *very-low, low, medium*, and *high*. We grow a tree for each one of the values of SD, however for lack of space we only show the ones concerning to SD = *medium* and SD = *low* although we discuss all of them. When SD = *high* we obtained a tree with several impure populated leaves. In particular there is a path that discriminates 209 out of 324 examples having IQV high. In fact, when SD = *high* and IR = *very-high* the tree classifies 236 examples with IQV high and also 4 having IQV medium. The tree has some branches of depth 4 trying

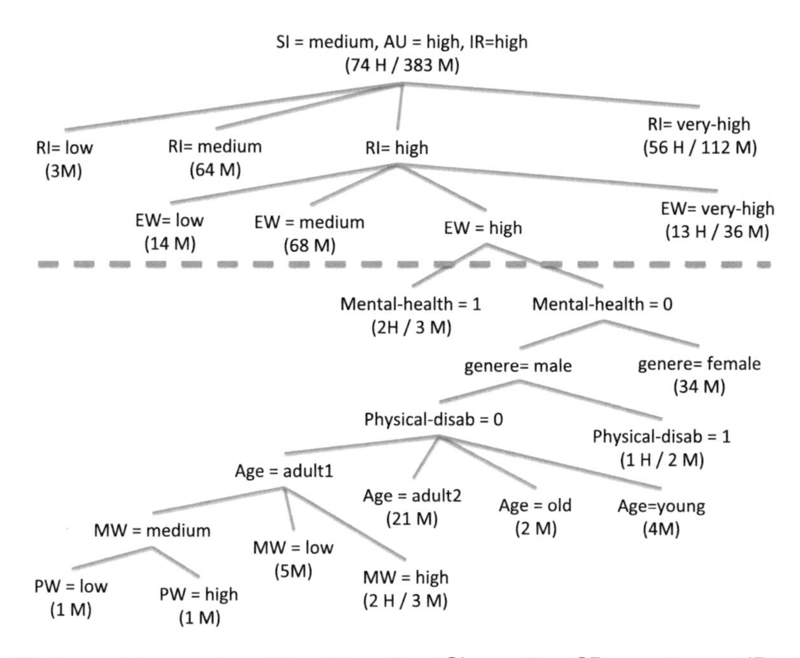

Fig. 11 A complete expansion of the subtree of root SI = *medium*, SD = *medium* and IR = *high*

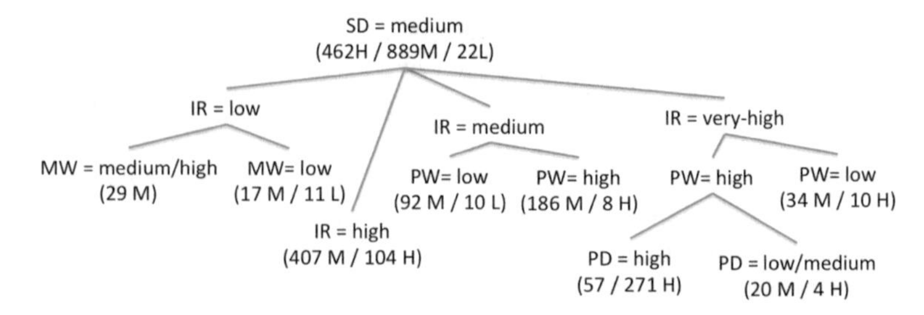

Fig. 12 Ftree classifying the examples with SD = *medium*

to discriminate between *medium* and *high* values of IQV but there is overfitting. A conclusion from this tree is that SD and IR are highly related.

When SD = *very-low* the tree has overfitting and it is not able to clearly discriminate between medium and low levels of IQV. This means that the frontier in defining both levels should be revised. The most interesting situations appear when SD = *low* and SD = *medium*. In both cases the subtrees have a maximum depth of 8 or 9 and high overfitting. This means that it is very difficult to separate the classes.

Figure 12 shows the tree built from the examples with SD = *medium*. We pruned the tree in order to avoid overfitting and to allow a better view of the tree. All those impure populated leaves lead to overfitting when the algorithm tries to separate the examples of the class with lower number of examples. From the tree we can extract the following rules:

- When SD = *medium*, IR = *low*, the IQV never is high (it classifies 11 out of 22 examples with IQV low).
- When SD = *medium*, IR = *high* or *very-high*, the IQV never is low (it classifies 285 out of 462 with IQV high).

When IR = *low*, there are not examples with IQV high. We do not show here the tree because it has not any path that clearly separates the examples of the classes low and medium. This tree has a high depth (some paths have length 7) and high overfitting. Therefore, the conclusion is that both classes are not separable, at least with the discretization intervals we used.

When IR = *medium*, most of examples have IQV medium and low. The tree has not a high depth if we want to separate the examples having IQV low from those having IQV high. This can be done using the attributes IR, PD and SD. Most of examples having IQV medium (212 out of 221) are classified under low and medium values of PD.

When IR = *high* the tree is deeper than in the previous cases we can extract the following rules:

- When IR = *high*, SD = *very-low*, the IQV never is high.
- When IR = *high*, SD = *medium* or *high*, the IQV never is low.
- When IR = *high*, SD = *low*, SI = *high*, the IQV never is low.

The analysis of the domain using decision trees allows us to extract some interesting aspects to be discussed with the expert. Because in this particular domain the attributes to be used to describe the examples are out of discussion, we see that the discretization has to be redone in a most accurate way since it seems to be the cause of obtaining deep trees with high overfitting and that are not able to completely separate the classes. Examples of the minor class of impure populated nodes could serve as basis to change this discretization.

With the current discretization, we focused only on subtrees that separate high values of IQV from low values. We see that trees mainly use the dimensions SI, SD and IR with which overfitting can be avoided and also most of leaves are impure populated. Other dimensions that sometimes are relevant are PD and PW although they appear in lower levels of the tree. We also see that rights (RI), material well-being (MW) or aspects related with a particular person such as the age or the genre do not seem to be relevant. Conversely, what seems to be very important is the self-determination (SD) of the person, i.e., the capability of deciding about its own life.

When analyzing the three most relevant attributes with their respective levels, we found trees with high overfitting, but when we purge them focussing on the separation of low and high values of IQV, we obtain, in general, trees with deep 3 or 4, meaning that may be with a more accurate discretization the classes could be well separated. There are two exceptions to this: one is the tree obtained using only those examples with SD = *low* and the other is the tree obtained using only those examples with IR = *high*. In both cases the depth of the tree grows due to one example with low IQV that has to be separated from the examples with IQV high. This example could be an error and, thus, it should be analyzed in depth by the expert.

We also see that high values of IQV are related with high values of the three mentioned dimensions. In other words, low values of SD are mainly produced when SI is high or when EW is high. This is an interesting aspect that needs to be analyzed in depth by the experts: the relation among these dimensions. Notice that IQV seems to be directly correlated with SD (and the same happens with IR): as higher the value of the dimension is, the highest is the value of IQV.

From our analysis we propose to the expert to analyze in more detail the following aspects:

- In which sense SI, SD and IR could be key factors in the definition of the quality of life of a person with intellectual disabilities.
- Why material well-being and rights appear in low levels of the trees, denoting a low degree of relevancy.
- Analyze the validity of the rules mentioned in the previous section, and propose a theory of relationship between dimensions that could be tested in a further analysis.

6 Discussion

The shortcomings of the kind of analysis we propose are the ones of the decision trees. In particular, this analysis could not be possible on huge data sets because decision trees are not scalable. We plan to explore how to perform this analysis in such situations. We think that possible solutions could be the use of algorithms such as SPRINT [20] and RainForest [6] that are able to construct decision trees on huge data sets. Because experts commonly want to analyze subsets of data (for instance, only melanocytic lesions with typical basal cells) it is also possible that, in fact, the size of the data set is not a problem. However this is a restriction because then we could not have to analyze the entire set of data.

As we already mentioned, the attributes used to grow the tree depend on the selection criteria, therefore a first issue to take into account is whether or not the expert agrees the subset of features composing the tree. If he does not it would be necessary to use other criteria since it is fundamental that most of the attributes in the tree (especially those in the top levels) coincide with the ones commonly used by the expert. Nevertheless, when any of the trees grown uses the attributes considered by the experts as relevant we have to face with an interesting situation: according to the data at hand, the attributes that they assess as relevant actually they are not, and it would be necessary to reconsider these attributes.

We also have to take into account the presence of attributes with unknown values. In our experiments we deal with *unknown* as it was another value when partition the data set according to the values of an attribute an then assessing the most relevant attribute. However, once selected the most relevant attribute, the domain objects having unknown value in that attribute does not satisfy any path of the tree. For instance, in the domain of melanomas, there are some objects that have not value in the attribute typical-basal, and therefore they are not classified by the tree. As a way to obtain some information from these objects, we propose to construct another tree with them. This focuses the expert's attention on the actual importance of the values of the main tree. That is to say, if the expert considers that the attribute typical-basal is important, he must always give some value to it. In constructing a second tree with the remaining unclassified objects, the expert can also focus on alternative subsets of attributes that could be taken into account when the primarily important attributes (those in the first tree) have not values.

7 Related Work

Our work is related with approaches that use decision trees with background knowledge. In our case, the background knowledge is only used to obtain a subset of domain objects from which to grow the tree. In fact, our goal and the one of the other approaches using decision trees is different. Whereas these approaches use decision trees to obtain a domain model useful for prediction, our goal is the analysis of the available data. This is an important difference because to obtain a predictable model it is necessary to have as much examples as possible. Instead, the analysis we propose can be made even for small databases.

Therefore, most of works use background knowledge to improve the predictivity. The kind of background varies according to the domain and the concrete goal of the application. Thus [13] introduces the EG2 algorithm, an extension of the ID3 algorithm that uses an *is-a* hierarchy associated to the attributes and a cost criteria to select them.

An interesting paper that describes how to incorporate domain knowledge to improve a model is the one of [14]. The authors explain their experience when using C4.5 on the Reaction Control System (RCS) domain and obtained a tree that did not use the attributes commonly taken into account by domain experts. What they propose is to bias C4.5 by changing its criteria of selecting relevant attributes. Therefore, instead of using the information gain measure they use a kind of ranking taking into account the preferences of domain experts. Our motivation in using FTree is the same: attributes relevant according to the expert's criteria should be included in the tree. However, the procedure proposed by Ortega and Fisher is more dynamic than ours since in FTree the expert preferences are a filter of the original data and all the domain objects that do no satisfy them are excluded for tree growing. Instead, Ortega and Fisher's approach does not excludes any original domain object.

Sivagama introduces in [21] a formal analysis of how the Gini's coefficient splits a dataset. The relation between this work and FTree can be find in a final reflexion that the author performs in the sense that how the extracted rules are easily understood by domain experts. Consequently, he points out that these rules can be used to access the database to analyze it. Our approach goes beyond this idea since we can see FTree as this access to the database (since it filters objects) but then we propose to grow a tree in order to classify the domain objects satisfying the query.

Tsai et al. [22] also mentions the problem that the attributes included in the decision trees are not the ones commonly used by the expert. This is an important aspect since the consequence is that the expert has some kind of reluctance to use the obtained model. For this reason authors propose an approach that uses the expert experience in growing the tree. The idea is that experts analyze both the tree and the examples and determine possible paths that does not agree with their experience. These paths can be object of discussion in order to determine either if they correspond to invalid data or if they enlight some kind of new knowledge. This kind of analysis is similar to the one performed with FTree. However, in our approach we also propose a global analysis of the tree since we point out that the form of the tree (i.e., its depth, its width, etc.) is also important to extract general characteristics of the domain. The main difference is that Tsai et al., use this analysis to improve the classification accuracy whereas we propose FTree itself as a way to analyze the available data.

8 Conclusions

The work introduced in the present paper focuses on the analysis of a knowledge base. The common methods used for knowledge discovery are the clustering ones. They are unsupervised methods since they suppose that the classes of domain objects are not known in advance. The case where the classes are known is not commonly considered as knowledge discovery but as either classification or prediction. In that situation the goal is to find good descriptions of the classes and this goal is achieved by means of inductive learning methods such as, for instance, decision trees. The claim of the present paper is that decision trees can also be useful for knowledge discovery.

Our approach consists of constructing a decision tree classifying all the known examples and then to analyze this tree using the FTree methodology. Ftree gives some guidelines to interpret the structure of a tree, allowing to focus on conflictive parts of the dataset at hand. Thus, FTree allows to identify lacks of knowledge, possible errors in the data, classes that are too similar that could be merged, etc. This methodology has emerged from our experience in interacting with domain experts and has proved to be an useful tool. We used FTree to analyze two databases, one concerning the identification of malignant melanoma and the other concerning the assessment of life quality of people with intellectual disabilities. In addition, FTree allows growing trees focusing on subsets of attribute-values that experts consider relevant. With the resulting tree is possible to detect (as we have show in the data base of melanomas) that some attributes are not so relevant as experts expected.

Acknowledgments The authors thank Susana Puig their helpful comments and suggestions, and the Taller Jeroni de Moragas. This research is partially funded by the European Union's Horizon 2020 research and innovation programme under the Marie Sklodowska-Curie grant agreement No 689176 (SYSMICS project), the projects RASO (TIN2015-71799-C2-1-P) and RPREF (CSIC Intramural 201650E044) and the grants 2014-SGR-118 and 2014-SGR-788 from the Generalitat de Catalunya.

References

1. E. Armengol. Usages of generalization in CBR. In R.O. Weber and M. M. Richter, editors, *ICCBR-2007. Case-based Reasoning and Development*, number 4626 in Lecture Notes in Artificial Intelligence, pages 31–45. Springer-Verlag, 2007.
2. E. Armengol. Building partial domain theories from explanations. *Knowledge Intelligence*, 2/08:19–24, 2008.
3. E. Armengol and E. Plaza. Discovery of toxicological patterns with lazy learning. In V. Palade, R.J. Howlett, and L. Jain, editors, *KES-2003*, number 2774 in Lecture Notes in Artificial Intelligence, pages 919–926. Springer, 2003.
4. A. Asuncion and D.J. Newman. UCI machine learning repository, 2007.
5. L. Breiman, J. H. Friedman, R. A. Olshen, and C. J. Stone. *Classification and Regression Trees*. Wadsworth, 1984.
6. J. Gehrke, R. Ramakrishnan, and V. Ganti. RainForest - a framework for fast decision tree construction of large datasets. *Data Mining and Knowledge Discovery*, 4(2/3):127–162, 2000.

7. L.E. Gómez, M.A. Verdugo, B. Arias and R.L. Schalock. Formulari de l'escala gencat de qualitat de vida. manual d'aplicació de l'escala gencat de qualitat de vida. Technical report, Departament d'Acció Social i Ciutadania, Generalitat de Catalunya, Barcelona, 2008.
8. L.E. Gómez, M.A. Verdugo, B. Arias and R.L. Schalock. Informe sobre la creació d'una escala multidimensional per avaluar la qualitat de vida de les persones usuàries dels serveis socials a catalunya. Technical report, Departament d'Acció Social i Ciutadania, Generalitat de Catalunya, Barcelona, 2008.
9. A. K. Jain, M. N. Murty, and P. J. Flynn. Data clustering: a review. *ACM Comput. Surv.*, 31(3):264–323, September 1999.
10. T. Kohonen. The self-organizing map. *Neurocomputing*, 21(1-3):1–6, 1998.
11. R. López de Mántaras. A distance-based attribute selection measure for decision tree induction. *Machine Learning*, 6:81–92, 1991.
12. O. Maimon and L. Rokach, editors. *Data Mining and Knowledge Discovery Handbook, 2nd ed.* Springer, 2010.
13. M. Núñez. The use of background knowledge in decision tree induction. *Machine Learning*, 6:231–250, 1991.
14. J. Ortega and D. Fisher. Flexibly exploiting prior knowledge in empirical learning. In *Proceedings of the 14th international joint conference on Artificial intelligence - Volume 2*, IJCAI'95, pages 1041–1047, San Francisco, CA, USA, 1995. Morgan Kaufmann Publishers Inc.
15. M. J. Pazzani. Knowledge discovery from data? *IEEE Intelligent Systems*, 15(2):10–13, 2000.
16. J. R. Quinlan. Induction of decision trees. *Machine Learning*, 1(1):81–106, 1986.
17. J. R. Quinlan. *C4.5: Programs for Machine Learning*. Morgan Kaufmann, 1993.
18. J. R. Quinlan. Discovering rules by induction from large collection of examples. In *Expert Systems in the Microelectronic Age. D. Michie (Ed.)*, pages 168–201. Edimburg Eniversity Press, 1979.
19. R.L. Schalock and M.A. Verdugo. *Handbook of quality of life for human service practitioners*. Washington, DC, 2002.
20. J. C. Shafer, R. Agrawal, and M. Mehta. Sprint: A scalable parallel classifier for data mining. In *VLDB*, pages 544–555, 1996.
21. S. M. Sivagama. A knowledge discovery using decision tree by Gini coefficient. In *International Conference on Business, Engineering and Industrial Applications (ICBEIA)*, pages 232–235, 2011.
22. Y. Tsai, Paul H. King, Ph. D, Michael S. Higgins, Ph. D, and Nimesh P. Patel. An expert-guided decision tree construction strategy: An application in knowledge discovery with medical databases. In *AMIA Annual Fall Symposium*, pages 208–212, 1997.

Part II
Bags, Fuzzy Bags, and Some Other Fuzzy Extensions

L-Fuzzy Bags

Fateme Kouchakinejad, Mashaallah Mashinchi and Radko Mesiar

Abstract This chapter studies L-fuzzy bags and some of its applications in which L is a complete lattice. Furthermore, the concepts of α-cuts, (L-fuzzy) bag relations and related theorems are given. The chapter ends with the characterization of the algebraic structure of bags and L-fuzzy bags.

1 Introduction

The theory of bags, an alternative name for multisets, as a natural extension of the set theory was introduced by Yager [19]. So far, bags have been employed in practice; for example, in flexible querying [16], representation of relational information [19], decision problem analysis [2], criminal career analysis [8], and in biology [13]. As another example, bags can play the role of primary data bases in the real world problems. As a matter of fact, all of information should be considered in the data mining tasks [6], and in particular in the fuzzy clustering where each data point has a membership degree in each cluster. So, from the mathematical point of view, each cluster should be considered as a fuzzy bag, see [18]. Some other applications can be found in [3, 11, 14–17]. However, due to some existing drawbacks in the first definition of bags [19], the necessity of a revision of this notion has grown. The definitions proposed by Delgado et al. [5] for bags and fuzzy bags have improved these drawbacks. As it is shown in [9], there is some incompatibility with the nature of fuzziness in the

F. Kouchakinejad (✉)
Department of Mathematics, Graduate University of Advanced Technology, Kerman, Iran
e-mail: kouchakinezhad@gmail.com

M. Mashinchi
Department of Statistics, Faculty of Mathematics and Computer Sciences, Shahid Bahonar University of Kerman, Kerman, Iran
e-mail: mashinchi@uk.ac.ir

R. Mesiar
Faculty of Civil Engineering, Slovak University of Technology, Radlinského 11, 810 05 Bratislava, Slovak Republic
e-mail: radko.mesiar@stuba.sk

V. Torra et al. (eds.), *Fuzzy Sets, Rough Sets, Multisets and Clustering*,
Studies in Computational Intelligence 671, DOI 10.1007/978-3-319-47557-8_12

fuzzy bag's definition in [5]. The proposed definition for fuzzy bags in [10] resolved this problem. In this chapter, we summarize our recent results concerning bags and L-fuzzy bags from [9, 10] adding several examples and observations.

The chapter is structured as follows. In the next section, basic definitions and results concerning bags and L-fuzzy bags are reviewed. Section 3 deals with relations on bags and L-fuzzy bags. In Sect. 4, the α-cuts of L-fuzzy bags are studied. Section 5 brings the characterization of the algebraic structure of bags and L-fuzzy bags. Finally, some concluding remarks are added.

2 Definitions

It should be mentioned that, in general, non-empty sets P and O can be arbitrary (finite or infinite) but they are considered to be finite in this chapter. Throughout this chapter, $I_n = \{1, 2, \dots, n\}$, where $n \in \mathcal{N}$ and $\mathcal{N}$ is the set of natural numbers. Also, P and O are two finite universes (sets) called "properties" and "objects", respectively. We have the following definitions.

Definition 1 ([5]) A (crisp) bag $\mathcal{B}^f$ is a pair (f, B^f), where $f : P \to \mathcal{P}(O)$ is a function and B^f is the following subset of $P \times \mathcal{N}_0$

$$B^f = \{(p, card(f(p)))|p \in P\}.$$

Here, $\mathcal{P}(O)$ is the power set of O, $\mathcal{N}_0 = \mathcal{N} \cup \{0\}$, $card(X)$ is the cardinality of set X.

We will use the convention that $card(\emptyset) = 0$ if necessary. Also, we will not distinguish $\{(p, card(f(p))), p \in P\}$ and $\{(p, card(f(p))), p \in P, f(p) \neq \emptyset\}$.

Note 1 For the sake of simplicity, whenever $f(p) = \emptyset$ we may not write $(p, 0)$ in the set B^f.

In this characterization, a bag $\mathcal{B}^f$ consists of two parts. The first one is the function f that can be seen as an information source about the relation between objects and properties. The second part B^f is a summary of the information in f obtained by means of the count operation $card(.)$. This summary corresponds to the classical view of bags in the sense of [19]. Observe that, up to trivial cases, the knowledge of B^f is not enough to recover the original information source f (this was the main drawback of the original approach to bags in [19]). Obviously, f determines $\mathcal{B}^f$ univocally. However, we prefer to keep the notation (f, B^f) for bags as proposed in [4, 5] due to the higher transparency and link to the original notion of bags given in [19].

Notation 1 $\mathbf{B}(P, O)$ is the set of all bags $\mathcal{B}^f = (f, B^f)$ defined in Definition 1.

Table 1 Several functions: age-people

p	17	21	27	35
$f_1(p)$	{Bill, Sue}	{John, Tom}	∅	∅
$f_2(p)$	{Bill, Sue}	{John, Tom, Stan}	∅	{Ben}
$f_3(p)$	∅	{Stan}	{Ana}	{Ben}
$f_4(p)$	{Bill}	{John, Stan}	∅	∅
$f_5(p)$	{John, Tom}	{Ana, Stan}	∅	∅

Definition 2 We have $\mathcal{B}^0 = (0, B^0)$ and $\mathcal{B}^1 = (1, B^1)$ where, $0(p) = \emptyset$, $1(p) = O$ for all $p \in P$, $B^0 = \{(p, 0), p \in P\}$ and $B^1 = \{(p, card(O)), p \in P\}$. Clearly, $\mathcal{B}^0, \mathcal{B}^1 \in \mathbf{B}(P, O)$.

Example 1 ([4]) Let $O = \{$John, Ana, Bill, Tom, Sue, Stan, Ben$\}$ and $P = \{17, 21, 27, 35\}$ be the set of objects and the set of properties, respectively. Let $f_1, f_2, f_3, f_4, f_5 : P \to \mathcal{P}(O)$ be the functions in Table 1 with $f_i(p) \subseteq O$ for all $p \in P$. So, we can define bags $\mathcal{B}^{f_i} = (f_i, B^{f_i})$, $1 \leq i \leq 5$. Where,
$B^{f_1} = \{(17, 2), (21, 2)\}$,
$B^{f_2} = \{(17, 2), (21, 3), (35, 1)\}$,
$B^{f_3} = \{(21, 1), (27, 1), (35, 1)\}$,
$B^{f_4} = \{(17, 1), (21, 2)\}$ and
$B^{f_5} = \{(17, 2), (21, 2)\}$.

Now, we can define some binary operations between bags.

Definition 3 ([5]) Let $* \in \{\cup, \cap, \setminus\}$. Then

$$\mathcal{B}^f * \mathcal{B}^g = \mathcal{B}^{f*g} = (f * g, B^{f*g}),$$

where $f * g : P \to \mathcal{P}(O)$ such that $(f * g)(p) = f(p) * g(p)$ for all $p \in P$.

Example 2 ([4]) We can obtain some new bags from operations among bags in Example 1, where their functions are shown in Table 2 and the corresponding summaries are as follows
$B^{f_1 \cup f_2} = \{(17, 2), (21, 3), (35, 1)\}$,
$B^{f_2 \cap f_3} = \{(21, 1), (35, 1)\}$,
$B^{f_1 \setminus f_3} = \{(17, 2), (21, 2)\}$,
$B^{f_3 \setminus f_2} = \{(27, 1)\}$,
$B^{f_1 \cup f_5} = \{(17, 4), (21, 4)\}$,
$B^{f_1 \cap f_5} = \{(17, 0), (21, 0), (27, 0), (35, 0)\}$.

It should be noted that the values of function for different properties need not be disjoint. This means $f(p) \cap f(p')$ may be a non-empty set. As an example consider the bag $\mathcal{B}^{f_1 \cup f_5}$ in Example 2.

Table 2 Operations on functions from Example 1

p	17	21	27	35
$(f_1 \cup f_2)(p)$	{Bill, Sue}	{John, Tom, Stan}	∅	{Harry}
$(f_2 \cap f_3)(p)$	∅	{Stan}	∅	{Harry}
$(f_1 \setminus f_3)(p)$	{Bill, Sue}	{John, Tom}	∅	∅
$(f_3 \setminus f_2)(p)$	∅	∅	{Mary}	∅
$(f_1 \cup f_5)(p)$	{Bill, Sue, John, Tom}	{John, Tom, Ana, Stan}	∅	∅
$(f_1 \cap f_5)(p)$	∅	∅	∅	∅

From the point of view of the functions associated to a bag, we have the following definition.

Definition 4 ([5]) (i) A bag $\mathcal{B}^f$ is a sub bag of $\mathcal{B}^g$, denoted by $\mathcal{B}^f \sqsubseteq \mathcal{B}^g$, if $f(p) \subseteq g(p)$ for all $p \in P$.
(ii) Two bags $\mathcal{B}^f$ and $\mathcal{B}^g$ are equal, denoted by $\mathcal{B}^f = \mathcal{B}^g$ if $\mathcal{B}^f \sqsubseteq \mathcal{B}^g$ and $\mathcal{B}^g \sqsubseteq \mathcal{B}^f$ that means if $f = g$.

Remark 1 ([5]) Operations $\cap$ and $\cup$ in $\mathbf{B}(P, O)$ satisfy the laws of idempotency, commutativity, associativity, monotonicity and distributivity. Moreover, $\mathcal{B}^0$ is neutral for operation $\cup$ and $\mathcal{B}^1$ is neutral for operation $\cap$.

Definition 5 ([5]) Let $\mathcal{B}^f = (f, B^f)$. Then, complement of $\mathcal{B}^f$ is $\mathcal{B}^{f^c} = (\mathcal{B}^f)^c = (f^c, B^{f^c})$, where $f^c : P \to \mathcal{P}(O)$ is such that $f^c(p) = O \setminus f(p)$ for all $p \in P$.

As an example, observe that $(\mathcal{B}^0)^c = \mathcal{B}^1$.

In what follows, L is a complete lattice and $\mathcal{F}_L(O) = \{A | A : O \to L\}$ is the set of all L-fuzzy subsets of O. In the case of $L = [0, 1]$, we write $\mathcal{F}(O)$.

Definition 6 ([10]) An L-fuzzy bag $\tilde{\mathcal{B}}^{\tilde{f}}$ is a pair $(\tilde{f}, B^{\tilde{f}})$, where $\tilde{f} : P \to \mathcal{F}_L(O)$ is a function and $B^{\tilde{f}}$ is the following subset of $P \times L \times \mathcal{N}_0$

$$B^{\tilde{f}} = \{(p, \delta, card(O^p_\delta)) | p \in P, \delta \in L\}.$$

where, $O^p_\delta = \{o \in O | \tilde{f}(p)(o) = \delta\}$.

Obviously, a bag is a particular case of L-fuzzy bag where, for all $p \in P$, $\tilde{f}(p)$ is a crisp subset of O. Similar to bags, an L-fuzzy bag $\tilde{\mathcal{B}}^{\tilde{f}}$ consists of two parts. The first one is the function $\tilde{f}$ that can be seen as an information source about the relation between objects and properties. The second part $B^{\tilde{f}}$ is a summary of the information in $\tilde{f}$ obtained by means of the count operation $card(.)$.

Note 2 ([10]) In the case that $L = [0, 1]$, the defined bag in Definition 6 is called fuzzy bag.

Table 3 The degrees of memberships for Example 3

p	o								
	Ben	Sue	Tom	John	Stan	Bill	Kim	Ana	Sara
Young	0.7	0.2	0.4	0.0	0.7	0.4	0.2	0.7	0.1
Middle age	0.3	0.8	0.7	0.3	0.3	0.7	0.8	0.3	0.5
Old	0.1	0.2	0.1	0.9	0.1	0.1	0.2	0.1	0.5

Table 4 The degrees of memberships for Example 4

p	o								
	Ben	Sue	Tom	John	Stan	Bill	Kim	Ana	Sara
Tall	0.8	0.6	0.0	0.1	0.8	0.6	0.5	0.7	0.5
Medium	0.3	0.1	0.1	0.6	0.3	0.1	0.8	0.1	0.5
Short	0.1	0.0	0.9	0.4	0.1	0.0	0.2	0.0	0.1

Here, the concept of L-fuzzy bag is illustrated by two examples.

Example 3 ([10]) Let $L = [0, 1]$, $O = \{\text{Ben, Sue, Tom, John, Stan, Bill, Kim, Ana, Sara}\}$ and $P = \{\text{young, middle age, old}\}$ is the set of some linguistic descriptions of age. Let the degrees of membership of all $o \in O$ in the set of each property $p \in P$ are given as in Table 3.

So, by Definition 6, we can define fuzzy bag $\tilde{\mathcal{B}}^{\tilde{f}} = (\tilde{f}, B^{\tilde{f}})$ where,

$$\tilde{f}(\text{young}) = \{\frac{0.7}{\text{Ben}}, \frac{0.2}{\text{Sue}}, \frac{0.4}{\text{Tom}}, \frac{0.7}{\text{Stan}}, \frac{0.4}{\text{Bill}}, \frac{0.2}{\text{Kim}}, \frac{0.7}{\text{Ana}}, \frac{0.1}{\text{Sara}}\},$$
$$\tilde{f}(\text{middle age}) = \{\frac{0.3}{\text{Ben}}, \frac{0.8}{\text{Sue}}, \frac{0.7}{\text{Tom}}, \frac{0.3}{\text{John}}, \frac{0.3}{\text{Stan}}, \frac{0.7}{\text{Bill}}, \frac{0.8}{\text{Kim}}, \frac{0.3}{\text{Ana}}, \frac{0.5}{\text{Sara}}\},$$
$$\tilde{f}(\text{old}) = \{\frac{0.1}{\text{Ben}}, \frac{0.2}{\text{Sue}}, \frac{0.1}{\text{Tom}}, \frac{0.9}{\text{John}}, \frac{0.1}{\text{Stan}}, \frac{0.1}{\text{Bill}}, \frac{0.2}{\text{Kim}}, \frac{0.1}{\text{Ana}}, \frac{0.5}{\text{Sara}}\},$$

and

$$\begin{aligned} B^{\tilde{f}} = \{&(\text{young}, 0.7, 3), (\text{young}, 0.4, 2), (\text{young}, 0.2, 2), (\text{young}, 0.1, 1),\\ &(\text{middle age}, 0.8, 2), (\text{middle age}, 0.7, 2), (\text{middle age}, 0.5, 1),\\ &(\text{middle age}, 0.3, 4), (\text{old}, 0.9, 1), (\text{old}, 0.5, 1), (\text{old}, 0.2, 2), (\text{old}, 0.1, 5)\}.\end{aligned}$$

Example 4 ([10]) Let $L = [0, 1]$, O be as in Example 3 and $P = \{\text{tall, medium, short}\}$ is the set of some linguistic descriptions of height. Let the degrees of membership of all $o \in O$ in the set of each property $p \in P$ be given as in Table 4.

So, by Definition 6, we can define fuzzy bag $\tilde{\mathcal{B}}^{\tilde{g}} = (\tilde{g}, B^{\tilde{g}})$ where,

Table 5 The membership fuzzy sets for Example 5

p	o				
	Ben	Sue	Tom	John	Stan
Young	$\{\frac{0.4}{0.6}, \frac{0.5}{0.7}, \frac{0.6}{0.8}\}$	$\{\frac{0.9}{0.2}, \frac{0.8}{0.3}\}$	$\{\frac{0.7}{0.4}, \frac{0.8}{0.5}\}$	$\{\frac{1}{0.0}\}$	$\{\frac{0.5}{0.6}, \frac{0.5}{0.7}, \frac{0.6}{0.8}\}$
Middle age	$\{\frac{0.7}{0.2}, \frac{0.8}{0.3}, \frac{0.7}{0.4}\}$	$\{\frac{0.7}{0.8}, \frac{0.8}{0.9}\}$	$\{\frac{0.7}{0.7}, \frac{0.8}{0.8}\}$	$\{\frac{0.8}{0.3}, \frac{0.9}{0.4}, \frac{0.8}{0.5}\}$	$\{\frac{0.7}{0.2}, \frac{0.8}{0.3}, \frac{0.7}{0.4}\}$
Old	$\{\frac{1}{0.1}\}$	$\{\frac{0.8}{0.1}, \frac{0.9}{0.2}\}$	$\{\frac{0.8}{0.1}, \frac{0.9}{0.2}, \frac{0.8}{0.3}\}$	$\{\frac{1}{0.9}\}$	$\{\frac{0.9}{0.1}\}$

$$\tilde{g}(\text{tall}) = \{\frac{0.8}{\text{Ben}}, \frac{0.6}{\text{Sue}}, \frac{0.1}{\text{John}}, \frac{0.8}{\text{Stan}}, \frac{0.6}{\text{Bill}}, \frac{0.5}{\text{Kim}}, \frac{0.7}{\text{Ana}}, \frac{0.5}{\text{Sara}}\},$$
$$\tilde{g}(\text{medium}) = \{\frac{0.3}{\text{Ben}}, \frac{0.1}{\text{Sue}}, \frac{0.1}{\text{Tom}}, \frac{0.6}{\text{John}}, \frac{0.3}{\text{Stan}}, \frac{0.1}{\text{Bill}}, \frac{0.8}{\text{Kim}}, \frac{0.1}{\text{Ana}}, \frac{0.5}{\text{Sara}}\},$$
$$\tilde{g}(\text{short}) = \{\frac{0.1}{\text{Ben}}, \frac{0.9}{\text{Tom}}, \frac{0.4}{\text{John}}, \frac{0.1}{\text{Stan}}, \frac{0.2}{\text{Kim}}, \frac{0.1}{\text{Sara}}\},$$

and

$$\begin{aligned} B^{\tilde{g}} = \{&(\text{tall}, 0.8, 2), (\text{tall}, 0.7, 1), (\text{tall}, 0.6, 2), (\text{tall}, 0.5, 2), (\text{tall}, 0.1, 1),\\ &(\text{medium}, 0.8, 1), (\text{medium}, 0.6, 1), (\text{medium}, 0.5, 1), (\text{medium}, 0.3, 2),\\ &(\text{medium}, 0.1, 4), (\text{short}, 0.9, 1), (\text{short}, 0.4, 1), (\text{short}, 0.2, 1), (\text{short}, 0.1, 3)\}. \end{aligned}$$

Remark 2 Let in Definition 6, the lattice is $\mathcal{F}_L(L)$. Then, we have type-2 L-fuzzy bag $\tilde{\mathcal{B}}^{\tilde{f}^2} = (\tilde{f}^2, B^{\tilde{f}^2})$.

Example 5 Let $L = \mathcal{F}([0, 1])$, $O = \{\text{Ben, Sue, Tom, John, Stan}\}$ and P be as in the Example 3. Let the membership of each $o \in O$ in the set of each property $p \in P$ be given as in Table 5.

So, by Definition 6 and Remark 2, we can define type-2 L-fuzzy bag $\tilde{\mathcal{B}}^{\tilde{f}^2} = (\tilde{f}^2, B^{\tilde{f}^2})$ where,

$$\tilde{f}^2(\text{Young}) = \{\frac{\{\frac{0.4}{0.6}, \frac{0.5}{0.7}, \frac{0.6}{0.8}\}}{\text{Ben}}, \frac{\{\frac{0.9}{0.2}, \frac{0.8}{0.3}\}}{\text{Sue}}, \frac{\{\frac{0.7}{0.4}, \frac{0.8}{0.5}\}}{\text{Tom}}, \frac{\{\frac{1}{0.0}\}}{\text{John}}, \frac{\{\frac{0.5}{0.6}, \frac{0.5}{0.7}, \frac{0.6}{0.8}\}}{\text{Stan}}\},$$
$$\tilde{f}^2(\text{Middle age}) = \{\frac{\{\frac{0.7}{0.2}, \frac{0.8}{0.3}, \frac{0.7}{0.4}\}}{\text{Ben}}, \frac{\{\frac{0.7}{0.8}, \frac{0.8}{0.9}\}}{\text{Sue}}, \frac{\{\frac{0.7}{0.7}, \frac{0.8}{0.8}\}}{\text{Tom}}, \frac{\{\frac{0.8}{0.3}, \frac{0.9}{0.4}, \frac{0.8}{0.5}\}}{\text{John}}, \frac{\{\frac{0.7}{0.2}, \frac{0.8}{0.3}, \frac{0.7}{0.4}\}}{\text{Stan}}\},$$
$$\tilde{f}^2(\text{Old}) = \{\frac{\{\frac{1}{0.1}\}}{\text{Ben}}, \frac{\{\frac{0.8}{0.1}, \frac{0.9}{0.2}\}}{\text{Sue}}, \frac{\{\frac{0.8}{0.1}, \frac{0.9}{0.2}, \frac{0.8}{0.3}\}}{\text{Tom}}, \frac{\{\frac{1}{0.9}\}}{\text{John}}, \frac{\{\frac{0.9}{0.1}\}}{\text{Stan}}\},$$

and

$$B^{\tilde{f}^2} = \{(\text{young}, \{\frac{0.6}{0.8}\}, 2), (\text{young}, \{\frac{0.5}{0.7}\}, 2), (\text{young}, \{\frac{0.5}{0.6}\}, 1), (\text{young}, \{\frac{0.4}{0.6}\}, 1),$$
$$(\text{young}, \{\frac{0.8}{0.5}\}, 1), (\text{young}, \{\frac{0.7}{0.4}\}, 1), (\text{young}, \{\frac{0.8}{0.3}\}, 1), (\text{young}, \{\frac{0.9}{0.2}\}, 1),$$
$$(\text{young}, \{\frac{1.0}{0.0}\}, 1), (\text{middle age}, \{\frac{0.8}{0.9}\}, 1), (\text{middle age}, \{\frac{0.8}{0.8}\}, 1),$$
$$(\text{middle age}, \{\frac{0.7}{0.8}\}, 1), (\text{middle age}, \{\frac{0.7}{0.7}\}, 1), (\text{middle age}, \{\frac{0.8}{0.5}\}, 1),$$
$$(\text{middle age}, \{\frac{0.9}{0.4}\}, 1), (\text{middle age}, \{\frac{0.7}{0.4}\}, 2), (\text{middle age}, \{\frac{0.8}{0.3}\}, 3),$$
$$(\text{middle age}, \{\frac{0.7}{0.2}\}, 2), (\text{old}, \{\frac{1.0}{0.9}\}, 1), (\text{old}, \{\frac{0.8}{0.3}\}, 1), (\text{old}, \{\frac{0.9}{0.2}\}, 2),$$
$$(\text{old}, \{\frac{1.0}{0.1}\}, 1), (\text{old}, \{\frac{0.9}{0.1}\}, 1), (\text{old}, \{\frac{0.8}{0.1}\}, 2)\}.$$

Remark 3 ([10]) As it can be seen, the more important part of an L-fuzzy bag is information function $\tilde{f}$. Therefore, it is possible to study the properties of L-fuzzy bags just by considering their information functions.

Notation 2 ([10]) We set $\tilde{\mathbf{B}}_L(P, O)$ as the set of all L-fuzzy bags $\tilde{\mathcal{B}}^{\tilde{f}} = (\tilde{f}, B^{\tilde{f}})$. Where, $\tilde{f} : P \to \mathcal{F}_L(O)$ and $B^{\tilde{f}}$ are as defined in Definition 6. Also, we set $\tilde{\mathbf{B}}(P, O)$ as the set of all fuzzy bags.

The following theorem gives the relation among bags, fuzzy bags and L-fuzzy bags.

Theorem 1 ([10]) *Let a complete lattice L_1 be a sub lattice of a complete lattice L_2. Then, $\tilde{\boldsymbol{B}}_{L_1}(P, O) \subseteq \tilde{\boldsymbol{B}}_{L_2}(P, O)$. In particular, $\boldsymbol{B}(P, O) = \tilde{\boldsymbol{B}}_{\{0,1\}}(P, O) \subseteq \tilde{\boldsymbol{B}}_{[0,1]}(P, O) = \tilde{\boldsymbol{B}}(P, O)$.*

Here, we define the binary operations among L-fuzzy bags.

Definition 7 Let $\tilde{\mathcal{B}}^{\tilde{f}_i} \in \tilde{\mathbf{B}}_L(P_i, O_i)$ for all $i \in I_n$ be given L-fuzzy bags, $\overline{O} = \cap_{i\in I_n} O_i \neq \emptyset$ and $\overline{P} = \cap_{i\in I_n} P_i \neq \emptyset$. Then, their intersection is L-fuzzy bag

$$\cap_{i\in I_n} \tilde{\mathcal{B}}^{\tilde{f}_i} = (\cap_{i\in I_n} \tilde{f}_i, B^{\cap_{i\in I_n} \tilde{f}_i}), \tag{1}$$

where $\cap_{i\in I_n} \tilde{f}_i : \overline{P} \to \mathcal{F}_L(\overline{O})$ such that $(\cap_{i\in I_n} \tilde{f}_i)(p) = \cap_{i\in I_n} \tilde{f}_i(p)$. Also,

$$B^{\cap_{i\in I_n} \tilde{f}_i} = \{(p, \delta, card(O_\delta^p)) | p \in \overline{P}, \delta \in L\},$$

where $O_\delta^p = \{o \in \overline{O} | (\cap_{i\in I_n} \tilde{f}_i)(p)(o) = \delta\}$.

Note that by Definition 6, $\cap_{i\in I_n} \tilde{\mathcal{B}}^{\tilde{f}_i} = \tilde{\mathcal{B}}^{\cap_{i\in I_n} \tilde{f}_i}$.

Table 6 Values of $\tilde{f}_1(p)(o)$

p	o				
	Nancy	Lia	Elena	Suzi	Sam
Tall	0.6	0.8	0.3	0.0	0.6
Medium	0.8	0.4	0.6	0.2	0.4
Short	0.0	0.0	0.8	1.0	0.3

Table 7 Values of $\tilde{f}_2(p)(o)$

p	o			
	Liu	Sam	Bob	Suzi
Extremly tall	0.9	0.2	0.4	0.0
Tall	1.0	0.7	0.7	0.0
Medium	0.1	0.3	0.2	0.3
Short	0.0	0.2	0.1	0.9

Definition 8 Let $\tilde{\mathcal{B}}^{\tilde{f}_i} \in \tilde{\mathbf{B}}_L(P_i, O_i)$ for all $i \in I_n$ be given L-fuzzy bags, $\overline{O} = \cup_{i \in I_n} O_i$ and $\overline{P} = \cup_{i \in I_n} P_i$. Then, their union is L-fuzzy bag

$$\cup_{i \in I_n} \tilde{\mathcal{B}}^{\tilde{f}_i} = (\cup_{i \in I_n} \tilde{f}_i, B^{\cup_{i \in I_n} \tilde{f}_i}), \tag{2}$$

where $\cup_{i \in I_n} \tilde{f}_i : \overline{P} \to \mathcal{F}_L(\overline{O})$ such that $(\cup_{i \in I_n} \tilde{f}_i)(p) = \cup_{i \in I_n} \tilde{f}_i(p)$. Also,

$$B^{\cup_{i \in I_n} \tilde{f}_i} = \{(p, \delta, card(O^p_\delta)) | p \in \overline{P}, \delta \in L\},$$

where $O^p_\delta = \{o \in \overline{O} | (\dot{\cup}_{i \in I_n} \tilde{f}_i)(p)(o) = \delta\}$.

Note that by Definition 6, $\cup_{i \in I_n} \tilde{\mathcal{B}}^{\tilde{f}_i} = \tilde{\mathcal{B}}^{\cup_{i \in I_n} \tilde{f}_i}$.

Example 6 Let $O_1 = \{\text{Nancy, Lia, Sam, Elena, Suzi}\}$, $O_2 = \{\text{Liu, Sam, Bob, Suzi}\}$, $P_1 = \{\text{tall, medium, short}\}$, $P_2 = \{\text{extremely tall, tall, medium, short}\}$ and $L = [0, 1]$. Consider $\tilde{\mathcal{B}}^{\tilde{f}_1} \in \tilde{\mathbf{B}}(P_1, O_1)$ and $\tilde{\mathcal{B}}^{\tilde{f}_2} \in \tilde{\mathbf{B}}(P_2, O_2)$ in which the values of $\tilde{f}_1$ and $\tilde{f}_2$ are as in Tables 6 and 7.

So, the intersection is $\tilde{\mathcal{B}}^{\tilde{f}_1 \cap \tilde{f}_2} = (\tilde{f}_1 \cap \tilde{f}_2, B^{\tilde{f}_1 \cap \tilde{f}_2})$ where,

$$(\tilde{f}_1 \cap \tilde{f}_2)(\text{tall}) = \{\frac{0.6}{\text{Sam}}\},$$
$$(\tilde{f}_1 \cap \tilde{f}_2)(\text{medium}) = \{\frac{0.2}{\text{Suzi}}, \frac{0.3}{\text{Sam}}\},$$
$$(\tilde{f}_1 \cap \tilde{f}_2)(\text{short}) = \{\frac{0.9}{\text{Suzi}}, \frac{0.2}{\text{Sam}}\}.$$

and

$$B^{\tilde{f}_1 \cap \tilde{f}_2} = \{(\text{tall}, 0.6, 1), (\text{medium}, 0.2, 1), (\text{medium}, 0.3, 1), (\text{short}, 0.9, 1), (\text{short}, 0.2, 1)\}.$$

And the union is $\tilde{\mathcal{B}}^{\tilde{f}_1 \cup \tilde{f}_2} = (\tilde{f}_1 \cup \tilde{f}_2, B^{\tilde{f}_1 \cup \tilde{f}_2})$ where,

$$(\tilde{f}_1 \cup \tilde{f}_2)(\text{extremely tall}) = \{\frac{0.9}{\text{Liu}}, \frac{0.2}{\text{Sam}}, \frac{0.4}{\text{Bob}}\},$$
$$(\tilde{f}_1 \cup \tilde{f}_2)(\text{tall}) = \{\frac{1.0}{\text{Liu}}, \frac{0.7}{\text{Sam}}, \frac{0.7}{\text{Bob}}, \frac{0.6}{\text{Nancy}}, \frac{0.8}{\text{Lia}}, \frac{0.3}{\text{Elena}}\},$$
$$(\tilde{f}_1 \cup \tilde{f}_2)(\text{medium}) = \{\frac{0.8}{\text{Nancy}}, \frac{0.4}{\text{Lia}}, \frac{0.6}{\text{Elena}}, \frac{0.3}{\text{Suzi}}, \frac{0.1}{\text{Liu}}, \frac{0.4}{\text{Sam}}, \frac{0.2}{\text{Bob}}\},$$
$$(\tilde{f}_1 \cup \tilde{f}_2)(\text{short}) = \{\frac{0.8}{\text{Elena}}, \frac{1.0}{\text{Suzi}}, \frac{0.3}{\text{Sam}}, \frac{0.1}{\text{Bob}}\}.$$

and

$$\begin{aligned} B^{\tilde{f}_1 \cup \tilde{f}_2} =& \{(\text{extremely tall}, 0.2, 1), (\text{extremely tall}, 0.4, 1), (\text{extremely tall}, 0.9, 1), \\ & (\text{tall}, 0.3, 1), (\text{tall}, 0.6, 1), (\text{tall}, 0.7, 2), (\text{tall}, 0.8, 1), (\text{tall}, 1.0, 1), \\ & (\text{medium}, 0.1, 1), (\text{medium}, 0.2, 1), (\text{medium}, 0.3, 1), (\text{medium}, 0.4, 2), \\ & (\text{medium}, 0.6, 1), (\text{medium}, 0.8, 1), (\text{short}, 0.1, 1), (\text{short}, 0.3, 1), \\ & (\text{short}, 0.8, 1), (\text{short}, 1.0, 1)\}. \end{aligned}$$

The following definition equips the set of all L-fuzzy bags with an order.

Definition 9 ([10]) (i) An L-fuzzy bag $\tilde{\mathcal{B}}^{\tilde{f}}$ is an L-fuzzy sub bag of $\tilde{\mathcal{B}}^{\tilde{g}}$, denoted by $\tilde{\mathcal{B}}^{\tilde{f}} \tilde{\sqsubseteq} \tilde{\mathcal{B}}^{\tilde{g}}$ if and only if $\tilde{f}(p) \tilde{\subseteq} \tilde{g}(p)$ for all $p \in P$. That means $\tilde{\mathcal{B}}^{\tilde{f}} \tilde{\sqsubseteq} \tilde{\mathcal{B}}^{\tilde{g}}$ if and only if for all $p \in P$, $\tilde{f}(p)$ be an L-fuzzy subset of $\tilde{g}(p)$.
(ii) Two L-fuzzy bags $\tilde{\mathcal{B}}^{\tilde{f}}$ and $\tilde{\mathcal{B}}^{\tilde{g}}$ are equal, denoted by $\tilde{\mathcal{B}}^{\tilde{f}} \cong \tilde{\mathcal{B}}^{\tilde{g}}$ if $\tilde{\mathcal{B}}^{\tilde{f}} \tilde{\sqsubseteq} \tilde{\mathcal{B}}^{\tilde{g}}$ and $\tilde{\mathcal{B}}^{\tilde{g}} \tilde{\sqsubseteq} \tilde{\mathcal{B}}^{\tilde{f}}$ that means if $\tilde{f} = \tilde{g}$.

The next theorem gives some useful results about L-fuzzy bags.

Theorem 2 ([10]) *Operations $\cup$ and $\cap$ in $\tilde{\boldsymbol{B}}_L(P, O)$ satisfy the laws of idempotency, commutativity, associativity, monotonicity and distributivity. Moreover, $\mathcal{B}^0$ is neutral for operation $\cup$ and $\mathcal{B}^1$ is neutral for operation $\cap$.*

In the following definition, we review the concept of the complement of an L-fuzzy bag.

Definition 10 ([10]) Let $\eta : L \to L$ be a fixed strong negation [1], this means an involutive decreasing bijection. Consider $\tilde{\mathcal{B}}^{\tilde{f}} = (\tilde{f}, B^{\tilde{f}})$. Then, the $\eta-$complement of $\tilde{\mathcal{B}}^{\tilde{f}}$ is L-fuzzy bag $(\tilde{\mathcal{B}}^{\tilde{f}})^c = (\tilde{f}^c, B^{\tilde{f}^c})$, where $\tilde{f}^c : P \to \mathcal{F}_L(O)$ such that $\tilde{f}^c(p)(o) = \eta(\tilde{f}(p)(o))$ for all $p \in P$ and $o \in O$.

Note that by Definition 6, $(\tilde{\mathcal{B}}^{\tilde{f}})^c = \tilde{\mathcal{B}}^{\tilde{f}^c}$.

Note 3 ([10]) In Definition 10, if $L = [0, 1]$ and η is the standard negation, $\eta(x) = 1 - x$ for all $x \in [0, 1]$ [1], then $\tilde{\mathcal{B}}^{\tilde{f}^c}$ is called the complement of $\tilde{\mathcal{B}}^{\tilde{f}}$.

Example 7 ([10]) The complement of the fuzzy bag in Example 4 is $\tilde{\mathcal{B}}^{\tilde{g}^c} = (\tilde{g}^c, B^{\tilde{g}^c})$ where,

$$\tilde{g}^c(\text{tall}) = \{\frac{0.2}{\text{Ben}}, \frac{0.4}{\text{Sue}}, \frac{1.0}{\text{Tom}}, \frac{0.9}{\text{John}}, \frac{0.2}{\text{Stan}}, \frac{0.4}{\text{Bill}}, \frac{0.5}{\text{Kim}}, \frac{0.3}{\text{Ana}}, \frac{0.5}{\text{Sara}}\},$$
$$\tilde{g}^c(\text{medium}) = \{\frac{0.7}{\text{Ben}}, \frac{0.9}{\text{Sue}}, \frac{0.9}{\text{Tom}}, \frac{0.4}{\text{John}}, \frac{0.7}{\text{Stan}}, \frac{0.9}{\text{Bill}}, \frac{0.2}{\text{Kim}}, \frac{0.9}{\text{Ana}}, \frac{0.5}{\text{Sara}}\},$$
$$\tilde{g}^c(\text{short}) = \{\frac{0.9}{\text{Ben}}, \frac{1.0}{\text{Sue}}, \frac{0.1}{\text{Tom}}, \frac{0.6}{\text{John}}, \frac{0.9}{\text{Stan}}, \frac{1.0}{\text{Bill}}, \frac{0.8}{\text{Kim}}, \frac{1.0}{\text{Ana}}, \frac{0.9}{\text{Sara}}\},$$

and

$$\begin{aligned} B^{\tilde{g}^c} = \{&(\text{tall}, 1.0, 1), (\text{tall}, 0.9, 1), (\text{tall}, 0.5, 2), (\text{tall}, 0.4, 2), (\text{tall}, 0.3, 1),\\ &(\text{tall}, 0.2, 2), (\text{medium}, 0.9, 4), (\text{medium}, 0.7, 2), (\text{medium}, 0.5, 1),\\ &(\text{medium}, 0.4, 1), (\text{medium}, 0.2, 1), (\text{short}, 1.0, 3), (\text{short}, 0.9, 3),\\ &(\text{short}, 0.8, 1), (\text{short}, 0.6, 1), (\text{short}, 0.1, 1)\}. \end{aligned}$$

Note 4 In the process of determining the degrees of membership in Definition 6, some degrees are very close to each other and may be they are not different in the decision maker's point of view. This situation appears specially when the cardinality of O is big. In this case, we can cluster the objects based on their degrees of membership. For example consider Example 5 in the case that we have $card(O) = 100$.

Table 8 Clusters

Cluster head	Cluster members
$\frac{0.06}{70}$	$\frac{0.08}{12}, \frac{0.08}{19}, \frac{0.00}{23}, \frac{0.08}{27}, \frac{0.08}{46}, \frac{0.05}{52}, \frac{0.06}{70}, \frac{0.02}{74}, \frac{0.04}{75}, \frac{0.08}{90}$
$\frac{0.11}{21}$	$\frac{0.11}{21}, \frac{0.12}{48}, \frac{0.11}{60}, \frac{0.10}{65}, \frac{0.13}{66}$
$\frac{0.15}{15}$	$\frac{0.16}{1}, \frac{0.15}{15}, \frac{0.15}{35}, \frac{0.14}{36}, \frac{0.14}{40}$
$\frac{0.18}{33}$	$\frac{0.17}{5}, \frac{0.18}{33}, \frac{0.18}{49}, \frac{0.17}{76}, \frac{0.19}{84}, \frac{0.18}{86}$
$\frac{0.24}{47}$	$\frac{0.26}{7}, \frac{0.23}{13}, \frac{0.26}{29}, \frac{0.26}{34}, \frac{0.24}{47}, \frac{0.24}{50}, \frac{0.24}{63}, \frac{0.23}{71}$
$\frac{0.34}{57}$	$\frac{0.31}{3}, \frac{0.35}{43}, \frac{0.34}{57}, \frac{0.37}{59}, \frac{0.35}{72}, \frac{0.30}{82}, \frac{0.37}{87}, \frac{0.31}{96}$
$\frac{0.42}{51}$	$\frac{0.45}{11}, \frac{0.44}{20}, \frac{0.40}{28}, \frac{0.43}{31}, \frac{0.40}{45}, \frac{0.42}{51}, \frac{0.39}{62}, \frac{0.40}{64}, \frac{0.45}{80}, \frac{0.44}{94}, \frac{0.45}{95}$
$\frac{0.51}{44}$	$\frac{0.53}{4}, \frac{0.54}{17}, \frac{0.51}{44}, \frac{0.49}{55}, \frac{0.49}{56}, \frac{0.49}{93}, \frac{0.51}{97}, \frac{0.51}{98}$
$\frac{0.58}{38}$	$\frac{0.60}{6}, \frac{0.58}{38}, \frac{0.55}{39}, \frac{0.62}{42}, \frac{0.58}{69}, \frac{0.55}{81}, \frac{0.63}{88}$
$\frac{0.69}{9}$	$\frac{0.65}{8}, \frac{0.69}{9}, \frac{0.65}{77}, \frac{0.73}{78}, \frac{0.65}{79}, \frac{0.74}{83}, \frac{0.69}{85}$
$\frac{0.79}{2}$	$\frac{0.79}{2}, \frac{0.75}{10}, \frac{0.83}{16}, \frac{0.77}{24}, \frac{0.82}{25}, \frac{0.80}{30}, \frac{0.78}{61} \frac{0.82}{73}, \frac{0.78}{89}, \frac{0.78}{92}, \frac{0.82}{99}, \frac{0.79}{100}$
$\frac{0.90}{53}$	$\frac{0.91}{14}, \frac{0.87}{26}, \frac{0.91}{32}, \frac{0.87}{37}, \frac{0.85}{41}, \frac{0.90}{53}, \frac{0.90}{58}, \frac{0.93}{91}$
$\frac{0.96}{22}$	$\frac{1.00}{18}, \frac{0.96}{22}, \frac{0.94}{54}, \frac{0.94}{67}, \frac{0.96}{68}$

Let us have the following fuzzy set for the property "young".

$$\tilde{f}(\text{young}) = \{\frac{0.16}{o_1}, \frac{0.79}{o_2}, \frac{0.31}{o_3}, \frac{0.53}{o_4}, \frac{0.17}{o_5}, \frac{0.60}{o_6}, \frac{0.26}{o_7}, \frac{0.65}{o_8}, \frac{0.69}{o_9}, \frac{0.75}{o_{10}},$$
$$\frac{0.45}{o_{11}}, \frac{0.08}{o_{12}}, \frac{0.23}{o_{13}}, \frac{0.91}{o_{14}}, \frac{0.15}{o_{15}}, \frac{0.83}{o_{16}}, \frac{0.54}{o_{17}}, \frac{1.00}{o_{18}}, \frac{0.08}{o_{19}}, \frac{0.44}{o_{20}}, \frac{0.11}{o_{21}},$$
$$\frac{0.96}{o_{22}}, \frac{0.00}{o_{23}}, \frac{0.77}{o_{24}}, \frac{0.82}{o_{25}}, \frac{0.87}{o_{26}}, \frac{0.08}{o_{27}}, \frac{0.40}{o_{28}}, \frac{0.26}{o_{29}}, \frac{0.80}{o_{30}}, \frac{0.43}{o_{31}}, \frac{0.91}{o_{32}},$$
$$\frac{0.18}{o_{33}}, \frac{0.26}{o_{34}}, \frac{0.15}{o_{35}}, \frac{0.14}{o_{36}}, \frac{0.87}{o_{37}}, \frac{0.58}{o_{38}}, \frac{0.55}{o_{39}}, \frac{0.14}{o_{40}}, \frac{0.85}{o_{41}}, \frac{0.62}{o_{42}}, \frac{0.35}{o_{43}},$$
$$\frac{0.51}{o_{44}}, \frac{0.40}{o_{45}}, \frac{0.08}{o_{46}}, \frac{0.24}{o_{47}}, \frac{0.12}{o_{48}}, \frac{0.18}{o_{49}}, \frac{0.24}{o_{50}}, \frac{0.42}{o_{51}}, \frac{0.05}{o_{52}}, \frac{0.90}{o_{53}}, \frac{0.94}{o_{54}},$$
$$\frac{0.49}{o_{55}}, \frac{0.49}{o_{56}}, \frac{0.34}{o_{57}}, \frac{0.90}{o_{58}}, \frac{0.37}{o_{59}}, \frac{0.11}{o_{60}}, \frac{0.78}{o_{61}}, \frac{0.39}{o_{62}}, \frac{0.24}{o_{63}}, \frac{0.40}{o_{64}}, \frac{0.10}{o_{65}},$$
$$\frac{0.13}{o_{66}}, \frac{0.94}{o_{67}}, \frac{0.96}{o_{68}}, \frac{0.58}{o_{69}}, \frac{0.06}{o_{70}}, \frac{0.23}{o_{71}}, \frac{0.35}{o_{72}}, \frac{0.82}{o_{73}}, \frac{0.02}{o_{74}}, \frac{0.04}{o_{75}}, \frac{0.17}{o_{76}},$$
$$\frac{0.65}{o_{77}}, \frac{0.73}{o_{78}}, \frac{0.65}{o_{79}}, \frac{0.45}{o_{80}}, \frac{0.55}{o_{81}}, \frac{0.30}{o_{82}}, \frac{0.74}{o_{83}}, \frac{0.19}{o_{84}}, \frac{0.69}{o_{85}}, \frac{0.18}{o_{86}}, \frac{0.37}{o_{87}},$$
$$\frac{0.63}{o_{88}}, \frac{0.78}{o_{89}}, \frac{0.08}{o_{90}}, \frac{0.93}{o_{91}}, \frac{0.78}{o_{92}}, \frac{0.49}{o_{93}}, \frac{0.44}{o_{94}}, \frac{0.45}{o_{95}}, \frac{0.31}{o_{96}}, \frac{0.51}{o_{97}}, \frac{0.51}{o_{98}},$$
$$\frac{0.82}{o_{99}}, \frac{0.79}{o_{100}}\}$$

By K-medoids method and choosing 13 clusters, we have the results of Table 8.

3 Relations on Bags and Fuzzy Bags

Let P_i and O_i be the sets of properties and objectives for all $i \in I_n$, respectively. We have the following results.

Definition 11 ([10]) An n-dimensional bag is the pair $\mathcal{B}^l = (l, B^l)$ where,

$$l : \Pi_{i\in I_n} P_i \to \Pi_{i\in I_n} \mathcal{P}(O_i)$$

and

$$B^l = \{((p_1, \ldots, p_n), card(l(p_1, \ldots, p_n)))|p_i \in P_i,\ i \in I_n\}.$$

It should be mentioned that in what follows, for convenience, we use both notations Π and $\times$ for Cartesian product.

Table 9 Values of $(f_1 \times f_2)((p_1, p_2))$

(p_1, p_2)	$(f_1 \times f_2)((p_1, p_2))$
(male, 18)	{A, C, D, E, G, H, I, J, K} × {B, J}
(male, 19)	{A, C, D, E, G, H, I, J, K} × {E, K, L}
(male, 20)	{A, C, D, E, G, H, I, J, K} × {A, D}
(male, 21)	{A, C, D, E, G, H, I, J, K} × {I}
(male, 22)	{A, C, D, E, G, H, I, J, K} × {C, F, G, H, M}
(female, 18)	{B, F, L, M} × {B, J}
(female, 19)	{B, F, L, M} × {E, K, L}
(female, 20)	{B, F, L, M} × {A, D}
(female, 21)	{B, F, L, M} × {I}
(female, 22)	{B, F, L, M} × {C, F, G, H, M}

Definition 12 ([10]) Let $\mathcal{B}^{f_i} \in \mathbf{B}(P_i, O_i)$ for all $i \in I_n$. Define bag $\Pi_{i \in I_n} \mathcal{B}^{f_i} = (\Pi_{i \in I_n} f_i, B^{\Pi_{i \in I_n} f_i})$ as the Cartesian product of $\{\mathcal{B}^{f_i}\}_{i \in I_n}$ which is called C_n-bag. Where, $(\Pi_{i \in I_n} f_i)((p_1, \ldots, p_n)) = \Pi_{i \in I_n} f_i(p_i)$ as the Cartesian product of n sets and

$$B^{\Pi_{i \in I_n} f_i} = \{((p_1, \ldots, p_n), card(\Pi_{i \in I_n} f_i(p_i)) | p_i \in P_i, \text{for all } i \in I_n\}.$$

Note that by Definition 1, $\mathcal{B}^{\Pi_{i \in I_n} f_i} = \Pi_{i \in I_n} \mathcal{B}^{f_i}$.

Theorem 3 ([10]) *C_n-bag $\mathcal{B}^{\Pi_{i \in I_n} f_i}$ is an n-dimensional bag.*

Example 8 ([10]) Let $O = \{\text{A, B, C, D, E, F, G, H, I, J, K, L, M}\}$, $P_1 = \{\text{male, female}\}$ and $P_2 = \{18, 19, 20, 21, 22\}$. Let $\mathcal{B}^{f_1} \in \mathbf{B}(P_1, O)$ and $\mathcal{B}^{f_2} \in \mathbf{B}(P_2, O)$, where

$$f_1(\text{male}) = \{\text{A, C, D, E, G, H, I, J, K}\}, \quad f_1(\text{female}) = \{\text{B, F, L, M}\},$$
$$f_2(18) = \{\text{B, J}\}, \quad f_2(19) = \{\text{E, K, L}\}, \quad f_2(20) = \{\text{A, D}\},$$
$$f_2(21) = \{\text{I}\}, \quad f_2(22) = \{\text{C, F, G, H, M}\}.$$

Hence, the C_2-bag of $\mathcal{B}^{f_1}$ and $\mathcal{B}^{f_2}$ is $\mathcal{B}^{f_1 \times f_2} = (f_1 \times f_2, B^{f_1 \times f_2})$, where the values of $(f_1 \times f_2)((p_1, p_2))$ are as in Table 9. So, according to Table 9, $B^{f_1 \times f_2}$ is as follows.

$$\begin{aligned} B^{f_1 \times f_2} = & \{((\text{male}, 18), 18), ((\text{male}, 19), 27), ((\text{male}, 20), 18), ((\text{male}, 21), 9), \\ & ((\text{male}, 22), 45), ((\text{female}, 18), 8), ((\text{female}, 19), 12), ((\text{female}, 20), 8), \\ & ((\text{female}, 21), 4), ((\text{female}, 22), 20)\} \end{aligned}$$

Definition 13 Let $\mathcal{B}^{f_i} \in \mathbf{B}(P_i, O_i)$ for all $i \in I_n$ and $\overline{O} = \cup_{i \in I_n} O_i$. Define bag conjunctive Cartesian product of $\{\mathcal{B}^{f_i}\}_{i \in I_n}$, which is called C_n^c-bag, by

Table 10 Values of $(f_1 \times^c f_2)((p_1, p_2))$

(p_1, p_2)	$(f_1 \times^c f_2)((p_1, p_2))$
(male, 18)	{J}
(male, 19)	{E, K}
(male, 20)	{A, D}
(male, 21)	{I}
(male, 22)	{C, G, H}
(female, 18)	{B}
(female, 19)	{L}
(female, 20)	∅
(female, 21)	∅
(female, 22)	{F, M}

$$\Pi^c_{i\in I_n} \mathcal{B}^{f_i} = (\Pi^c_{i\in I_n} f_i, B^{\Pi^c_{i\in I_n} f_i}), \tag{3}$$

where $\Pi^c_{i\in I_n} f_i : \Pi_{i\in I_n} P_i \to P(\overline{O})$ such that $(\Pi^c_{i\in I_n} f_i)((p_1, p_2, \ldots, p_n)) = \cap_{i\in I_n} f_i(p_i)$ for all $p_i \in P_i$. Also,

$$B^{\Pi^c_{i\in I_n} f_i} = \{((p_1, p_2, \ldots, p_n), card(\Pi_{i\in I_n} f_i(p_i)))|p_i \in P_i \text{ for all } i \in I_n\}.$$

Note that by Definition 1, $\Pi^c_{i\in I_n} \mathcal{B}^{f_i} = \mathcal{B}^{\Pi^c_{i\in I_n} f_i}$.

Definition 14 Let $\mathcal{B}^{f_i} \in \mathbf{B}(P_i, O_i)$ for all $i \in I_n$ and $\overline{O} = \cup_{i\in I_n} O_i$. Define bag disjunctive Cartesian product of $\{\mathcal{B}^{f_i}\}_{i\in I_n}$, which is called C^d_n-bag, by

$$\Pi^d_{i\in I_n} \mathcal{B}^{f_i} = (\Pi^d_{i\in I_n} f_i, B^{\Pi^d_{i\in I_n} f_i}), \tag{4}$$

where $\Pi^d_{i\in I_n} f_i : \Pi_{i\in I_n} P_i \to P(\overline{O})$ such that $(\Pi^d_{i\in I_n} f_i)((p_1, p_2, \ldots, p_n)) = \cup_{i\in I_n} f_i(p_i)$ for all $p_i \in P_i$. Also,

$$B^{\Pi^d_{i\in I_n} f_i} = \{((p_1, p_2, \ldots, p_n), card(\Pi_{i\in I_n} f_i(p_i)))|p_i \in P_i \text{ for all } i \in I_n\}.$$

Note that by Definition 1, $\Pi^d_{i\in I_n} \mathcal{B}^{f_i} = \mathcal{B}^{\Pi^d_{i\in I_n} f_i}$. Moreover, $f^i(p_i) \subseteq O_i \subseteq \overline{O}$ for all $i \in I_n$ and thus, $\cap_{i\in I_n}$ is well defined.

Example 9 Let $\mathcal{B}^{f_1} \in \mathbf{B}(P_1, O)$ and $\mathcal{B}^{f_2} \in \mathbf{B}(P_2, O)$ be as in Example 8. The C^c_n-bag of $\mathcal{B}^{f_1}$ and $\mathcal{B}^{f_2}$ is $\mathcal{B}^{f_1 \times^c f_2} = (f_1 \times^c f_2, B^{f_1 \times^c f_2})$, where the values of $(f_1 \times^c f_2)((p_1, p_2))$ are as in Table 10.

So, according to Table 10, $B^{f_1 \times^c f_2}$ is as follows.

$$B^{f_1 \times^c f_2} = \{((\text{male}, 18), 1), ((\text{male}, 19), 2), ((\text{male}, 20), 2), ((\text{male}, 21), 1), ((\text{male}, 22), 3), ((\text{female}, 18), 1), ((\text{female}, 19), 1), ((\text{female}, 20), 0), ((\text{female}, 21), 0), ((\text{female}, 22), 2)\}.$$

Definition 15 ([10]) Fix C_n−bag $\mathcal{B}^{\Pi_{i=1}^n f_i}$. An n−ary bag relation is a sub bag of C_n−bag $\mathcal{B}^{\Pi_{i=1}^n f_i}$ which is denoted by $\mathcal{B}^{f_R} = (f_R, B^{f_R})$.

Example 10 ([10]) Consider C_2−bag of Example 8. The bag, $\mathcal{B}^{f_R} = (f_R, B^{f_R})$, is a 2−ary bag relation which introduces people who are older than twenty, where

$$\begin{aligned}
f_R((\text{male}, 20)) &= \{\text{A, C, D, E, G, H, I, J, K}\} \times \{\text{A, D}\},\\
f_R((\text{male}, 21)) &= \{\text{A, C, D, E, G, H, I, J, K}\} \times \{\text{I}\},\\
f_R((\text{male}, 22)) &= \{\text{A, C, D, E, G, H, I, J, K}\} \times \{\text{C, F, G, H, M}\},\\
f_R((\text{female}, 20)) &= \{\text{B, F, L, M}\} \times \{\text{A, D}\},\\
f_R((\text{female}, 21)) &= \{\text{B, F, L, M}\} \times \{\text{I}\},\\
f_R((\text{female}, 22)) &= \{\text{B, F, L, M}\} \times \{\text{C, F, G, H, M}\}
\end{aligned}$$

and

$$B^{f_R} = \{((\text{male}, 20), 18), ((\text{male}, 21), 9), ((\text{male}, 22), 45), ((\text{female}, 20), 8), ((\text{female}, 21), 4), ((\text{female}, 22), 20)\}.$$

Definition 16 ([10]) Let $R_n \subseteq \Pi_{i \in I_n} \mathcal{P}(O_i)$ and $l_n : \Pi_{i \in I_n} P_i \to \Pi_{i \in I_n} \mathcal{P}(O_i)$. Then, $\mathcal{B}_{R_n}^{l_n} = (l_n, B_{R_n}^{l_n})$, where

$$B_{R_n}^{l_n} = \{((p_1, \ldots, p_n), card(l_n(p_1, \ldots, p_n)))|\, l_n(p_1, \ldots, p_n) \in R_n\}$$

is called the bag induced by R_n.

Example 11 ([10]) Let $O_1 = \{m_i | i \in I_{40}\}$ and $O_2 = \{w_i | i \in I_{40}\}$ where, m_i, w_i for all $i \in I_{40}$ are man and woman, respectively. Let $R_2 = \{(m_i, w_i) | i \in I_{40}\}$ shows the relation of "spouse". Now, Table 11 gives $l_2 : P_1 \times P_2 \to R_2 \subseteq \mathcal{P}(O_1) \times \mathcal{P}(O_2)$, where $P_1 = P_2 = \{A, B, AB, O\}$ is the set of all blood groups.

Thus, by Definition 16, we can present this information by the bag $\mathcal{B}_{R_2}^{l_2} = (l_2, B_{R_2}^{l_2})$, where $B_{R_2}^{l_2}$ is as follows.

$$B_{R_2}^{l_2} = \{((A, A), 3), ((A, B), 3), ((A, AB), 3), ((B, A), 5), ((B, B), 1), ((B, AB), 2), ((B, O), 3), ((AB, A), 4), ((AB, B), 2), ((AB, AB), 2), ((AB, O), 1), ((O, A), 3), ((O, B), 2), ((O, AB), 2), ((O, O), 4)\}.$$

Table 11 function l_2

(p_1, p_2)	$l_2(p_1, p_2)(o_1, o_2)$
(A, A)	$\{(m_5, w_5), (m_{11}, w_{11}), (m_{26}, w_{26})\}$
(A, B)	$\{(m_4, w_4), (m_{18}, w_{18}), (m_{36}, w_{36})\}$
(A, AB)	$\{(m_{12}, w_{12}), (m_{27}, w_{27}), (m_{39}, w_{39})\}$
(A, O)	$\emptyset$
(B, A)	$\{(m_8, w_8), (m_{25}, w_{25}), (m_{28}, w_{28}), (m_{31}, w_{31}), (m_{35}, w_{35})\}$
(B, B)	$\{(m_{14}, w_{14})\}$
(B, AB)	$\{(m_{17}, w_{17}), (m_{23}, w_{23})\}$
(B, O)	$\{(m_9, w_9), (m_{20}, w_{20}), (m_{32}, w_{32})\}$
(AB, A)	$\{(m_3, w_3), (m_{22}, w_{22}), (m_{38}, w_{38}), (m_{40}, w_{40})\}$
(AB, B)	$\{(m_6, w_6), (m_{29}, w_{29})\}$
(AB, AB)	$\{(m_2, w_2), (m_{15}, w_{15})\}$
(AB, O)	$\{(m_{37}, w_{37})\}$
(O, A)	$\{(m_{10}, w_{10}), (m_{24}, w_{24}), (m_{30}, w_{30})\}$
(O, B)	$\{(m_1, w_1), (m_{19}, w_{19})\}$
(O, AB)	$\{(m_{13}, w_{13}), (m_{21}, w_{21})\}$
(O, O)	$\{(m_7, w_7), (m_{16}, w_{16}), (m_{33}, w_{33}), (m_{34}, w_{34})\}$

Remark 4 ([10]) As a matter of fact, if people eat food that is not compatible with their blood type, they will experience many health problems. On the other hand, if a person eats food that is compatible, he/she will be healthier [20]. Since an appropriate diet can affect the unborn child's health, giving a proposal of a special diet to spouses can be helpful. Using the concept of relations on bags, one can screen all spouses with the similar blood groups.

Now, we study relations on L-fuzzy bags and give some results about them. First, we should review the concept of n-dimensional L-fuzzy bag.

Definition 17 ([10]) An n-dimensional L-fuzzy bag is the pair $\tilde{\mathcal{B}}^{\tilde{l}} = (\tilde{l}, B^{\tilde{l}})$ where,

$$\tilde{l} : \Pi_{i\in I_n} P_i \to \Pi_{i\in I_n} \mathcal{F}_L(O_i)$$

and

$$B^{\tilde{l}} = \{((p_1, \ldots, p_n), \delta, card(O_\delta^{p_1,\ldots,p_n}))|p_i \in P_i,\ i \in I_n, \delta \in L,\ O_\delta^{p_1,\ldots,p_n}\}.$$

where, $O_\delta^{p_1,\ldots,p_n} = \{(o_1, \ldots, o_n) \in \Pi_{i\in I_n} O_i | \tilde{l}(p_1, \ldots, p_n)(o_1, \ldots, o_n) = \delta\}$.

Notation 3 ([10]) In the sequel, we use notation $\mathcal{F}_L(O)^n$ for $\underbrace{\mathcal{F}_L(O) \times \mathcal{F}_L(O) \times \cdots \times \mathcal{F}_L(O)}_{n-times}$.

Definition 18 ([10]) Let $\tilde{\mathcal{B}}^{\tilde{f}_i} \in \tilde{\mathbf{B}}_L(P_i, O_i)$ for all $i \in I_n$. Define L-fuzzy bag Cartesian product of $\{\tilde{\mathcal{B}}^{\tilde{f}_i}\}_{i\in I_n}$, which is called C_n-L-fuzzy bag, by

$$\Pi_{i=1}^n \tilde{\mathcal{B}}^{\tilde{f}_i} = (\Pi_{i=1}^n \tilde{f}_i, B^{\Pi_{i=1}^n \tilde{f}_i}).$$

where, $(\Pi_{i=1}^n \tilde{f}_i)((p_1, \ldots, p_n)) = \Pi_{i=1}^n \tilde{f}_i(p_i)$ is the Cartesian product of n L-fuzzy sets and

$$B^{\Pi_{i=1}^n \tilde{f}_i} = \{((p_1, \ldots, p_n), \delta, card(O_\delta^{p_1,\ldots,p_n}))|p_i \in P_i,\ i \in I_n, \delta \in L, O_\delta^{p_1,\ldots,p_n}\},$$

and $O_\delta^{p_1,\ldots,p_n} = \{(o_1, \ldots, o_n) \in \Pi_{i=1}^n O_i |\min\{\tilde{f}_1(p_1)(o_1), \ldots, \tilde{f}_n(p_n)(o_n)\} = \delta\}$.

Note that by Definition 6, $\mathcal{B}^{\Pi_{i=1}^n \tilde{f}_i} = \Pi_{i=1}^n \mathcal{B}^{\tilde{f}_i}$.

Theorem 4 ([10]) *C_n-L-fuzzy bag is an n-dimensional L-fuzzy bag.*

An example of a 2-dimensional L-fuzzy bag is given in the following example.

Example 12 ([10]) Consider the fuzzy bags of Examples 3 and 4. The C_2-fuzzy bag of $\tilde{\mathcal{B}}^{\tilde{f}}$ and $\tilde{\mathcal{B}}^{\tilde{g}}$ is $\tilde{\mathcal{B}}^{\tilde{f}\times\tilde{g}} = (\tilde{f} \times \tilde{g}, B^{\tilde{f}\times\tilde{g}})$, where the values of $(\tilde{f} \times \tilde{g})((p_1, p_2))$ are as in Table 12. According to Table 12, $B^{\tilde{f}\times\tilde{g}}$ can be easily given.

Definition 19 Let $\tilde{\mathcal{B}}^{\tilde{f}_i} \in \tilde{\mathbf{B}}_L(P_i, O_i)$ for all $i \in I_n$ and $\overline{O} = \cup_{i\in I_n} O_i$. Define L-fuzzy bag conjunctive Cartesian product of $\{\tilde{\mathcal{B}}^{\tilde{f}_i}\}_{i\in I_n}$, which is called C_n^c-L-fuzzy bag, by

$$\Pi_{i\in I_n}^c \tilde{\mathcal{B}}^{\tilde{f}_i} = (\Pi_{i\in I_n}^c \tilde{f}_i, B^{\Pi_{i\in I_n}^c \tilde{f}_i}), \tag{5}$$

where $\Pi_{i\in I_n}^c \tilde{f}_i : \Pi_{i\in I_n} P_i \to \mathcal{F}_L(\overline{O})$ is such that $(\Pi_{i\in I_n}^c \tilde{f}_i)((p_1, p_2, \ldots, p_n)) = \cap_{i\in I_n} \tilde{f}_i(p_i)$ for all $p_i \in P_i$. Also,

$$B^{\Pi_{i\in I_n}^c \tilde{f}_i} = \{((p_1, p_2, \ldots, p_n), \delta, card(O_\delta^{p_1,p_2,\ldots,p_n}))|p_i \in P_i, \delta \in L\},$$

where $O_\delta^{p_1,p_2,\ldots,p_n} = \{o \in \overline{O}|(\Pi_{i\in I_n}^d \tilde{f}_i)((p_1, p_2, \ldots, p_n))(o) = \delta\}$.

Note that by Definition 6, $\Pi_{i\in I_n}^d \tilde{\mathcal{B}}^{\tilde{f}_i} = \tilde{\mathcal{B}}^{\Pi_{i\in I_n}^d \tilde{f}_i}$.

Definition 20 Let $\tilde{\mathcal{B}}^{\tilde{f}_i} \in \tilde{\mathbf{B}}_L(P_i, O_i)$ for all $i \in I_n$ and $\overline{O} = \cup_{i\in I_n} O_i$. Define L-fuzzy bag disjunctive Cartesian product of $\{\tilde{\mathcal{B}}^{\tilde{f}_i}\}_{i\in I_n}$, which is called C_n^d-L-fuzzy bag, by

$$\Pi_{i\in I_n}^d \tilde{\mathcal{B}}^{\tilde{f}_i} = (\Pi_{i\in I_n}^d \tilde{f}_i, B^{\Pi_{i\in I_n}^d \tilde{f}_i}), \tag{6}$$

where $\Pi_{i\in I_n}^d \tilde{f}_i : \Pi_{i\in I_n} P_i \to \mathcal{F}_L(\overline{O})$ such that $(\Pi_{i\in I_n}^d \tilde{f}_i)((p_1, p_2, \ldots, p_n)) = \cup_{i\in I_n} \tilde{f}_i(p_i)$ for all $p_i \in P_i$. Also,

Table 12 Values of $(\tilde{f} \times \tilde{g})((p_1, p_2))$

(p_1, p_2)	$(\tilde{f} \times \tilde{g})((p_1, p_2))$
(young, tall)	$\{\frac{0.7}{\text{Ben}}, \frac{0.2}{\text{Sue}}, \frac{0.4}{\text{Tom}}, \frac{0}{\text{John}}, \frac{0.7}{\text{Stan}}, \frac{0.4}{\text{Bill}}, \frac{0.2}{\text{Kim}}, \frac{0.7}{\text{Ana}}, \frac{0.1}{\text{Sara}}\} \times \{\frac{0.8}{\text{Ben}}, \frac{0.6}{\text{Sue}}, \frac{0}{\text{Tom}}, \frac{0.1}{\text{John}}, \frac{0.8}{\text{Stan}}, \frac{0.6}{\text{Bill}}, \frac{0.5}{\text{Kim}}, \frac{0.7}{\text{Ana}}, \frac{0.5}{\text{Sara}}\}$
(young, medium)	$\{\frac{0.7}{\text{Ben}}, \frac{0.2}{\text{Sue}}, \frac{0.4}{\text{Tom}}, \frac{0}{\text{John}}, \frac{0.7}{\text{Stan}}, \frac{0.4}{\text{Bill}}, \frac{0.2}{\text{Kim}}, \frac{0.7}{\text{Ana}}, \frac{0.1}{\text{Sara}}\} \times \{\frac{0.3}{\text{Ben}}, \frac{0.1}{\text{Sue}}, \frac{0.1}{\text{Tom}}, \frac{0.6}{\text{John}}, \frac{0.3}{\text{Stan}}, \frac{0.1}{\text{Bill}}, \frac{0.8}{\text{Kim}}, \frac{0.1}{\text{Ana}}, \frac{0.5}{\text{Sara}}\}$
(young, short)	$\{\frac{0.7}{\text{Ben}}, \frac{0.2}{\text{Sue}}, \frac{0.4}{\text{Tom}}, \frac{0}{\text{John}}, \frac{0.7}{\text{Stan}}, \frac{0.4}{\text{Bill}}, \frac{0.2}{\text{Kim}}, \frac{0.7}{\text{Ana}}, \frac{0.1}{\text{Sara}}\} \times \{\frac{0.1}{\text{Ben}}, \frac{0}{\text{Sue}}, \frac{0.9}{\text{Tom}}, \frac{0.4}{\text{John}}, \frac{0.1}{\text{Stan}}, \frac{0}{\text{Bill}}, \frac{0.2}{\text{Kim}}, \frac{0}{\text{Ana}}, \frac{0.1}{\text{Sara}}\}$
(middle age, tall)	$\{\frac{0.3}{\text{Ben}}, \frac{0.8}{\text{Sue}}, \frac{0.7}{\text{Tom}}, \frac{0.3}{\text{John}}, \frac{0.3}{\text{Stan}}, \frac{0.7}{\text{Bill}}, \frac{0.8}{\text{Kim}}, \frac{0.3}{\text{Ana}}, \frac{0.5}{\text{Sara}}\} \times \{\frac{0.8}{\text{Ben}}, \frac{0.6}{\text{Sue}}, \frac{0}{\text{Tom}}, \frac{0.1}{\text{John}}, \frac{0.8}{\text{Stan}}, \frac{0.6}{\text{Bill}}, \frac{0.5}{\text{Kim}}, \frac{0.7}{\text{Ana}}, \frac{0.5}{\text{Sara}}\}$
(middle age, medium)	$\{\frac{0.3}{\text{Ben}}, \frac{0.8}{\text{Sue}}, \frac{0.7}{\text{Tom}}, \frac{0.3}{\text{John}}, \frac{0.3}{\text{Stan}}, \frac{0.7}{\text{Bill}}, \frac{0.8}{\text{Kim}}, \frac{0.3}{\text{Ana}}, \frac{0.5}{\text{Sara}}\} \times \{\frac{0.3}{\text{Ben}}, \frac{0.1}{\text{Sue}}, \frac{0.1}{\text{Tom}}, \frac{0.6}{\text{John}}, \frac{0.3}{\text{Stan}}, \frac{0.1}{\text{Bill}}, \frac{0.8}{\text{Kim}}, \frac{0.1}{\text{Ana}}, \frac{0.5}{\text{Sara}}\}$
(middle age, short)	$\{\frac{0.3}{\text{Ben}}, \frac{0.8}{\text{Sue}}, \frac{0.7}{\text{Tom}}, \frac{0.3}{\text{John}}, \frac{0.3}{\text{Stan}}, \frac{0.7}{\text{Bill}}, \frac{0.8}{\text{Kim}}, \frac{0.3}{\text{Ana}}, \frac{0.5}{\text{Sara}}\} \times \{\frac{0.1}{\text{Ben}}, \frac{0}{\text{Sue}}, \frac{0.9}{\text{Tom}}, \frac{0.4}{\text{John}}, \frac{0.1}{\text{Stan}}, \frac{0}{\text{Bill}}, \frac{0.2}{\text{Kim}}, \frac{0}{\text{Ana}}, \frac{0.1}{\text{Sara}}\}$
(old, tall)	$\{\frac{0.1}{\text{Ben}}, \frac{0.2}{\text{Sue}}, \frac{0.1}{\text{Tom}}, \frac{0.9}{\text{John}}, \frac{0.1}{\text{Stan}}, \frac{0.1}{\text{Bill}}, \frac{0.2}{\text{Kim}}, \frac{0.1}{\text{Ana}}, \frac{0.5}{\text{Sara}}\} \times \{\frac{0.8}{\text{Ben}}, \frac{0.6}{\text{Sue}}, \frac{0}{\text{Tom}}, \frac{0.1}{\text{John}}, \frac{0.8}{\text{Stan}}, \frac{0.6}{\text{Bill}}, \frac{0.5}{\text{Kim}}, \frac{0.7}{\text{Ana}}, \frac{0.5}{\text{Sara}}\}$
(old, medium)	$\{\frac{0.1}{\text{Ben}}, \frac{0.2}{\text{Sue}}, \frac{0.1}{\text{Tom}}, \frac{0.9}{\text{John}}, \frac{0.1}{\text{Stan}}, \frac{0.1}{\text{Bill}}, \frac{0.2}{\text{Kim}}, \frac{0.1}{\text{Ana}}, \frac{0.5}{\text{Sara}}\} \times \{\frac{0.3}{\text{Ben}}, \frac{0.1}{\text{Sue}}, \frac{0.1}{\text{Tom}}, \frac{0.6}{\text{John}}, \frac{0.3}{\text{Stan}}, \frac{0.1}{\text{Bill}}, \frac{0.8}{\text{Kim}}, \frac{0.1}{\text{Ana}}, \frac{0.5}{\text{Sara}}\}$
(old, short)	$\{\frac{0.1}{\text{Ben}}, \frac{0.2}{\text{Sue}}, \frac{0.1}{\text{Tom}}, \frac{0.9}{\text{John}}, \frac{0.1}{\text{Stan}}, \frac{0.1}{\text{Bill}}, \frac{0.2}{\text{Kim}}, \frac{0.1}{\text{Ana}}, \frac{0.5}{\text{Sara}}\} \times \{\frac{0.1}{\text{Ben}}, \frac{0}{\text{Sue}}, \frac{0.9}{\text{Tom}}, \frac{0.4}{\text{John}}, \frac{0.1}{\text{Stan}}, \frac{0}{\text{Bill}}, \frac{0.2}{\text{Kim}}, \frac{0}{\text{Ana}}, \frac{0.1}{\text{Sara}}\}$

Table 13 The values of $(\tilde{f} \times^c \tilde{g})((p_1, p_2))(o)$ with minimum

(p_1, p_2)	o								
	Ben	Sue	Tom	John	Stan	Bill	Kim	Ana	Sara
(young, tall)	0.8	0.6	0.4	0.1	0.8	0.6	0.5	0.7	0.5
(young, medium)	0.7	0.2	0.4	0.6	0.7	0.4	0.8	0.7	0.5
(young, short)	0.7	0.2	0.9	0.4	0.7	0.4	0.2	0.7	0.1
(middle age, tall)	0.8	0.8	0.7	0.3	0.8	0.7	0.8	0.7	0.5
(middle age, medium)	0.3	0.8	0.7	0.6	0.3	0.7	0.8	0.3	0.5
(middle age, short)	0.3	0.8	0.9	0.4	0.3	0.7	0.8	0.3	0.5
(old, tall)	0.8	0.6	0.1	0.9	0.8	0.6	0.5	0.7	0.5
(old, medium)	0.3	0.2	0.1	0.9	0.3	0.1	0.8	0.1	0.5
(old, short)	0.1	0.2	0.9	0.9	0.1	0.1	0.2	0.1	0.5

Table 14 The values of $(\tilde{f} \times^c \tilde{g})((p_1, p_2))(o)$ with product

(p_1, p_2)	o								
	Ben	Sue	Tom	John	Stan	Bill	Kim	Ana	Sara
(young, tall)	0.56	0.12	0.0	0.0	0.56	0.24	0.1	0.49	0.05
(young, medium)	0.21	0.02	0.04	0.0	0.21	0.04	0.01	0.49	0.05
(young, short)	0.07	0.0	0.36	0.0	0.07	0.0	0.04	0.0	0.01
(middle age, tall)	0.21	0.24	0.0	0.03	0.24	0.42	0.4	0.21	0.25
(middle age, medium)	0.09	0.08	0.07	0.18	0.09	0.07	0.16	0.03	0.25
(middle age, short)	0.03	0.0	0.63	0.12	0.03	0.0	0.16	0.0	0.05
(old, tall)	0.08	0.12	0.0	0.09	0.08	0.06	0.10	0.07	0.25
(old, medium)	0.03	0.02	0.01	0.54	0.03	0.01	0.16	0.01	0.25
(old, short)	0.1	0.0	0.09	0.36	0.01	0.0	0.04	0.0	0.05

$$B^{\Pi^d_{i\in I_n}\tilde{f}_i} = \{((p_1, p_2, \ldots, p_n), \delta, card(O_\delta^{p_1,p_2,\ldots,p_n}))|p_i \in P_i, \delta \in L\},$$

where $O_\delta^{p_1,p_2,\ldots,p_n} = \{o \in \overline{O}|(\Pi^d_{i\in I_n}\tilde{f}_i)((p_1, p_2, \ldots, p_n))(o) = \delta\}$.

Note that by Definition 6, $\Pi^d_{i\in I_n}\tilde{\mathcal{B}}^{\tilde{f}_i} = \tilde{\mathcal{B}}^{\Pi^d_{i\in I_n}\tilde{f}_i}$.

Remark 5 Definitions 19 and 20 can be defined with t-norm T or t-conorm S, see [7], instead of minimum or maximum, respectively, i.e. we can consider $T(\tilde{f}_1(p_1)(o), \ldots, \tilde{f}_n(p_n)(o))$ and $S(\tilde{f}_1(p_1)(o), \ldots, \tilde{f}_n(p_n)(o))$ for all $o \in O$, respectively.

Example 13 Consider the fuzzy bags of Examples 3 and 4. The C_n^c-L-fuzzy bag of $\tilde{\mathcal{B}}^{\tilde{f}}$ and $\tilde{\mathcal{B}}^{\tilde{g}}$ is $\tilde{\mathcal{B}}^{\tilde{f}\times^c\tilde{g}} = (\tilde{f} \times^c \tilde{g}, B^{\tilde{f}\times^c\tilde{g}})$, where the values of $(\tilde{f} \times^c \tilde{g})((p_1, p_2))$ for three different t-norms are given in Tables 13, 14 and 15. It is easy to write $\tilde{B}^{\tilde{f}\times^d\tilde{g}}$ using the tables.

Definition 21 ([10]) Fix C_n- L-fuzzy bag $\mathcal{B}^{\Pi_{i\in I_n}\tilde{f}_i}$. An $n-$ary L-fuzzy bag relation is a L-fuzzy sub bag of C_n- L-fuzzy bag $\mathcal{B}^{\Pi_{i\in I_n}\tilde{f}_i}$ which is denoted by $\tilde{\mathcal{B}}^{\tilde{f}_R} = (\tilde{f}_R, B^{\tilde{f}_R})$.

Table 15 The values of $(\tilde{f} \times^c \tilde{g})((p_1, p_2))(o)$ with Lukasiewicz t-norm

(p_1, p_2)	o								
	Ben	Sue	Tom	John	Stan	Bill	Kim	Ana	Sara
(young, tall)	0.5	0.0	0.0	0.0	0.5	0.0	0.0	0.4	0.0
(young, medium)	0.0	0.0	0.0	0.0	0.0	0.0	0.0	0.0	0.0
(young, short)	0.0	0.0	0.3	0.0	0.0	0.0	0.0	0.0	0.0
(middle age, tall)	0.1	0.4	0.0	0.0	0.1	0.3	0.3	0.0	0.0
(middle age, medium)	0.0	0.0	0.0	0.0	0.0	0.0	0.6	0.0	0.0
(middle age, short)	0.0	0.0	0.6	0.0	0.0	0.0	0.0	0.0	0.0
(old, tall)	0.0	0.0	0.0	0.0	0.0	0.0	0.0	0.0	0.0
(old, medium)	0.0	0.0	0.0	0.5	0.0	0.0	0.0	0.0	0.0
(old, short)	0.0	0.0	0.0	0.3	0.0	0.0	0.0	0.0	0.0

Definition 22 ([10]) Let $\tilde{R}_n \subseteq \Pi_{i\in I_n}\mathcal{F}_L(O_i)$ and $\tilde{l}_n : \Pi_{i\in I_n} P_i \to \Pi_{i\in I_n}\mathcal{F}_L(O_i)$. Then, $\mathcal{B}^{\tilde{l}_n}_{\tilde{R}_n} = (\tilde{l}_n, B^{\tilde{l}_n}_{\tilde{R}_n})$ is the fuzzy bag induced by $\tilde{R}_n$ and $\tilde{l}_n$. Where,

$$B^{\tilde{l}_n}_{\tilde{R}_n} = \{((p_1, \ldots, p_n), \delta, card(O^{p_1,\ldots,p_n}_{\delta,\tilde{l}_n}))|\ p_i \in P_i,\ i \in I_n, \delta \in L\}$$

and $O^{p_1,\ldots,p_n}_{\delta,\tilde{l}_n} = \{(o_1, \ldots, o_n) \in \Pi_{i\in I_n} O_i|\tilde{l}((p_1, \ldots, p_n))(o_1, \ldots, o_n) = \delta, \tilde{l}_n$ $((p_1, \ldots, p_n)) \in \tilde{R}_n\}$.

Example 14 ([10]) Let $L = [0, 1]$, O be as in Example 4 and $R_2 = \{(o_1, o_2)|o_1, o_2 \in O, o_1 = o_2\}$. Now, let Table 16 gives $\tilde{l}_2 : P_1 \times P_2 \to \tilde{R}_2 \subseteq \mathcal{F}(O)^2$, where $P_1 = $ {young, middle age, old} and $P_2 = $ {tall, medium, short}.

Thus, by Definition 22, we can present this information by the fuzzy bag $\mathcal{B}^{\tilde{l}_2}_{\tilde{R}_2} = (\tilde{l}_2, B^{\tilde{l}_2}_{\tilde{R}_2})$, where $B^{\tilde{l}_2}_{\tilde{R}_2}$ can be easily given using information of Table 16.

4 Alpha-Cuts of L-Fuzzy Bags

The notion of α-cut plays a fairly big role in the fuzzy theory. So, in this section, we define this notion for the bags. Here are some notations.

Notation 4 ([12]) If $\alpha \in L$, then $\uparrow \alpha = \{c \in L|c \geq \alpha\}$. Thus, $\uparrow$ is a mapping from L into $\mathcal{P}(L)$ and $\uparrow \alpha$ is called the up set of α.

Definition 23 ([10]) The α-cut of an L-fuzzy bag $\tilde{\mathcal{B}}^{\tilde{f}} = (\tilde{f}, B^{\tilde{f}}) \in \tilde{\mathbf{B}}_L(P, O)$ is defined as the crisp bag $(\tilde{\mathcal{B}}^{\tilde{f}})_\alpha = (\tilde{f}_\alpha, B^{\tilde{f}_\alpha})$, where for all $p \in P$

$$\tilde{f}_\alpha(p) = \tilde{f}(p)^{-1}(\uparrow \alpha),$$

for all $\alpha \in L$.

Table 16 Function $\tilde{l}_2$

(p_1, p_2)	$\tilde{l}_2((p_1, p_2))(o_1, o_2)$
(young, tall)	$\{\frac{0.7}{(\text{Ben,Ben})}, \frac{0.2}{(\text{Sue,Sue})}, \frac{0}{(\text{Tom,Tom})}, \frac{0}{(\text{John,John})}, \frac{0.7}{(\text{Stan,Stan})}, \frac{0.4}{(\text{Bill,Bill})}, \frac{0.2}{(\text{Kim,Kim})}, \frac{0.7}{(\text{Ana,Ana})}, \frac{0.1}{(\text{Sara,Sara})}\}$
(young, medium)	$\{\frac{0.3}{(\text{Ben,Ben})}, \frac{0.1}{(\text{Sue,Sue})}, \frac{0.1}{(\text{Tom,Tom})}, \frac{0}{(\text{John,John})}, \frac{0.3}{(\text{Stan,Stan})}, \frac{0.1}{(\text{Bill,Bill})}, \frac{0.2}{(\text{Kim,Kim})}, \frac{0.1}{(\text{Ana,Ana})}, \frac{0.1}{(\text{Sara,Sara})}\}$
(young, short)	$\{\frac{0.1}{(\text{Ben,Ben})}, \frac{0}{(\text{Sue,Sue})}, \frac{0.4}{(\text{Tom,Tom})}, \frac{0}{(\text{John,John})}, \frac{0.1}{(\text{Stan,Stan})}, \frac{0}{(\text{Bill,Bill})}, \frac{0.2}{(\text{Kim,Kim})}, \frac{0}{(\text{Ana,Ana})}, \frac{0.1}{(\text{Sara,Sara})}\}$
(middle age, tall)	$\{\frac{0.3}{(\text{Ben,Ben})}, \frac{0.6}{(\text{Sue,Sue})}, \frac{0}{(\text{Tom,Tom})}, \frac{0.1}{(\text{John,John})}, \frac{0.3}{(\text{Stan,Stan})}, \frac{0.6}{(\text{Bill,Bill})}, \frac{0.5}{(\text{Kim,Kim})}, \frac{0.3}{(\text{Ana,Ana})}, \frac{0.5}{(\text{Sara,Sara})}\}$
(middle age, medium)	$\{\frac{0.3}{(\text{Ben,Ben})}, \frac{0.1}{(\text{Sue,Sue})}, \frac{0.1}{(\text{Tom,Tom})}, \frac{0.3}{(\text{John,John})}, \frac{0.3}{(\text{Stan,Stan})}, \frac{0.1}{(\text{Bill,Bill})}, \frac{0.8}{(\text{Kim,Kim})}, \frac{0.1}{(\text{Ana,Ana})}, \frac{0.5}{(\text{Sara,Sara})}\}$
(middle age, short)	$\{\frac{0.1}{(\text{Ben,Ben})}, \frac{0}{(\text{Sue,Sue})}, \frac{0.7}{(\text{Tom,Tom})}, \frac{0.3}{(\text{John,John})}, \frac{0.1}{(\text{Stan,Stan})}, \frac{0}{(\text{Bill,Bill})}, \frac{0.2}{(\text{Kim,Kim})}, \frac{0}{(\text{Ana,Ana})}, \frac{0.1}{(\text{Sara,Sara})}\}$
(old, tall)	$\{\frac{0.1}{(\text{Ben,Ben})}, \frac{0.2}{(\text{Sue,Sue})}, \frac{0}{(\text{Tom,Tom})}, \frac{0.1}{(\text{John,John})}, \frac{0.1}{(\text{Stan,Stan})}, \frac{0.1}{(\text{Bill,Bill})}, \frac{0.2}{(\text{Kim,Kim})}, \frac{0.1}{(\text{Ana,Ana})}, \frac{0.5}{(\text{Sara,Sara})}\}$
(old, medium)	$\{\frac{0.1}{(\text{Ben,Ben})}, \frac{0.1}{(\text{Sue,Sue})}, \frac{0.1}{(\text{Tom,Tom})}, \frac{0.6}{(\text{John,John})}, \frac{0.1}{(\text{Stan,Stan})}, \frac{0.1}{(\text{Bill,Bill})}, \frac{0.2}{(\text{Kim,Kim})}, \frac{0.1}{(\text{Ana,Ana})}, \frac{0.5}{(\text{Sara,Sara})}\}$
(old, short)	$\{\frac{0.1}{(\text{Ben,Ben})}, \frac{0}{(\text{Sue,Sue})}, \frac{0.1}{(\text{Tom,Tom})}, \frac{0.4}{(\text{John,John})}, \frac{0.1}{(\text{Stan,Stan})}, \frac{0}{(\text{Bill,Bill})}, \frac{0.2}{(\text{Kim,Kim})}, \frac{0}{(\text{Ana,Ana})}, \frac{0.1}{(\text{Sara,Sara})}\}$

Theorem 5 ([10]) *Let* $\tilde{\mathcal{B}}^{\tilde{f}}$ *and* $\tilde{\mathcal{B}}^{\tilde{g}} \in \tilde{\boldsymbol{B}}_L(P, O)$. *If* $\mathcal{B}^{\tilde{f}_\alpha} = \mathcal{B}^{\tilde{g}_\alpha}$ *for all* $\alpha \in L$, *then* $\tilde{\mathcal{B}}^{\tilde{f}} = \tilde{\mathcal{B}}^{\tilde{g}}$.

Thus, we have the following situation. A function $\tilde{f}(p) : O \to L$ induces a function $\tilde{f}(p)^{-1} \uparrow : L \to \mathcal{P}(O)$. We already know from the Theorem 5 that associating $\tilde{f}(p)$ with the function $\tilde{f}(p)^{-1} \uparrow$ is an injection.

Theorem 6 ([10]) *Let L be a complete lattice,* $\mathcal{F}_L(O)$ *be the set of all mappings from O to L, and* $\mathcal{L}(O)$ *be the set of all mappings* $g : L \to \mathcal{P}(O)$ *such that for all subsets D of L,*

$$g(\vee D) = \cap_{d \in D} g(d).$$

Then, the mapping $\Phi : \mathcal{F}_L(O) \to \mathcal{L}(O)$ *given by* $\Phi(\tilde{f}(p)) = \tilde{f}(p)^{-1} \uparrow$ *is a bijection.*

In the case of fuzzy bags, we study them more specifically see the following.

Definition 24 ([10]) Let $\alpha \in [0, 1]$. Then, α-cut of fuzzy bag $\tilde{\mathcal{B}}^{\tilde{f}} \in \tilde{\mathbf{B}}(P, O)$ is a crisp bag $(\tilde{\mathcal{B}}^{\tilde{f}})_\alpha = (\tilde{f}_\alpha, B^{\tilde{f}_\alpha})$ where, $\tilde{f}_\alpha : P \to \mathcal{P}(O)$ is a function in which for all $p \in P$, $\tilde{f}_\alpha(p) = \{o \in O | \tilde{f}(p)(o) \geqslant \alpha\}$ and

$$B^{\tilde{f}_\alpha} = \{(p, card(\tilde{f}_\alpha(p))) | p \in P\}.$$

Definition 25 ([10]) Let $\alpha \in [0, 1]$. Then, strong α-cut of fuzzy bag $\tilde{\mathcal{B}}^{\tilde{f}} \in \tilde{\mathbf{B}}(P, O)$ is the crisp bag $(\tilde{\mathcal{B}}^{\tilde{f}})_{\alpha^\cdot} = (\tilde{f}_{\alpha^\cdot}, B^{\tilde{f}_{\alpha^\cdot}})$ where, $\tilde{f}_{\alpha^\cdot} : P \to \mathcal{P}(O)$ is a function which for all $p \in P$, $\tilde{f}_{\alpha^\cdot}(p) = \{o \in O | \tilde{f}(p)(o) > \alpha\}$ and

$$B^{\tilde{f}_{\alpha^\cdot}} = \{(p, card(\tilde{f}_{\alpha^\cdot}(p))) | p \in P\}.$$

Note that by Definition 1, we have $\mathcal{B}^{\tilde{f}_\alpha} = (\tilde{\mathcal{B}}^{\tilde{f}})_\alpha$ and $\mathcal{B}^{\tilde{f}_{\alpha^\cdot}} = (\tilde{\mathcal{B}}^{\tilde{f}})_{\alpha^\cdot}$.

Notation 5 ([10]) For all $p \in P$, we set
$\tilde{f}_{[\alpha,\beta)}(p) = \{o \in O | \alpha \leq \tilde{f}(p)(o) < \beta\}$ and
$\tilde{f}_{(\alpha,\beta]}(p) = \{o \in O | \alpha < \tilde{f}(p)(o) \leq \beta\}$.

Some useful results for the fuzzy bags are given in the next theorem.

Theorem 7 ([10]) *Let* $\tilde{\mathcal{B}}^{\tilde{f}}, \tilde{\mathcal{B}}^{\tilde{g}} \in \tilde{\boldsymbol{B}}(P, O)$, $\alpha, \beta \in [0, 1]$ *and* $\alpha \leqslant \beta$.

(i) $\mathcal{B}^{\tilde{f}_{\beta^\cdot}} \tilde{\sqsubseteq} \mathcal{B}^{\tilde{f}_\beta} \tilde{\sqsubseteq} \mathcal{B}^{\tilde{f}_{\alpha^\cdot}} \tilde{\sqsubseteq} \mathcal{B}^{\tilde{f}_\alpha}$,
(ii) $\mathcal{B}^{\tilde{f}_\alpha} = \mathcal{B}^{\tilde{f}_\beta}$ *if and only if* $\mathcal{B}^{\tilde{f}_{[\alpha,\beta)}} = \mathcal{B}^0$,
(iii) $\mathcal{B}^{\tilde{f}_{\alpha^\cdot}} = \mathcal{B}^{\tilde{f}_{\beta^\cdot}}$ *if and only if* $\mathcal{B}^{\tilde{f}_{(\alpha,\beta]}} = \mathcal{B}^0$,
(iv) $(\tilde{\mathcal{B}}^{\tilde{f}} \cup \tilde{\mathcal{B}}^{\tilde{g}})_\alpha = \mathcal{B}^{\tilde{f}_\alpha} \cup \mathcal{B}^{\tilde{g}_\alpha}$ *and* $(\tilde{\mathcal{B}}^{\tilde{f}} \cup \tilde{\mathcal{B}}^{\tilde{g}})_{\alpha^\cdot} = \mathcal{B}^{\tilde{f}_{\alpha^\cdot}} \cup \mathcal{B}^{\tilde{g}_{\alpha^\cdot}}$,
(v) $(\tilde{\mathcal{B}}^{\tilde{f}} \cap \tilde{\mathcal{B}}^{\tilde{g}})_\alpha = \mathcal{B}^{\tilde{f}_\alpha} \cap \mathcal{B}^{\tilde{g}_\alpha}$ *and* $(\tilde{\mathcal{B}}^{\tilde{f}} \cap \tilde{\mathcal{B}}^{\tilde{g}})_{\alpha^\cdot} = \mathcal{B}^{\tilde{f}_{\alpha^\cdot}} \cap \mathcal{B}^{\tilde{g}_{\alpha^\cdot}}$.

In the following example, we compute α-cuts of a fuzzy bag.

Table 17 The values of $\tilde{f}_\alpha(p)$ for Example 15

α	p		
	Young	Middle age	Old
0.0	O	O	O
0.1	O \ {John}	O	O
0.2	O \ {John, Sara}	O	{Sue, John, Kim, Sara}
0.3	O \ {Sue, John, Kim, Sara}	O	{John, Sara}
0.4	O \ {Sue, John, Kim, Sara}	{Sue, Tom, Bill, Kim, Sara}	{John, Sara}
0.5	{Ben, Stan, Ana}	{Sue, Tom, Bill, Kim, Sara}	{John, Sara}
0.7	{Ben, Stan, Ana}	{Sue, Tom, Bill, Kim}	{John}
0.8	∅	{Sue, Kim}	{John}
0.9	∅	∅	{John}

Example 15 ([10]) Consider the fuzzy bag of Example 3. We compute α-cuts, $\mathcal{B}^{\tilde{f}_\alpha} = (\tilde{f}_\alpha, B^{\tilde{f}_\alpha})$. Where, $\tilde{f}_\alpha(p)$ is presented in Table 17 and $B^{\tilde{f}_\alpha}$ is as follows

$$B^{\tilde{f}_0} = \{(\text{young}, 9), (\text{middle age}, 9), (\text{old}, 9)\}, \quad B^{\tilde{f}_{0.1}} = \{(\text{young}, 8), (\text{middle age}, 9), (\text{old}, 9)\},$$
$$B^{\tilde{f}_{0.2}} = \{(\text{young}, 7), (\text{middle age}, 9), (\text{old}, 4)\}, \quad B^{\tilde{f}_{0.3}} = \{(\text{young}, 5), (\text{middle age}, 9), (\text{old}, 2)\},$$
$$B^{\tilde{f}_{0.4}} = \{(\text{young}, 5), (\text{middle age}, 5), (\text{old}, 2)\}, \quad B^{\tilde{f}_{0.5}} = \{(\text{young}, 3), (\text{middle age}, 5), (\text{old}, 2)\},$$
$$B^{\tilde{f}_{0.7}} = \{(\text{young}, 3), (\text{middle age}, 4), (\text{old}, 1)\}, \quad B^{\tilde{f}_{0.8}} = \{(\text{middle age}, 2), (\text{old}, 1)\},$$
$$B^{\tilde{f}_{0.9}} = \{(\text{old}, 1)\}.$$

Definition 26 ([10]) Let $\mathcal{B}^f \in \mathbf{B}(P, O)$ and $\alpha \in [0, 1]$. We define fuzzy bag $\widetilde{\alpha\mathcal{B}^f} = \tilde{\mathcal{B}}^{\widetilde{\alpha f}} = (\widetilde{\alpha f}, B^{\widetilde{\alpha f}})$. Where,

$$\widetilde{\alpha f}(p)(o) = \min(\alpha, \chi_{f(p)}(o)) = \alpha\chi_{f(p)}(o),$$

for all $o \in O$ and $p \in P$.

Theorem 8 ([10]) *Let $\tilde{\mathcal{B}}^{\tilde{f}}$ be a fuzzy bag and let $\mathcal{B}^{\tilde{f}_\alpha}$ be α-cut of $\tilde{\mathcal{B}}^{\tilde{f}}$. Then,*

$$\tilde{\mathcal{B}}^{\tilde{f}} = \bigcup_{\alpha\in[0,1]} \widetilde{\alpha\mathcal{B}^{\tilde{f}_\alpha}}.$$

Theorem 9 ([10]) *Let $\tilde{\mathcal{B}}^{\tilde{f}}$ be a fuzzy bag and let $\mathcal{B}^{\tilde{f}_{\alpha^\cdot}}$ be a strong α-cut of $\tilde{\mathcal{B}}^{\tilde{f}}$. Then,*

$$\tilde{\mathcal{B}}^{\tilde{f}} = \bigcup_{\alpha\in[0,1]} \widetilde{\alpha\mathcal{B}^{\tilde{f}_{\alpha^\cdot}}}.$$

Theorem 10 ([10]) *Let* $\tilde{\mathcal{B}}^{\tilde{f}} \in \tilde{\boldsymbol{B}}(P, O)$ *and* $\{\mathcal{B}^{\tilde{g}_\alpha} | \alpha \in [0, 1]\}$ *be a class of elements of* $\boldsymbol{B}(P, O)$ *such that* $\mathcal{B}^{\tilde{f}_{\alpha^\cdot}} \sqsubseteq \mathcal{B}^{\tilde{g}_\alpha} \sqsubseteq \mathcal{B}^{\tilde{f}_\alpha}$. *Then,*

$$\tilde{\mathcal{B}}^{\tilde{f}} = \bigcup_{\alpha \in [0,1]} \widetilde{\alpha \mathcal{B}^{\tilde{g}_\alpha}}.$$

Theorem 11 ([10]) *Let* $\{\mathcal{B}^{g_\alpha} | \alpha \in [0, 1]\}$ *be a class of elements of* $\boldsymbol{B}(P, O)$. *There exists* $\tilde{\mathcal{B}}^{\tilde{f}} \in \tilde{\boldsymbol{B}}(P, O)$ *such that for all* $\alpha \in [0, 1]$, $\mathcal{B}^{\tilde{f}_\alpha} = \mathcal{B}^{g_\alpha}$ *if and only if for all* $\alpha, \beta \in [0, 1]$ *such that* $\alpha \leq \beta$, $\mathcal{B}^{g_\beta} \sqsubseteq \mathcal{B}^{g_\alpha}$ *and* $\mathcal{B}^{g_0} = \mathcal{B}^1$.

So far, L-fuzzy bags and some basic concepts relevant to them are given. In the next section, the algebraic structure of the L-fuzzy bags is studied.

5 Algebraic Structure of Bags and L-Fuzzy Bags

In this section, we study the algebraic structure of bags and L-fuzzy bags. Let $\sqsubseteq$ and $\mathbf{B}(P, O)$ be as in Notation 1 and Definition 4. We have the following results. For terminology of this section, see [12].

Corollary 1 ([9]) $(\boldsymbol{B}(P, O), \cup, \cap, ^c, \mathcal{B}^0, \mathcal{B}^1)$ *is a De Morgan algebra.*

Theorem 12 ([9]) $(\boldsymbol{B}(P, O), \cup, \cap, ^c, \mathcal{B}^0, \mathcal{B}^1)$ *is a complete Boolean algebra.*

Hence, the set of all bags equipped with the proposed order is a complete Boolean algebra. Now, let $\tilde{\mathbf{B}}_L(P, O)$ and $\tilde{\sqsubseteq}$ be as in Notation 2 and Definition 9.

Theorem 13 ([10]) $(\tilde{\boldsymbol{B}}_L(P, O), \tilde{\sqsubseteq})$ *is a bounded distributive lattice.*

Definition 27 ([12]) Let X be a bounded lattice and let $x \in X$. Then, an element x^* is a pseudocomplement of x if $x \wedge x^* = 0$ and $y \leq x^*$ whenever $x \wedge y = 0$. That is, for each $x \in X$, there is a unique largest element whose meet with x is 0.

Theorem 14 ([10]) $(\tilde{\boldsymbol{B}}_L(P, O), \cup, \cap, \mathcal{B}^0, \mathcal{B}^1)$ *is pseudocomplemented.*

An element in a bounded lattice has at most one pseudocomplement since two pseudocomplements must each be less or equal to the other and hence equal. If every element has a pseudocomplement then the bounded lattice is pseudocomplemented and the unary operation * is called a pseudocomplement. The equation $x^* \vee x^{**} = 1$ is called Stone's identity and a Stone algebra is a pseudocomplemented distributive lattice satisfying this identity [12].

Definition 28 ([12]) If $(S, \vee, \wedge, ^*, 0, 1)$ is a Stone algebra, then for $S^* = \{s^* \in S | s \in S\}$, $(S^*, \vee, \wedge, ^*, 0, 1)$ is a Boolean algebra and it is called center of S.

Theorem 15 ([10]) $(\tilde{\boldsymbol{B}}_L(P, O), \cup, \cap, ^*, \mathcal{B}^0, \mathcal{B}^1)$ *is a Stone algebra whose center consists of the bags in* $\boldsymbol{B}(P, O)$.

Here, consider $F_L(O)^n$ as in Notation 3. We have the following order on it.

Definition 29 ([10]) Let $(\tilde{A}_1, \tilde{A}_2, \ldots, \tilde{A}_n), (\tilde{B}_1, \tilde{B}_2, \ldots, \tilde{B}_n) \in F_L(O)^n$. Then, $(\tilde{A}_1, \tilde{A}_2, \ldots, \tilde{A}_n) \tilde{\preceq} (\tilde{B}_1, \tilde{B}_2, \ldots, \tilde{B}_n)$ if and only if $\tilde{A}_i \tilde{\subseteq} \tilde{B}_i$ for all $i \in I_n$.

Note that since $\mathcal{F}_L(O)$ is a Stone algebra, $\mathcal{F}_L(O)^n$ is so as well [12]. The next theorem shows that the lattice of all L-fuzzy bags is isomorphic to the lattice of n-Cartesian product of $F_L(O)$.

Theorem 16 ([10]) *$\tilde{\boldsymbol{B}}_L(P, O)$ is isomorphic to $F_L(O)^n$, where $n = card(P)$.*

Now, let $\tilde{\mathcal{B}}^{\tilde{f}}, \tilde{\mathcal{B}}^{\tilde{g}} \in \tilde{\mathbf{B}}_L(P, O)$ and $n = card(P)$. Then, by Definition 9, $\tilde{\mathcal{B}}^{\tilde{f}} \tilde{\sqsubseteq} \tilde{\mathcal{B}}^{\tilde{g}}$ if and only if $\tilde{f}(p) \tilde{\subseteq} \tilde{g}(p)$ for all $p \in P$. That means $\tilde{\mathcal{B}}^{\tilde{f}} \tilde{\sqsubseteq} \tilde{\mathcal{B}}^{\tilde{g}}$ if and only if $\tilde{f}(p_i) \tilde{\subseteq} \tilde{g}(p_i)$ for all $i \in I_n$ if and only if $\Psi(\tilde{\mathcal{B}}^{\tilde{f}}) \tilde{\preceq} \Psi(\tilde{\mathcal{B}}^{\tilde{g}})$. Since Ψ is one to one and onto, Ψ is an order isomorphism and thus a lattice isomorphism. So, $\tilde{\mathbf{B}}_L(P, O)$ is a Stone algebra as already observed in Theorem 15.

6 Concluding Remarks

The notions of bags, L-fuzzy bags and some of their applications in which L is a complete lattice have been given. Furthermore, the concepts of α-cuts, (L-fuzzy) bag relations and related theorems were given and by some examples, these definitions have been illustrated. Finally, the algebraic structure of bags and L-fuzzy bags have been studied.

Acknowledgements This chapter is dedicated to Prof. Miyamoto. The major part was completed during reward stay of F. Kouchakinejad at Department of Mathematics and Descriptive Geometry, Faculty of Civil Engineering of Slovak University of Technology in Bratislava on the supervision of R. Mesiar. The first author acknowledges the financial support received from the Ministry of Science, Research and Technology of the Islamic Republic of Iran. The work of the third author was supported by the grant APVV-14-0013.

References

1. Beliakov G, Pradera A, Calvo T. Aggregation Functions: A Guide for Practitioners. Berlin, Heidelberg: Springer; 2007.
2. Biswas R. An application of Yager's bag theory in multicriteria based decision making problems. Int J Intell Syst 1999; 14:1231–1238.
3. Chakrabarty K. IF-Bags and Knowledge Representation in Soft Decision Analysis. Int J Uncertain Fuzz 2014; 22:783–790.
4. Delgado M, Martin Bautista MJ, Sanchez D, Vila MA. An extended characterization of fuzzy bags. Int J Intell Syst 2009; 24:706–721.
5. Delgado M, Ruiz MD, Sanchez D. RL-bags: A conceptual, level-based approach to fuzzy bags. Fuzzy Sets Syst 2012; 208:111–128.

6. Han J, Kamber M, Pei J. Data Mining: Concepts and Techniques. 3rd edition. Morgan Kaufmann; 2011.
7. Klement EP, Mesiar R, Pap E. Triangular Norms. Kluwer, Dordrecht; 2000.
8. Kosters WA, Laros JFJ. Metrics for mining multisets. in: 27th SGAI International Conference on Innovative Techniques and Applications of Artificial Intelligence. 2007; 293–303.
9. Kouchakinejad F, Mashinchi M. Algebraic structure of bags and fuzzy bags. The 46th Annual Iranian Mathematics Conference: Yazd University. Yazd. Iran. 2015; 1295–1298.
10. Kouchakinejad F, Mashinchi M, Mesiar R. On algebraic structure of L-fuzzy bags. Int J Intell Syst 2016; in press.
11. Meghabghab G. Crisp bags, one-dimensional fuzzy bags, multidimensional fuzzy bags: Comparing successful and unsuccessful user's Web behavior. Int J Intell Syst 2009; 649–676.
12. Nguyen HT, Walker EA. A first course in fuzzy logic. Boca Raton, Florida: CRC Press; 2006.
13. Paun G, Perez-Jimenez MJ. Membrane computing: brief introduction, recent results and applications. Bio Systems 2006; 85:11–22.
14. Rebai A. BBTOPSIS: A bag based technique for order preference by similarity to ideal solution. Fuzzy Sets Syst 1993; 60:143–162.
15. Rebai A, Martel J. A fuzzy bag approach to choosing the "best" multiattributed potential actions in a multiple judgement and non cardinal data context. Fuzzy Sets Syst 1997; 87:159–166.
16. Rocacher D. On fuzzy bags and their application to flexible querying. Fuzzy Sets Syst 2003; 140:93–110.
17. Syropoulos A. On generalized fuzzy multisets and their use in computation. Iranian J Fuzzy Syst 2012; 9:113–125.
18. Torra V, Miyamoto S. On the consistency of a Fuzzy C-Means algorithm for multisets. Proc. CCIA 2005.
19. Yager RR. On the theory of bags. Int J General Syst 1986; 13:23–37.
20. Zeratsky K. "Blood type diet: What is it? Does it work?". http://web.archive.org/web/20110612124046/http://www.mayoclinic.com/health/blood-type-diet/AN01415.

A Perspective on Differences Between Atanassov's Intuitionistic Fuzzy Sets and Interval-Valued Fuzzy Sets

Eulalia Szmidt and Janusz Kacprzyk

Abstract In the paper we show our perspective on some differences between Atanassov's intuitionistic fuzzy sets (A-IFSs, for short) and Interval-valued fuzzy sets (IVFSs, for short). First, we present some standard operators and extensions for the A-IFSs which have no counterparts for IVFSs. Next, we show on an example a practical application based on one of such operators. We also revisit, and further analyze, the concepts of two possible representations of A-IFSs: the two term one, in which the degrees of membership and non-membership are only involved, and the three term one, in which in addition to the above degrees of membership and non-membership the so called hesitation margin is explicitly accounted for. Though both representations are mathematically correct and may be considered equivalent, the second one involves explicitly an additional, conceptually different information than the degree of membership and non-membership only even if it directly results from these two degrees. We then show on some examples of decision making type problems its intuitive appeal and usefulness for reflecting more sophisticated intentions and preferences of the user which cannot be fully reflected via their counterpart IVFSs based models. Finally, we recall different measures that are important from the point of view of applications. We consider the measures for both types representations of the A-IFSs pointing out some further differences in comparison to the case of the IVFSs.

Keywords Intuitionistic fuzzy sets · Interval-valued fuzzy sets · Operators · Representation · Measures

E. Szmidt (✉) · J. Kacprzyk
Systems Research Institute, Polish Academy of Sciences,
ul. Newelska 6, 01-447 Warsaw, Poland
e-mail: szmidt@ibspan.waw.pl

E. Szmidt · J. Kacprzyk
Warsaw School of Information Technology, ul. Newelska 6, 01-447 Warsaw, Poland
e-mail: kacprzyk@ibspan.waw.pl

V. Torra et al. (eds.), *Fuzzy Sets, Rough Sets, Multisets and Clustering*,
Studies in Computational Intelligence 671, DOI 10.1007/978-3-319-47557-8_13

1 Introduction

Atanassov's intuitionistic fuzzy sets (Atanassov [1, 5, 7]), which are a generalization of fuzzy sets (Zadeh [57]), have found numerous applications in various areas. Their ability to better reflect human intentions, preferences, constraints, etc. is due to an ability to reflect an additional type of uncertainty related to a "pro" and "con" type representation both via a degree of membership, as in the case of traditional fuzzy sets, and also via a degree of non-membership. Such an approach follows other modern approaches based on some sort of bipolarity in judgments, evaluations, etc. (cf. Dubois and Prade [16], Grabisch, Greco and Pirlot [18]. The introduction of both a degree of membership and non-membership – which, by definition, do not need to sum up to 1 – results in the existence of an inherent degree of a lack of knowledge which is equal to how much the sum of the degree of membership and non-membership differs from 1, and which is called a hesitation margin. It is worth emphasizing that though the hesitation margin is a direct result of the degree of membership and non-membership, and can be viewed by a pure mathematical argument as irrelevant, it conveys a very relevant information that can be used in all kind soft human centric applications, especially decision making.

It is noteworthy that the A-IFSs can be considered to be a tool for grasping many aspects of uncertainty that are relevant for our purposes. First, they can represent the fuzziness because of the degree of membership and non-membership. Second, the hesitation margin can represent a lack of knowledge. Finally, the three degrees combined, that is, the membership, non-membership and hesitation, can indirectly represent uncertainty related to repetitive random events (cf. Szmidt and Kacprzyk [28, 29]).

In the literature there are two ways of constructing models using the A-IFS tools and techniques: using the two terms only, i.e. the degree of membership and non-membership (e.g., Atanassov [5]) or all three terms, i.e. the membership and non-membership degree and the hesitation margin (e.g., Szmidt and Kacprzyk [30, 32], Tasseva et al. [56], Atanassov et al. [9], Szmidt and Baldwin [24, 25], Deng and Feng [15], Tan and Zhang [55], Narukawa and Torra [21]). Both ways are equally correct from the mathematical point of view. Just to give an example, while considering distances, which are crucial in many theoretical analyses and applications, no matter if we use two or three terms describing the A-IFSs, all necessary and sufficient conditions of the distance are fulfilled. The reason can be stated as to be related to the fact that these analyses are related more to the syntactic than semantic aspects. However, the situation changes when we are concerned with applications because the use of the three term representation may easily be shown to provide a new insight and hence results; this can be implied by the fact that semantic aspects are more relevant in that case.

In spite of the fact that in the models based on the use of the A-IFSs almost always the degree of membership and non-membership are applied, sometimes the A-IFSs are said to be equal to the interval-valued fuzzy sets (IVFSs) as formally A-IFSs can be presented via the intervals. This may be equivalent from a formal point of view, but differs in terms of the very meaning. This is the motivation of our paper – to show some differences between A-IFSs and IVFSs going beyond a purely formal analysis.

We start from the presentation of some operators and extensions of the A-IFSs which have not counterparts for the IVFSs.

Next, we consider intervals which formally can be employed to represent the A-IFSs. We consider the two term and the three term representations of the A-IFSs with their corresponding interval interpretations. Simple examples show that the three term A-IFS representation, as opposed to the two term A-IFS representation, makes it possible to explicitly consider different "scenarios" of an uncertain future which is not possible for IVFSs. We show that the two term representation of A-IFSs can be expressed by one interval, whereas the three term representation of A-IFSs, while looking for it counterpart in terms of intervals, should be expressed by two intervals (which is obviously different than in the case of the IVFSs).

Next, we examine some selected measures that are important from the decision making point of view. The distance measures were our first object of interest due to their general relevance. The calculation of the distances, in particular by examining the correctness of some Hausdorff distances using the two approaches to represent the A-IFSs, clearly indicates that the three term A-IFS representation is a more reliable choice. Remember that the three term A-IFS representation can be expressed via two intervals, i.e., in a different way than for the IVFSs approach for which only one interval is considered.

Similar conclusions about advantages of the three term representation of the A-IFSs and the influence of each of the three terms while considering the distances, remain valid for the Pearson correlation coefficient (Szmidt and Kacprzyk [39]), Spearman correlation coefficient (Szmidt and Kacprzyk [40]), Kendall correlation coefficient (Szmidt and Kacprzyk [43, 44]), Principal Component Analysis (Szmidt and Kacprzyk [46], Szmidt et al. [48]), ranking procedures (Szmidt and Kacprzyk [35, 36, 38]), construction of a classifier for imbalanced and overlapping classes (cf. Szmidt and Kukier [52–54]), construction of the intuitionistic fuzzy trees (Bujnowski et al. [11]), etc.

It is worth stressing again that not only the three type representation of A-IFSs but also the two type representation differs from IVFSs because: we collect data in a different way, look for different answers, and use different types of operators for both kinds of models.

2 A Brief Introduction to A-IFSs

One of the possible generalizations of a fuzzy set in X (Zadeh [57]) given by

$$A^{'} = \{< x, \mu_{A'}(x) > |x \in X\} \tag{1}$$

where $\mu_{A'}(x) \in [0, 1]$ is the membership function of the fuzzy set $A^{'}$, is an A-IFS (Atanassov [1, 5, 7]) A is given by

$$A = \{< x, \mu_A(x), \nu_A(x) > | x \in X\} \tag{2}$$

where: $\mu_A : X \to [0, 1]$ and $\nu_A : X \to [0, 1]$ such that

$$0 \leq \mu_A(x) + \nu_A(x) \leq 1 \tag{3}$$

and $\mu_A(x)$, $\nu_A(x) \in [0, 1]$ denote a degree of membership and a degree of non-membership of $x \in A$, respectively. (An approach to the assigning memberships and non-memberships for A-IFSs from data is proposed by Szmidt and Baldwin [26]).

Obviously, each fuzzy set may be represented by the following A-IFS: $A = \{< x, \mu_{A'}(x), 1 - \mu_{A'}(x) > | x \in X\}$.

An additional concept for each A-IFS in X, that is not only an obvious result of (2) and (3) but which is also relevant for applications, we will call (Atanassov [5])

$$\pi_A(x) = 1 - \mu_A(x) - \nu_A(x) \tag{4}$$

a *hesitation margin* of $x \in A$ which expresses a lack of knowledge of whether x belongs to A or not (cf. Atanassov [5]). It is obvious that $0 \leq \pi_A(x) \leq 1$, for each $x \in X$.

The hesitation margin turns out to be important while considering the distances (Szmidt and Kacprzyk [27, 30, 32], entropy (Szmidt and Kacprzyk [31, 33]), similarity (Szmidt and Kacprzyk [34]) for the A-IFSs, etc. i.e., the measures that play a crucial role in virtually all information processing tasks (Szmidt [23]).

The hesitation margin turns out to be relevant for applications – in image processing (cf. Bustince et al. [12, 13]), the classification of imbalanced and overlapping classes (cf. Szmidt and Kukier [52–54]), the classification applying intuitionistic fuzzy trees (cf. Bujnowski [11]), group decision making (e.g., [10]), genetic algorithms [22], negotiations, voting and other situations (cf. Szmidt and Kacprzyk papers).

2.1 Two and Three Term Representations of A-IFSs

In our previous works (e.g., Szmidt [23], Szmidt and Kacprzyk [30, 47, 50]) we have analyzed in detail both types of representations of the A-IFSs, i.e., using two and three terms. We have also presented geometrical representations for both approaches. Here, because of limited space, we only remind briefly the idea.

The two term representation of the A-IFSs means that only the membership value μ and non-membership value ν are taken into account assuming (correctly) that we know the value of the third term π from (4) as $\pi(.) = 1 - \mu(.) - \nu(.)$. However, the fact that π can be directly calculated does not mean that it should be omitted in the formulas. It will be elaborated upon later.

The three term representation of the A-IFSs means that the membership value μ, non-membership value ν, and the value of the hesitation margin π are explicitly taken into account.

3 Operations and Extensions Specific to the A-IFSs but Not IVFSs

The A-IFSs can be viewed as a direct generalization of the fuzzy sets, but their theory comprises some modal operators that have no equivalents in the traditional fuzzy logic (cf. Atanassov [5–7]). Over the A-IFSs (and only for them, in particular not for the IVFSs) the modal operators of *necessity* and *possibility* are introduced.

Foe each A-IFS A Atanassov [5–7] introduced:

- the *necessity operator*:

$$\square A = \{\langle x, \mu_A(x), 1 - \mu_A(x)\rangle | x \in X\}, \tag{5}$$

- the *possibility operator*:

$$\diamond A = \{\langle x, 1 - \nu_A(x), \nu_A(x)\rangle | x \in X\}. \tag{6}$$

However, if A is an ordinary fuzzy set, then due to Atanassov [5–7]:

$$\square A = A = \diamond A. \tag{7}$$

The equalities (7) show that these operators do not have analogues in the case of fuzzy sets. On the other hand, for the A-IFSs, (7) is not fulfilled, i.e., the operators do not give the same results. Moreover, these two operators are extended (cf. Atanassov [5]) to many other modal-type operators over the IFSs which have no analogues either in the IVFS theory or in modal logic (cf. Atanassov [5–7]). The first group of the extended modal operators are the following:

$$D_\alpha(A) = \{\langle x, \mu_A(x) + \alpha.\pi_A(x), \nu_A(x) + (1-\alpha).\pi_A(x)\rangle | x \in X\}, \tag{8}$$
$$F_{\alpha,\beta}(A) = \{\langle x, \mu_A(x) + \alpha.\pi_A(x), \nu_A(x) + \beta.\pi_A(x)\rangle | x \in X\} \tag{9}$$
$$G_{\alpha,\beta}(A) = \{\langle x, \alpha.\mu_A(x), \beta.\nu_A(x)\rangle | x \in X\}, \tag{10}$$
$$H_{\alpha,\beta}(A) = \{\langle x, \alpha.\mu_A(x), \nu_A(x) + \beta.\pi_A(x)\rangle | x \in X\}, \tag{11}$$
$$H^*_{\alpha,\beta}(A) = \{\langle x, \alpha.\mu_A(x), \nu_A(x) + \beta.(1 - \alpha.\mu_A(x) - \nu_A(x))\rangle | x \in X\}, \tag{12}$$
$$J_{\alpha,\beta}(A) = \{\langle x, \mu_A(x) + \alpha.\pi_A(x), \beta.\nu_A(x)\rangle | x \in X\}, \tag{13}$$
$$J^*_{\alpha,\beta}(A) = \{\langle x, \mu_A(x) + \alpha.(1 - \mu_A(x) - \beta.\nu_A(x)), \beta.\nu_A(x)\rangle | x \in X\}, \tag{14}$$

where $\alpha, \beta \in [0, 1]$ and $\alpha + \beta \leq 1$.

The above operators are extended to the operators (cf. Atanassov [5, 7]):

$$d_{\alpha}(A) = \{\langle x, \nu_A(x) + \alpha.\pi_A(x), \mu_A(x) + (1-\alpha).\pi_A(x)\rangle | x \in X\}, \quad (15)$$

$$f_{\alpha,\beta}(A) = \{\langle x, \nu_A(x) + \alpha.\pi_A(x), \mu_A(x) + \beta.\pi_A(x)\rangle | x \in X\}, \quad (16)$$

$$g_{\alpha,\beta}(A) = \{\langle x, \alpha.\nu_A(x), \beta.\mu_A(x)\rangle | x \in X\}, \quad (17)$$

$$h_{\alpha,\beta}(A) = \{\langle x, \alpha.\nu_A(x), \mu_A(x) + \beta.\pi_A(x)\rangle | x \in X\}, \quad (18)$$

$$h^*_{\alpha,\beta}(A) = \{\langle x, \alpha.\nu_A(x), \mu_A(x) + \beta.(1 - \alpha.\nu_A(x) - \mu_A(x))\rangle | x \in X\}, \quad (19)$$

$$j_{\alpha,\beta}(A) = \{\langle x, \nu_A(x) + \alpha.\pi_A(x), \beta.\mu_A(x)\rangle | x \in X\}, \quad (20)$$

$$j^*_{\alpha,\beta}(A) = \{\langle x, \nu_A(x) + \alpha.(1 - \nu_A(x) - \beta.\mu_A(x)), \beta.\mu_A(x)\rangle \quad (21)$$

$$P_{\alpha,\beta}(A) = \{\langle x, \max(\alpha, \mu_A(x)), \min(\beta, \nu_A(x))\rangle | x \in X\}, \quad (22)$$

$$Q_{\alpha,\beta}(A) = \{\langle x, \min(\alpha, \mu_A(x)), \max(\beta, \nu_A(x))\rangle | x \in X\}, \quad (23)$$

where $\alpha + \beta \leq 1$.

For other extensions of the modal operators we refer an interested reader to Atanassov's [5, 7] works. What is worth stressing, the presented operators are not considered for the IVFSs.

Another kind of operators defined over the A-IFSs are the topological operators (cf. Atanassov [2]) which are analogous of the operators of *closure* C and *interior* I from topology (e.g. [19]). In [3] relations between the modal and the topological operators over the A-IFSs were studied.

Definition 1 ([2]) For each A-IFS A, we define:

$$C(A) = \{\langle x, K, L\rangle | x \in E\} \quad (24)$$

where:

$$K = \max_{y \in E} \mu_A(y)$$

$$L = \min_{y \in E} \nu_A(y)$$

and

$$I(A) = \{\langle x, k, l\rangle | x \in E\} \quad (25)$$

where:

$$k = \min_{y \in E} \mu_A(y)$$

$$l = \max_{y \in E} \nu_A(y)$$

In the area of IVFSs neither modal nor topological operators exist.

The A-IFSs being a generalization of fuzzy sets are, in turn, generalized, too. One of the generalizations of the A-IFSs are the intuitionistic L-fuzzy sets in which the values of functions μ and ν are elements of some fixed lattice L. Another extension of the A-IFSs are the A-IFSs of type 2, for which (3) is replaced by

$$0 \leq \mu_A(x)^2 + \nu_A(x)^2 \leq 1 \tag{26}$$

Another generalization of the A-IFS are the so called *temporal A-IFSs* (cf. [4]).

It was not our goal to study here either all operations or extensions of the A-IFSs but to stress that the A-IFSs are different from the IVFSs just from this point of view as the respective operators and extensions do not exist for the IVFSs.

Now we will show an example of using one of the A-IFS operators discussed above. It is worth stressing that a counterpart IVFS model would be difficult to derived.

3.1 Using an Extended Modal Operator for the A-IFS as a Natural Way to Obtaining an Intuitionistic Fuzzy Classifier for the Imbalanced Classes

The recognition of imbalanced classes is well known as an important and tough task. Here we will describe briefly an intuitionistic fuzzy classifier making use of one of the extended modal operators for the A-IFSs and data presented via the A-IFSs. After expressing the data by relative frequency distributions, the algorithm presented by Szmidt and Baldwin [24–26] is applied to describe smaller and bigger classes in the space of all attributes. As a result, a data instance e is described as an intuitionistic fuzzy element (all three terms are taken into account: the membership value μ, non-membership value ν, and hesitation margin π), i.e.

$$e : (\mu_e, \nu_e, \pi_e) \tag{27}$$

To enhance the possibility of a proper classification of instances belonging to a smaller class, while training the intuitionistic fuzzy classifier, the hesitation margins are used which assign the (width of) intervals where the unknown values of memberships lie. Namely, the $D_\alpha(A)$ operator (8) is applied making it possible to see as well as possible the elements of the class we are interested in. To be more precise, the values of the hesitation margins are divided so as to better "see" the smaller class, i.e. each instance e (27) is expressed as

$$e : (\mu_e + \alpha\pi_e, \nu_e + (1-\alpha)\pi_e) \tag{28}$$

where $\alpha \in (0.5, 1)$ is a parameter.

To guarantee the best behavior of the fuzzy classifier, the parameter α is chosen separately for each attribute, and next, the results are aggregated (see Szmidt and Kukier [52–54]).

The above model is built for each attribute separately, and then the obtained results are aggregated. A full description of the classifier and very promising results of its use are presented in Szmidt and Kukier [52–54].

Here we only wish to emphasize that the idea of the presented classifier making use of the operator $D_\alpha(A)$ (8) is rather natural within the A-IFSs, while it would be difficult both to invent and implement it within the IVFSs.

Now we will consider an interpretation of the A-IFSs as intervals which seems to prove that the A-IFSs and IVFSs are "the same". However, a deeper analysis shows something different.

4 Representation of the A-IFSs by Intervals

We will consider here the fact with its consequences that the A-IFSs can be represented in the form of intervals.

Imagine a very simple situation with a degenerated, one element, A-IFS A consisting of x only. Assuming the two term representation of A-IFSs, i.e., $A = \{< x, \mu, \nu >\}$ we may consider A in the form of an interval, namely:

$$[\mu(x), 1 - \nu(x)] \tag{29}$$

which, from (4), is equivalent to

$$[\mu(x), \mu(x) + \pi(x)] \tag{30}$$

The use of (29)–(30) makes it possible to express A in terms of the upper and lower membership values (instead of in terms of the membership and non-membership degrees). In other words, the two term representation is equivalent to the representation of the A-IFSs by one interval.

Example 1 Let x be a house with its advantages expressed by the membership value μ, and disadvantages expressed by the non-membership value ν. Using the two term representation of the respective A-IFS implies that the advantage of the house lies in the interval $[\mu(x), \mu(x) + \pi(x)]$.

In turn, consider the same situation with a degenerated A-IFS consisting of one element x only but by making use of the three term representation of A, i.e., $A = < x, \mu, \nu, \pi >$. Using all three terms means that the membership degree μ lies in the interval

$$[\mu_{min}, \mu_{max}] = [\mu, \mu + \pi] \tag{31}$$

and the non-membership degree ν lies in the interval

$$[\nu_{min}, \nu_{max}] = [\nu, \nu + \pi] \tag{32}$$

Next, for (31) and (32) the condition (3) must be fulfilled which means that it is not allowed, for example, that $< x, \mu + \pi, \nu + \pi >$.

Other possibilities of making use of the hesitation margin π to construct models are considered, e.g., in Montero et al. [20] but in this paper we stick to the above interpretation.

Example 2 Consider again the house x described above by using the two term representation of the A-IFSs. Now, we will employ the three term description $< x, \mu, \nu, \pi >$ with advantages given by $\mu(x)$, disadvantages expressed by $\nu(x)$, and lack of knowledge considering both the advantages, and the disadvantages given by $\pi(x)$. Assume $< x, 0.4, 0.2, 0.4 >$ which means that $\mu(x)$ is from the interval $[0.4, 0.8]$, and $\nu(x)$ is from the interval $[0.2, 0.6]$. In other words, we have the following possibilities:
– in the best situation the home assessment is
$\{< x, 0.8, 0.2, 0 >\}$, i.e. $\mu = 0.8$, $\nu = 0.2$, and $\pi = 0$,
– in the worst situation the home assessment is
$\{< x, 0.4, 0.6, 0 >\}$, i.e. $\mu = 0.4$, $\nu = 0.6$, and $\pi = 0$,
– in an average situation the home assessment is
$\{< x, 0.6, 0.4, 0 >\}$, i.e. $\mu = 0.6$, $\nu = 0.4$, and $\pi = 0$.

We may notice that the three term representation of A-IFSs makes it possible to use more information, and obtain more insight in a more transparent way than while using the two term representation. For instance, it is easy to consider different "scenarios" (cf. Example 2) taking into account the values of the hesitation margins. Considering such scenarios might be useful while decisions are to be made about some future events which can be described to some extent only, like, e.g., in voting processes, introducing a new product into a market, looking for a new job or buying a house. The success or defeat depend then strongly on the values of the hesitation margins. It is why market analysis and pre-voting analysis are performed, to reduce as much as possible a lack of knowledge (hesitation margin) concerning a phenomenon in question.

It is worth stressing that the three term representation is different from the representation used in the IVFSs.

In other words, the A-IFSs expressed by the two term representation and the IVFSs are, like Atanassov and Gargov [8] said, "equipollent", that is, deducible from each other, but still not "the same" (e.g. because of different operators), whereas the A-IFSs expressed via three term representation are equivalent to considering two intervals which for sure means that they are not "the same" as the IVFSs.

5 Two and Three Term A-IFSs Measures

Almost all important measures used in the area of A-IFSs, as well as in other areas, are non-linear. For example distances which play a decisive role in many models, and are a basis for many other measures, are usually non-linear. So it seems natural to expect that quite different qualitative results will be obtained while using the two term representation of the A-IFSs than the results obtained while using the three term representation of the A-IFSs. Szmidt and Kacprzyk [30, 32], Szmidt and Baldwin [24, 25] discuss the results obtained for the most often used distances when the two and the three term representations of the A-IFS are used. Examples of the distances between any two A-IFSs A and B in $X = \{x_1, x_2, \ldots, x_n\}$ while using the three term representation (Szmidt and Kacprzyk [30], Szmidt and Baldwin [24, 25]) may be as follows:

- the normalized Hamming distance:

$$l_{IFS}(A,B) = \frac{1}{2n}\sum_{i=1}^{n}\Big(\big|\mu_A(x_i) - \mu_B(x_i)\big| + \big|\nu_A(x_i) - \nu_B(x_i)\big| + \big|\pi_A(x_i) - \pi_B(x_i)\big|\Big) \quad (33)$$

- the normalized Euclidean distance:

$$e_{IFS}(A,B) = \left(\frac{1}{2n}\sum_{i=1}^{n}(\mu_A(x_i) - \mu_B(x_i))^2 + (\nu_A(x_i) - \nu_B(x_i))^2 + (\pi_A(x_i) - \pi_B(x_i))^2\right)^{\frac{1}{2}} \quad (34)$$

The values of both distances are from the interval [0, 1].

The corresponding distances to the above ones while using the two term representation of the A-IFSs are:

- the normalized Hamming distance:

$$l'(A,B) = \frac{1}{2n}\sum_{i=1}^{n}\Big(|\mu_A(x_i) - \mu_B(x_i)| + |\nu_A(x_i) - \nu_B(x_i)|\Big) \quad (35)$$

- the normalized Euclidean distance:

$$q'(A,B) = \left(\frac{1}{2n}\sum_{i=1}^{n}(\mu_A(x_i) - \mu_B(x_i))^2 + (\nu_A(x_i) - \nu_B(x_i))^2\right)^{\frac{1}{2}} \quad (36)$$

Both the two term distances and three term distances [23, 30] are correct from the mathematical point of view, i.e. all the needed properties are fulfilled, but practical problem solutions are more intuitively appealing while using the three term representations (cf. e.g., Szmidt and Kacprzyk [36]).

Even a more convincing argument for using the three term representation of the A-IFSs occurs in the case of the Hausdorff distance based on the Hamming metric.

The Hausdorff Distance

The Hausdorff distance (defined by Hausdorff in 1914) is the *maximum distance of a set to the nearest point in the other set*, i.e.

Definition 2 The Hausdorff distance $H(A, B)$ for two finite sets $A = \{a_1, \ldots, a_p\}$ and $B = \{b_1, \ldots, b_q\}$, is:

$$H(A, B) = \max\{h(A, B), h(B, A)\} \tag{37}$$

where

$$h(A, B) = \max_{a \in A} \min_{b \in B} d(a, b) \tag{38}$$

where:

- a and b mean elements of sets A and B respectively,
- $d(a, b)$ means any metric between elements a and b,
- $h(A, B)$ and $h(B, A)$ (38) are called the directed Hausdorff distances.

To assign the directed Hausdorff distance $h(A, B)$ from A to B each element of A is ranked based on its distance to the nearest element of B, and then the highest ranked element points out the value of the distance; The values $h(A, B)$ and $h(B, A)$ may be different (the directed distances are not symmetric).

Definition 2 implies that for A and B containing one element each (a_1 and b_1, respectively), the Hausdorff distance is just equal to $d(a_1, b_1)$. In other words, if a formula which is expected to express the Hausdorff distance gives for separate elements the results which are not consistent with the metric d used (e.g., the Hamming distance, the Euclidean distance, etc.), the formula considered is not a proper definition of the Hausdorff distance.

By applying the algorithm of calculating the directed Hausdorff distances, while using the two term type distance (35) for the A-IFSs, we obtain:

$$d_h(A, B) = \frac{1}{n} \sum_{i=1}^{n} max\{|\mu_A(x_i) - \mu_B(x_i)|, |\nu_A(x_i) - \nu_B(x_i)|\} \tag{39}$$

If the above distance is a properly calculated Hausdorff distance, in the case of the degenerated, i.e., one-element sets $A = \{< x, \mu_A(x), \nu_A(x) >\}$ and $B = \{< x, \mu_B(x), \nu_B(x) >\}$, it should give the same results as the the two term type Hamming distance. It means that while using the two term type Hamming distance, for the degenerated, one element A-IFSs, the following equations should give just the same results:

$$l'(A, B) = \frac{1}{2}(|\mu_A(x) - \mu_B(x)| + |\nu_A(x) - \nu_B(x)|) \tag{40}$$

$$d_h(A, B) = \max\{|\mu_A(x) - \mu_B(x)|, |\nu_A(x) - \nu_B(x)|\} \tag{41}$$

where (40) is the normalized two term type Hamming distance, and (41) should be its counterpart Hausdorff distance.

However, it is easy to show that in general the above condition does not work. For example, for the following one element A-IFSs [37]: $A, B, \in X = \{x\}$, $A = < x, 1, 0 >$, $B = \{< x, \frac{1}{4}, \frac{1}{4} >\}$, the result obtained from (41) is:

$$d_h(A, B) = \max\{|1 - 1/4|, |0 - 1/4|\} = 0.75$$

The corresponding Hamming distance calculated from (40) is:

$$l'(A, B) = 0.5(|1 - 1/4| + |0 - 1/4||) = 0.5$$

i.e. the value of the Hamming distances (40) used to propose the Hausdorff measure (41), and the value of the resulting Hausdorff distance (41) calculated for the separate elements are not consistent (as they should be).

Szmidt and Kacprzyk [37] have shown that the inconsistencies as shown above occur for an infinite number of other cases, and the conditions were formulated under which (40) and (41) produce the same results. The conclusion is that an attempt to construct the Hausdorff distance using the two term type Hamming distance between the A-IFSs seems to be an unjustified idea.

On the other hand, by applying the three term type Hamming distance for the A-IFSs, we obtain correct (in the sense of Definition 2) Hausdorff distance.

To be more precise, if we calculate the three term type Hamming distance between two degenerated, i.e. one-element A-IFSs, A and B in the spirit of Szmidt and Kacprzyk [30, 32], Szmidt and Baldwin [24, 25], i.e., in the following way:

$$l_{IFS}(A, B) = \frac{1}{2}(|\mu_A(x) - \mu_B(x)| + |\nu_A(x) - \nu_B(x)| + |\pi_A(x) - \pi_B(x)|) \tag{42}$$

then, it is possible to give a counterpart of the above distance in terms of the max function:

$$H_3(A, B) = \max\{|\mu_A(x) - \mu_B(x)|, |\nu_A(x) - \nu_B(x)|, |\pi_A(x) - \pi_B(x)|\} \tag{43}$$

If $H_3(A, B)$ (43) is a properly specified Hausdorff distance (in the sense that for two degenerated, one element A-IFSs the result is equal to the metric used), the following condition should be fulfilled:

$$\frac{1}{2}(|\mu_A(x) - \mu_B(x)| + |\nu_A(x) - \nu_B(x)|) + |\pi_A(x) - \pi_B(x)|) =$$
$$= \max\{|\mu_A(x) - \mu_B(x)|, |\nu_A(x) - \nu_B(x)|, |\pi_A(x) - \pi_B(x)|\} \quad (44)$$

Szmidt and Kacprzyk [37] have shown that the above condition is always valid. It means that, in general, the three term Hausdorff distance (43) (expressed using the memberships, non-memberships and hesitation margin degrees) and the Hamming distance (42) give for one element A-IFSs fully consistent results (as it should be).

The similar conclusions about the importance of the three term representation of the A-IFSs, from the perspective of the distances, remain valid for

- the Pearson correlation coefficient (Szmidt and Kacprzyk [39]),
- the Spearman correlation coefficient (Szmidt and Kacprzyk [40]),
- the Kendall correlation coefficient (Szmidt and Kacprzyk [43, 44]),
- the Principal Component Analysis (Szmidt and Kacprzyk [46], Szmidt et al. [48]),
- ranking procedures (Szmidt and Kacprzyk [35, 36, 38]),
- similarity measures (Szmidt and Kreinovich [51], Szmidt and Kacprzyk [34]),
- construction of a classifier for imbalanced and overlapping classes (cf. Szmidt and Kukier [52–54]),
- constructing intuitionistic fuzzy trees (Bujnowski et al. [11]).

For example, in [39, 40, 42] was shown that the third term (hesitation margin) could considerably enhance the Pearson correlation coefficient of the variables considered (in [49] as it was shown on the well-known benchmarks from the UC, Irvine). An analogical conclusion was drawn for the Spearman correlation coefficient [40] and the Kendall correlation coefficient. It seems to be quite natural to take into account the third term as for practical purposes (e.g., in decision making) it seems rather useful to know correlation concerning a lack of knowledge represented by the variables considered. The same approach – from theoretical considerations to a verification on benchmarks [45, 48], followed by the solution of real world examples that are planned – was used in the case of the Principal Component Analysis. In [51] it was shown that in order to produce similarity measures which were in a better accordance with common sense, we should take into account all three terms. Examples are presented in [34]. Next, without taking into account the third term, the classifier for the elements from imbalanced classes [52–54] would not be built. Because of the space limitation we are not able to discuss here in length all the obtained results. However, in general, analysis of the results obtained from the above mentioned models (Szmidt and Kacprzyk [35, 36, 38–40, 43, 44, 46, 48], Szmidt and Kukier [52–54], Bujnowski et al. [11]) have shown that each of the three terms plays an important role in data analysis and decision making.

It is worth stressing again that not only the three type representation of the A-IFSs but also the two type representation differ conceptually from the IVFSs because: we collect data in a different way, look for different answers, and use different types of operators while using the two kinds of models.

6 Conclusion

We have pointed out some differences between the A-IFSs and IVFSs mainly from the semantics focused perspective that is relevant for applications. First, we have paid attention to the fact that some operators and extensions typical for the A-IFSs do not exist for the IVFSs. The fact has been illustrated by a practical example of using one of such operators to build a classifier for imbalanced classes.

We have also revisited, and further analyzed, the concepts of two possible representations of the A-IFSs, namely, the two term one, in which the degrees of membership and non-membership are only involved, and the three term one, in which in addition to the above degrees of membership and non-membership the so called hesitation margin is explicitly accounted for. Making use of the two representations of the A-IFSs we have considered the representation of the A-IFS via intervals concluding that the three term representation differs considerably from the representation used for the IVFSs. In comparison to the IVFSs, the three term representation of the A-IFSs gives possibilities to analyze in a deeper way more sophisticated aspects of the problems, for instance in decision making.

Finally, we have recalled several measures that are important from the point of view of applications. It turns out that the three term representation of A-IFSs (which differs from the IVFSs) leads to view measures that are interesting from the theoretical and useful from the practical point of view.

This all can be viewed as a strong argument that, from the perspective of the existence of many operations, semantics and applications, the statement of an inherent equivalence between the A-IFSs and IVFSs may possibly be justified just from a narrow perspective of purely formal point of view but not generally, that is, in the perspective of looking for a general and applicable framework.

References

1. Atanassov K.T. (1983) Intuitionistic Fuzzy Sets. VII ITKR Session. Sofia (Deposed in Centr. Sci.-Techn. Library of Bulg. Acad. of Sci., 1697/84) (in Bulgarian).
2. Atanassov K.T. (1986) Intuitionistic Fuzzy Sets. Fuzzy Sets and Systems, 20, 87–96.
3. Atanassov K. T. (1989) More on intuitionistic fuzzy sets. Fuzzy Sets and Systems, 33(1), 37–46.
4. Atanassov K.T. (1991) Temporal intuitionistic fuzzy sets. Comptes Rendus de l'Academie Bulgare des Sciences, Tome 44, No. 7, 5–7.
5. Atanassov K.T. (1999) Intuitionistic Fuzzy Sets: Theory and Applications. Springer-Verlag.
6. Atanassov K.T. (2006) Intuitionistic fuzzy sets and interval valued fuzzy sets. First Int. Workshop on IFSs, GNs, KE, London, 6–7 Sept. 2006, 1–7.
7. Atanassov K.T. (2012) On Intuitionistic Fuzzy Sets Theory. Springer-Verlag.
8. Atanassov K.T. and Gargov G. (1989) Interval Valued Intuitionistic Fuzzy Sets. Fuzzy Sets and Systems, 31, 343–349.
9. Atanassov K., Tasseva V, Szmidt E. and Kacprzyk J. (2005): On the geometrical interpretations of the intuitionistic fuzzy sets. In: Issues in the Representation and Processing of Uncertain and Imprecise Information. Fuzzy Sets, Intuitionistic Fuzzy Sets, Generalized Nets, and Related Topics. (Eds. Atanassov K., Kacprzyk J., Krawczak M., Szmidt E.), EXIT, Warsaw, 11–24.

10. Atanassova V. (2004) Strategies for Decision Making in the Conditions of Intuitionistic Fuzziness. Int. Conf. 8th Fuzzy Days, Dortmund, Germany, 263–269.
11. Bujnowski P., Szmidt E., Kacprzyk J. (2014): Intuitionistic Fuzzy Decision Trees - a new Approach. In: Rutkowski L., Korytkowski M., Scherer R., Tadeusiewicz R., Zadeh L., urada J. (Eds.): Artificial Intelligence and Soft Computing, Part I. Springer, Switzerland, 181–192.
12. Bustince H., Mohedano V., Barrenechea E., and Pagola M. (2005): *Image thresholding using intuitionistic fuzzy sets.* In: Issues in the Representation and Processing of Uncertain and Imprecise Information. Fuzzy Sets, Intuitionistic Fuzzy Sets, Generalized Nets, and Related Topics. (Eds. Atanassov K., Kacprzyk J., Krawczak M., Szmidt E.), EXIT, Warsaw.
13. Bustince H., Mohedano V., Barrenechea E., and Pagola M. (2006): *An algorithm for calculating the threshold of an image representing uncertainty through A-IFSs.* IPMU'2006, 2383–2390.
14. Bustince H., Barrenechea E., Pagola M., Fernandez J., Xu Z., Bedregal B., Montero J., Hagras H., Herrera F., and De Baets B., A Historical Account of Types of Fuzzy Sets and Their Relationships. IEEE Transactions on Fuzzy Systems, 24 (1), 2016, 179–194.
15. Deng-Feng Li (2005): Multiattribute decision making models and methods using intuitionistic fuzzy sets. Journal of Computer and System Sciences, 70, 73–85.
16. Dubois D., Prade H.(2008): *An introduction to bipolar representations of information and preference.* Int. J. Intell. Syst. 23(8), 866–877.
17. Feys R. Modal logic. Foundation Universitaire de Belgique, Paris 1965.
18. Grabisch M., Greco S., Pirlot M. (2008): *Bipolar and bivariate models in multicriteria decision analysis: Descriptive and constructive approaches.* International Journal of Intelligent Systems, 23 (9), 930–969.
19. Kuratowski K. (1966) Topology, Vol. 1, New York, Acad. Press.
20. Montero J., Gmez D., Bustince Sola H. (2007): *On the relevance of some families of fuzzy sets.* Fuzzy Sets and Systems 158(22), 2429–2442.
21. Narukawa Y. and Torra V. (2006): Non-monotonic fuzzy measure and intuitionistic fuzzy set. MDAI 2006, LNAI 3885, 150–160, Springer-Verlag.
22. Roeva O. and Michalikova A. (2013) Generalized net model of intuitionistic fuzzy logic control of genetic algorithm parameters. In: Notes on Intuitionistic Fuzzy Sets. Academic Publishing House, Sofia, Bulgaria. Vol. 19, No. 2, 71–76. ISSN 1310-4926.
23. Szmidt E. (2014): Distances and Similarities in Intuitionistic Fuzzy Sets. Springer.
24. Szmidt E. and Baldwin J. (2003) New similarity measure for intuitionistic fuzzy set theory and mass assignment theory.
25. Szmidt E. and Baldwin J. (2004) Entropy for intuitionistic fuzzy set theory and mass assignment theory. Notes on IFSs, 10 (3), 15–28.
26. Szmidt E. and Baldwin J. (2006): *Intuitionistic Fuzzy Set Functions, Mass Assignment Theory, Possibility Theory and Histograms.* 2006 IEEE World Congress on Computational Intelligence, 237–243.
27. Szmidt E. and Kacprzyk J. (1997): *On measuring distances between intuitionistic fuzzy sets.* Notes on IFS, 3(4), 1–13.
28. Szmidt E. and Kacprzyk J. (1999): Probability of Intuitionistic Fuzzy Events and their Applications in Decision Making. EUSFLAT-ESTYLF, Palma de Mallorca, 457–460.
29. Szmidt E. and Kacprzyk J. (1999b): A Concept of a Probability of an Intuitionistic Fuzzy Event. FUZZ-IEEE'99, Seoul, Korea, III 1346–1349.
30. Szmidt E., Kacprzyk J. (2000): *Distances between intuitionistic fuzzy sets.* Fuzzy Sets and Systems, 114 (3), 505–518.
31. Szmidt E., Kacprzyk J. (2001): *Entropy for intuitionistic fuzzy sets.* Fuzzy Sets and Systems, 118, Elsevier, 467–477.
32. Szmidt E., Kacprzyk J. (2006): *Distances Between Intuitionistic Fuzzy Sets: Straightforward Approaches may not work.* 3rd International IEEE Conference Intelligent Systems IEEE IS'06, London, 716–721.
33. Szmidt E. and Kacprzyk J. (2007): *Some problems with entropy measures for the Atanassov intuitionistic fuzzy sets.* Applications of Fuzzy Sets Theory. LNAI 4578, Springer-Verlag, 291–297

34. Szmidt E. and Kacprzyk J. (2007a): *A New Similarity Measure for Intuitionistic Fuzzy Sets: Straightforward Approaches may not work.* 2007 IEEE Conf. on Fuzzy Systems, 481–486
35. Szmidt E. and Kacprzyk J. (2008) A new approach to ranking alternatives expressed via intuitionistic fuzzy sets. In: D. Ruan et al. (Eds.) Computational Intelligence in Decision and Control. World Scientific, 265–270.
36. Szmidt E. and Kacprzyk J.: Ranking of Intuitionistic Fuzzy Alternatives in a Multi-criteria Decision Making Problem. In: Proceedings of the conference: NAFIPS 2009, Cincinnati, USA, June 14–17, 2009, IEEE, ISBN: 978-1-4244-4577-6.
37. Szmidt E., Kacprzyk J. (2011) Intuitionistic fuzzy sets – Two and three term representations in the context of a Hausdorff distance. Acta Universitatis Matthiae Belii, Series Mathematics, available at http://ACTAMTH.SAVBB.SK, Vol.19, No. 19, 2011, 53–62.
38. Szmidt E. and Kacprzyk J. (2009) Amount of information and its reliability in the ranking of Atanassov's intuitionistic fuzzy alternatives. In: Recent Advances in decision Making, SCI 222. E. Rakus-Andersson, R. Yager, N. Ichalkaranje, L.C. Jain (Eds.), Springer-Verlag, 7–19.
39. Szmidt E. and Kacprzyk J. (2010): Correlation between intuitionistic fuzzy sets. LNAI 6178 (Computational Intelligence for Knowledge-Based Systems Design, Eds. E.Hullermeier, R. Kruse, F. Hoffmann), 169–177.
40. Szmidt E. and Kacprzyk J. (2010): The Spearman rank correlation coefficient between intuitionistic fuzzy sets. In: Proc. 2010 IEEE Int. Conf. on Intelligent Systems IEEE'IS 2010, London, 276–280.
41. Szmidt E., Kacprzyk J., Bujnowski P. (2011): *Measuring the Amount of Knowledge for Atanassov's Intuitionistic Fuzzy Sets.* Fuzzy Logic and Applications, Lecture Notes in Artificial Intelligence, Vol. 6857, 2011, 17–24.
42. Szmidt E., Kacprzyk J. and Bujnowski P. (2011) Pearson's coefficient between intuitionistic fuzzy sets. Notes on Intuitionistic Fuzzy Sets, 17 (2), 25–34.
43. Szmidt E. and Kacprzyk J. (2011) The Kendall Rank Correlation between Intuitionistic Fuzzy Sets. In: Proc.: World Conference on Soft Computing, San Francisco, CA, USA, 23/05/2011-26/05/2011.
44. Szmidt E. and Kacprzyk J. (2011) The Spearman and Kendall rank correlation coefficients between intuitionistic fuzzy sets. In: Proc. 7th conf. European Society for Fuzzy Logic and Technology, Aix-Les-Bains, France, Antantic Press, 521–528.
45. Szmidt E., Kacprzyk J., Bujnowski P. (2012): *Advances in Principal Component Analysis for Intuitionistic Fuzzy Data Sets.* 2012 IEEE 6th International Conference „Intelligent Systems", 194–199.
46. Szmidt E. and Kacprzyk J. (2012) A New Approach to Principal Component Analysis for Intuitionistic Fuzzy Data Sets. S. Greco et al. (Eds.): IPMU 2012, Part II, CCIS 298, 529–538, Springer-Verlag Berlin Heidelberg.
47. Szmidt E. and Kacprzyk J. (2015) Two and Three Term Representations of Intuitionistic Fuzzy Sets: Some Conceptual and Analytic Aspects. IEEE International Conference on Fuzzy Systems (FUZZ-IEEE 2015), Istanbul, Turkey, August 2–5, 2015, IEEE, ss. 1-8.
48. Szmidt E., Kacprzyk J. and Bujnowski P. (2012) Advances in Principal Component Analysis for Intuitionistic Fuzzy Data Sets. 2012 IEEE 6th International Conference "Intelligent Systems", 194–199.
49. Szmidt E., Kacprzyk J. and Bujnowski P. (2012): *Correlation between Intuitionistic Fuzzy Sets: Some Conceptual and Numerical Extensions.* WCCI 2012 IEEE World Congress on Computational Intelligence, Brisbane, Australia, 480–486.
50. Szmidt E., Kacprzyk J., Bujnowski P. (2014): *How to measure the amount of knowledge conveyed by Atanassov's intuitionistic fuzzy sets.* Information Sciences, 257, 276–285.
51. Szmidt E., Kreinovich V. (2009) Symmetry between true, false, and uncertain: An explanation. Notes on Intuitionistic Fuzzy Sets 15, No. 4, 1–8.
52. Szmidt E. and Kukier M. (2006): *Classification of Imbalanced and Overlapping Classes using Intuitionistic Fuzzy Sets.* IEEE IS'06, London, 722–727.
53. Szmidt E. and Kukier M. (2008): *A New Approach to Classification of Imbalanced Classes via Atanassov's Intuitionistic Fuzzy Sets.* In: Hsiao-Fan Wang (Ed.): Intelligent Data Analysis

: Developing New Methodologies Through Pattern Discovery and Recovery. Idea Group, 85–101.

54. Szmidt E., Kukier M. (2008): *Atanassov's intuitionistic fuzzy sets in classification of imbalanced and overlapping classes.* In: P. Chountas, I. Petrounias, J. Kacprzyk (Eds.): Intelligent Techniques and Tools for Novel System Architectures. Series: Studies in Computational Intelligence, Vol. 109, Springer, Berlin Heidelberg 2008, 455–471.
55. Tan Ch. and Zhang Q. (2005): Fuzzy multiple attribute TOPSIS decision making method based on intuitionistic fuzzy set theory. IFSA 2005, 1602–1605.
56. Tasseva V., Szmidt E. and Kacprzyk J. (2005): On one of the geometrical interpretations of the intuitionistic fuzzy sets. Notes on IFS, 11 (3), 21–27.
57. Zadeh L.A. (1965): *Fuzzy sets.* Information and Control, 8, 338–353.

Part III
Rough Sets

Attribute Importance Degrees Corresponding to Several Kinds of Attribute Reduction in the Setting of the Classical Rough Sets

Masahiro Inuiguchi

Abstract In this paper, we propose several attribute reduction concepts which are ordered linearly. For each attribute reduction, we give a discernibility matrix which enables to enumerate all reduced attribute sets. We define measures to evaluate the specificity of decision class and the retention ability of specificity corresponding to the proposed concepts of attribute reduction. Using those measures, attribute importance degrees are defined based on cooperative game theory. We show that the attribute importance degree is very different by the requirement to what extent we preserve the class information of objects. Finally, we describe the possible application of the attribute reduction to the group decision making and give modifications in case when decision classes are linearly ordered.

1 Introduction

Rough set [5, 6] is a useful tool for reasoning from data. Attribute reduction and rule induction are major applications of rough sets. The classical rough sets are generalized in many ways but we focus on the classical rough sets in this paper. More specifically, we investigate attribute reduction and attribute importance of the classical rough sets.

Several approaches [2–4, 8] to attribute reduction, or simply, reducts in rough set theory have been investigated. In classical rough sets, they are classified into two kinds: reducts preserving lower approximations and reducts preserving upper approximations (or equivalently, reducts preserving boundary regions) [2, 8]. The calculations of those reducts are well investigated. Corresponding to those reducts, set functions have been defined and degrees of attribute importance based on set functions have been investigated [1, 2].

In this paper, we show that other kinds of reducts can be defined. They are between the two kinds of reducts and linearly ordered. We present discernibility matrices to

M. Inuiguchi (✉)
Graduate School of Engineering Science, Osaka University, Toyonaka, Osaka 560-8531, Japan
e-mail: inuiguti@sys.es.osaka-u.ac.jp

V. Torra et al. (eds.), *Fuzzy Sets, Rough Sets, Multisets and Clustering*, Studies in Computational Intelligence 671, DOI 10.1007/978-3-319-47557-8_14

calculate all kinds of reducts. Moreover, we investigate the attribute importance and interaction indices. Set functions are defined corresponding to those reducts and the numerical attribute importance degrees are defined based on cooperative game theory. It is shown that the degrees of attribute importance are very different depending on the underlying reducts. A possible application of the proposed reducts to group decision making and modifications for the ordinal decision attribute are given.

This paper is organized as follows. In next section, we introduce decision tables and the rough set approach briefly. The proposed approach is described in Sect. 3. In Sect. 4, a numerical example is given to see the meaningfulness of the proposed approach. Application guide for the proposed approach is described in Sect. 5. In Sect. 6, a possible application of the proposed approach is described and modifications to a case when the decision attribute is ordinal is given. Concluding remarks are given in Sect. 6.

2 Decision Tables and Rough Set Approach

2.1 Decision Tables and Rough Sets

The classical rough sets are defined under an equivalence relation which is often called an indiscernibility relation. In this paper, we restrict ourselves to discussions of the classical rough sets under decision tables. A decision table is characterized by four-tuple $\mathcal{I} = \langle U, C \cup \{d\}, V, \rho \rangle$, where U is a finite set of objects, C is a finite set of condition attributes, d is a decision attribute, $V = \bigcup_{a \in C \cup \{d\}} V_a$ and V_a is a domain of the attribute a, and $\rho : U \times C \cup \{d\} \to V$ is an information function such that $\rho(x, a) \in V_a$ for every $a \in C \cup \{d\}$, $x \in U$.

Given a set of attributes $A \subseteq C \cup \{d\}$, we can define an equivalence relation I_A referred to as an indiscernibility relation by $I_A = \{(x, y) \in U \times U \mid \rho(x, a) = \rho(y, a),\ \forall a \in A\}$. From I_A, we have an equivalence class, $[x]_A = \{y \in U \mid (y, x) \in I_A\}$. When $A = \{d\}$, we define

$$\mathcal{D} = \{D_j,\ j = 1, 2, \ldots, p\} = \{[x]_{\{d\}} \mid x \in U\},\ D_i \neq D_j\ (i \neq j). \quad (1)$$

D_j is called a 'decision class'. There exists a unique $v_j \in V_d$ such that $\rho(x, d) = v_j$ for each $x \in D_j$, i.e., $D_j = \{x \in U \mid \rho(x, d) = v_j\}$. Moreover, since $D_i \cap D_j = \emptyset$ $(i \neq j)$ and $\bigcup \mathcal{D} = U$ hold, $\mathcal{D}$ forms a partition.

For a set of condition attributes $A \subseteq C$, the lower and upper approximations of an object set $X \subseteq U$ are defined as follows:

$$A_*(X) = \{x \mid [x]_A \subseteq X\}, \quad A^*(X) = \{x \mid [x]_A \cap X \neq \emptyset\}. \quad (2)$$

A pair $(A_*(X), A^*(X))$ is called a rough set of X. The boundary region of X is defined by $BN_A(X) = A^*(X) - A_*(X)$. Since $[x]_A$ can be seen as a set of objects indiscernible from $x \in U$ in view of condition attributes in A, $A_*(X)$ is interpreted as a collection of objects whose membership to X is noncontradictive in view of condition attributes in A. $BN_A(X)$ is interpreted as a collection of objects whose membership to X is doubtful in view of condition attributes in A. $A^*(X)$ is interpreted as a collection of possible members. For $x \in U$, the generalized decision class $\partial_A(x)$ of x with respect to a condition attribute set $A \subseteq C$ is defined by

$$\partial_A(x) = \{\rho(y, d) \mid y \in [x]_A\}. \tag{3}$$

Let $X, Y \subseteq U$. We have the following properties: (for other properties, see [6]):

$$A_*(X) \subseteq X \subseteq A^*(X), \tag{4}$$
$$A \subseteq B \Rightarrow A_*(X) \subseteq B_*(X),\ \ A^*(X) \supseteq B^*(X), \tag{5}$$
$$A_*(X \cap Y) = A_*(X) \cap A_*(Y),\ \ A^*(X \cup Y) = A^*(X) \cup A^*(Y), \tag{6}$$
$$A_*(X \cup Y) \supseteq A_*(X) \cup A_*(Y),\ \ A^*(X \cap Y) \subseteq A^*(X) \cap A^*(Y), \tag{7}$$
$$BN_A(X) = A^*(X) \cap A^*(U - X),\ \ A_*(X) = X - BN_A(X), \tag{8}$$
$$A^*(X) = X \cup BN_A(X) = U - A_*(U - X), \tag{9}$$
$$A_*(X) = A^*(X) - A^*(U - X) = U - A^*(U - X). \tag{10}$$

2.2 Attribute Reduction

Attribute reduction is one of the major topics in rough set approaches. It indicates minimally necessary attributes to classify objects and reveals important attributes. A set of minimally necessary attributes is called a reduct. In the classical rough set analysis, reducts preserving lower approximations are frequently used. Namely, a set of condition attributes, $A \subseteq C$ is called a reduct if and only if it satisfies (L1) $A_*(D_j) = C_*(D_j)$, $j = 1, 2, \ldots, p$ and (L2) $\nexists a \in A$, $(A - \{a\})_*(D_j) = C_*(D_j)$, $j = 1, 2, \ldots, p$. Since we discuss several kinds of reducts, we call this reduct, a 'reduct preserving lower approximations' or an 'L-reduct' for short. Let $\mathcal{R}^{\mathrm{L}}$ be a set of L-reducts. Then $\bigcap \mathcal{R}^{\mathrm{L}}$ is called the 'core preserving lower approximation' or the 'L-core'. Attributes in the L-core are important because without any of them, we cannot preserve all lower approximations of decision classes.

We consider reducts preserving upper approximations or equivalently, preserving boundary regions [2, 8]. A set of condition attributes, $A \subseteq C$ is called a 'reduct preserving upper approximations' or a 'U-reduct' for short if and only if it satisfies (U1) $A^*(D_j) = C^*(D_j)$, $j = 1, 2, \ldots, p$ and (U2) $\nexists a \in A$, $(A - \{a\})^*(D_j) = C^*(D_j)$, $j = 1, 2, \ldots, p$. On the other hand, a set of condition attributes, $A \subseteq C$ is called a 'reduct preserving boundary regions' or a 'B-reduct' for short if and only if it satisfies (B1) $BN_A(D_j) = BN_C(D_j)$, $j = 1, 2, \ldots, p$ and (B2) $\nexists a \in A$, $BN_{(A-\{a\})}(D_j) = BN_C(D_j)$, $j = 1, 2, \ldots, p$.

The following relations are known among L-reducts, U-reducts and B-reducts:

(R1) A U-reduct is also a B-reduct and vice versa.
(R2) There exists an L-reduct A for a U-reduct B such that $B \supseteq A$.
(R3) There exists an L-reduct A for a B-reduct B such that $B \supseteq A$.

Those relations can be proved easily from (6), (9) and (10). Since B-reduct is equivalent to U-reduct, we describe only U-reduct in what follows. Let $\mathcal{R}^{\mathrm{U}}$ be a set of U-reducts. Then $\bigcap \mathcal{R}^{\mathrm{U}}$ is called the 'core preserving upper approximation' or the 'U-core'. Attributes in the U-core are important because without any of them, we cannot preserve all upper approximations of decision classes.

To obtain a part or all of reducts, many approaches have been proposed in the literature [6, 7]. Among them, we mention an approach based on a discernibility matrix [7]. In this approach, we construct a Boolean function which characterizes the preservation of the lower approximations to obtain L-reducts. Each L-reduct is obtained as a prime implicant of the Boolean function. For the detailed discussion of the discernibility matrix for L-reducts, see Ref. [7]. However, the discernibility matrix is included in a special case of the discernibility matrix described later.

2.3 Game-Theoretic Approach to Attribute Importance

Corresponding to reducts, we obtain set functions. Under set functions, we can evaluate the importance of each member by cooperative game theory. Therefore, in order to measure the importance of condition attribute, we apply cooperative game theory. We introduce indices based on the Shapley function. Given a set function $\mu : 2^C \to \mathbf{R}$, the Shapley value of $a \in C$ is defined by

$$\phi_\mu^S(a) = \sum_{K \subseteq C - \{a\}} \frac{(|C| - |K| - 1)!|K|!}{|C|!}(\mu(K \cup \{a\}) - \mu(K)), \tag{11}$$

where $|Z|$ shows the cardinality of set Z. The Shapley value of $a \in C$ shows the contribution degree to μ of condition attribute a. The Shapley value is extended to measure the interaction among condition attributes. The Shapley interaction index of $A \subseteq C$ is given by

$$I_\mu^S(A) = \sum_{K \subseteq C - A} \frac{(|C| - |K| - |A|)!|K|!}{(|C| - |A| + 1)!} \sum_{L \subseteq A} (-1)^{|A| - |L|} \mu(K \cup L). \tag{12}$$

The meanings of the Shapley value and interaction index strongly depend on set function μ. When $|A| = 2$, $I_\mu^S(A) = I_\mu^S(\{a_1, a_2\})$ shows the interaction between a_1 and a_2. If $I_\mu^S(A) > 0$, the simultaneous existence of a_1 and a_2 creates a synergy effect. On the other hand, if $I_\mu^S(A) < 0$ the simultaneous existence of a_1 and a_2 creates a cancel effect.

We also introduce Möbius transform m_μ of μ defined by

$$m_\mu(A) = \sum_{B \subseteq A} (-1)^{|A-B|} \mu(B), \tag{13}$$

for each $A \subseteq C$. m_μ is also called as Harsanyi dividend or simply dividend of μ. Möbius transform $m_\mu(A)$ shows the value obtained by the creation of a set (coalition) $A \subseteq C$, i.e., additional value which cannot be obtained by sets B strictly smaller than A. Namely, $m_\mu(A)$ can be also seen as a kind of interaction index of A.

Corresponding to L-reducts, the following set function μ^{Q} is considered and the attribute importance and interaction indices have been calculated in Greco et al. [1]:

$$\mu^{\mathrm{Q}}(A) = \gamma_A(\mathcal{D}) = \frac{\sum_{i=1}^{p} |A_*(D_i)|}{|U|}, \tag{14}$$

where $\gamma_A(\mathcal{D})$ is called a 'quality of approximation' of partition $\mathcal{D}$. Moreover, corresponding to U-reducts, the following two set functions μ^{sp} and μ^{cl} are considered and the attribute importance and interaction indices have been calculated in Inuiguchi and Tsurumi [2]:

$$\mu^{\mathrm{sp}}(A) = \sigma_A(\mathcal{D}) = \frac{\sum_{i=1}^{p} |U - A^*(D_i)|}{(p-1)|U|} = \frac{\sum_{x \in U} (p - |\partial_A(x)|)}{(p-1)|U|}, \tag{15}$$

$$\mu^{\mathrm{cl}}(A) = \pi_A(\mathcal{D}) = \gamma_A(\mathcal{E}) = \frac{\sum_{i=1}^{q} |A_*(E_i)|}{|U|}, \tag{16}$$

where $\mathcal{E}$ is a partition with elementary sets $E_i = \{y \in U \mid \partial_C(y) = \partial_C(x)\}$ for any $x \in U$, $\partial_A(x) = \{D_i \mid y \in D_i, y \in [x]_A\}$ for $A \subseteq C$ and $q = |\mathcal{E}|$. $\sigma_A(\mathcal{D})$ and $\pi_A(\mathcal{D})$ are called a 'measure of specificity' and a 'classification power index' with respect to partition $\mathcal{D}$, respectively.

We note that L- and U-reducts can be characterized as follows by using measures μ^{Q}, μ^{sp} and μ^{cl}. Namely, a set of condition attributes $A \subseteq C$ is called an L-reduct if and only if it satisfies (L1′) $\mu^{\mathrm{Q}}(A) = \mu^{\mathrm{Q}}(C)$ and (L2′) $\nexists a \in A$, $\mu^{\mathrm{Q}}(A - \{a\}) = \mu^{\mathrm{Q}}(C)$. Similarly, a set of condition attributes $A \subseteq C$ is called an U-reduct if and only if it satisfies (U1′) $\mu^{\mathrm{sp}}(A) = \mu^{\mathrm{sp}}(C)$ and (U2′) $\nexists a \in A$, $\mu^{\mathrm{sp}}(A - \{a\}) = \mu^{\mathrm{sp}}(C)$, or a set of condition attributes $A \subseteq C$ is called an U-reduct if and only if it satisfies (U1″) $\mu^{\mathrm{cl}}(A) = \mu^{\mathrm{cl}}(C)$ and (U2″) $\nexists a \in A$, $\mu^{\mathrm{cl}}(A - \{a\}) = \mu^{\mathrm{cl}}(C)$. In the sense described above, μ^{Q} corresponds to L-reducts while μ^{sp} and μ^{cl} correspond to U-reducts.

3 Refinement of Attribute Reduction and Attribute Importance

3.1 Refinement of Attribute Reduction

Consider a cover $\mathcal{F}_k = \{D_{i_1} \cup D_{i_2} \cup \cdots \cup D_{i_k} \mid 1 \le i_1 < i_2 < \cdots < i_k \le p\}$ for $k \in \{1, 2, \ldots, p-1\}$. A condition attribute set A is called an $\mathcal{F}_k$-reduct if and only if (F1(k)) $A_*(F) = C_*(F)$ for all $F \in \mathcal{F}_k$ and (F2(k)) $\nexists a \in A, (A - \{a\})_*(F) = C_*(F)$ for all $F \in \mathcal{F}_k$.

From (9) and (10), we know that an $\mathcal{F}_k$-reduct A is a minimal set such that $A^*(F) = C^*(F)$ for all $F \in \mathcal{F}_{p-k}$. Moreover, from (6), an $\mathcal{F}_k$-reduct A satisfies (F$_l$1) for all $l \le k$ and therefore, from (9) and (10), it satisfies $A^*(F) = C^*(F)$ for all $F \in \mathcal{F}_{p-l}$ and for all $l \le k$. Note that $\mathcal{F}_1$-reducts are equivalent to L-reducts and $\mathcal{F}_{p-1}$-reducts are equivalent to U-reducts. From this observation the strong-weak relations among $\mathcal{F}_k$-reducts for $1 \le k \le p-1$ can be depicted as in Fig. 1. The reducts located on the upper side of Fig. 1 are strong, i.e., the condition to be the upper reduct is stronger than the lower. On the contrary, the reducts located on the lower side of Fig. 1 are weak, i.e., the condition to be the lower reduct is weaker than the upper. Therefore, for any reduct A located on the upper side, there exists a reduct B located on the lower side such that $B \subseteq A$.

Let $\mathcal{R}(k)$ be a set of $\mathcal{F}_k$-reducts. Then $\bigcap \mathcal{R}(k)$ is called the '$\mathcal{F}_k$-core'. Attributes in the L-core are important because without any of them, we cannot preserve $C_*(F)$ for all $F \in \mathcal{F}_k$.

As all L-reducts can be calculated using a discernibility matrix [7], all $\mathcal{F}_k$-reducts for $1 \le k \le p-1$ can be calculated by a discernibility matrix. The (i, j)-component $\mathcal{D}^k_{ij}$ of the discernibility matrix $\mathcal{D}^k$ for calculating $\mathcal{F}_k$-reducts is obtained as the following set of attributes:

$$\mathcal{D}^k_{ij} = \begin{cases} \{a \in C \mid \rho(x_i, a) \neq \rho(x_j, a)\}, & \text{if } \partial_C(x_i) \neq \partial_C(x_j) \text{ and } |\partial_C(x_i)| \le k, \\ C, & \text{otherwise.} \end{cases} \tag{17}$$

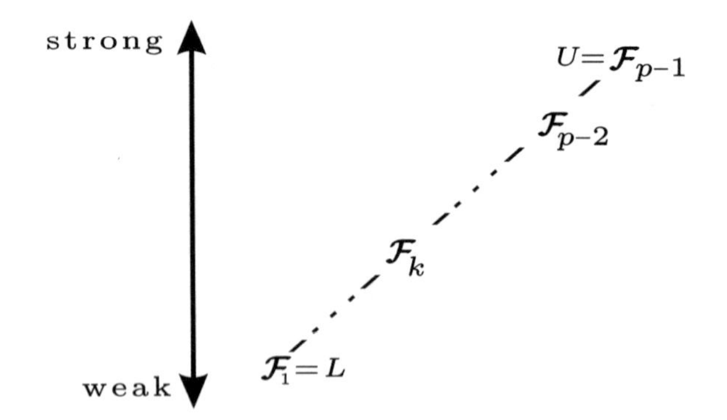

Fig. 1 The strong-weak relation among reducts

Then all $\mathcal{F}_k$-reducts are obtained as prime implicants of a Boolean function,

$$f^k = \bigwedge_{i,j:x_i,x_j \in U} \bigvee \mathcal{D}^k_{ij}, \tag{18}$$

where we regard $a \in \mathcal{D}^k_{ij}$ as a statement that 'the reduct includes a'.

Note that $\mathcal{D}^k_{ij}$ can be obtained from $\mathcal{D}^l_{ij}$ with $l > k$ by exchanging all components of i-th row such that $|\partial_C(x_i)| > k$ with C. Then, once $\mathcal{D}^{p-1}_{ij}$ is obtained, the other decision matrices can be obtained easily.

3.2 Attribute Importance

Attribute importance can be also considered with respect to $\mathcal{F}_k$. Corresponding to (15) and (16), we can define the following set functions with respect to $\mathcal{F}_k$ ($1 \leq k \leq p-1$);

$$\mu^{\mathrm{sp}}_k(A) = \frac{\displaystyle\sum_{\substack{i_1,\ldots,i_k \in \{1,\ldots,p\} \\ i_j \neq i_l}} |A_*(D_{i_1} \cup \cdots \cup D_{i_k})|}{\binom{p-1}{k-1}|U|}, \tag{19}$$

$$\mu^{\mathrm{cl}}_k(A) = \frac{\displaystyle\sum_{\substack{E_i \in \mathcal{E} \\ x \in E_i, |\partial_C(x)| \leq k}} |A_*(E_i)|}{|U|}. \tag{20}$$

Note that when $k = 1$, both $\mu^{\mathrm{sp}}_k(A)$ and $\mu^{\mathrm{cl}}_k(A)$ equal to $\mu^{\mathrm{Q}}(A)$ in (14). On the other hand, we have $\mu^{\mathrm{sp}}(A) = \mu^{\mathrm{sp}}_k(A)$ and $\mu^{\mathrm{cl}}(A) = \mu^{\mathrm{cl}}_k(A) + |\{x \in U \mid |\partial_C(x)| = p\}| / |U|$ when $k = p - 1$. Moreover, $\mu^{\mathrm{cl}}(A) = \mu^{\mathrm{cl}}_k(A)$ when $k = p$. When there is no $x \in U$ such that $|\partial_C(x)| = p$, we have $\mu^{\mathrm{cl}}(A) = \mu^{\mathrm{cl}}_{p-1}(A)$.

As μ^{Q} corresponds to L-reducts and both μ^{sp} and μ^{cl} correspond to U-reducts, both μ^{sp}_k and μ^{cl}_k correspond to $\mathcal{F}_k$-reducts. Namely, $A \subseteq C$ is an $\mathcal{F}_k$-reduct if and only if (F1$(k)'$) $\mu^{\mathrm{sp}}_k(A) = \mu^{\mathrm{sp}}_k(C)$ and (F2$(k)'$) $\nexists a \in A$, $\mu^{\mathrm{sp}}_k(A - \{a\}) = \mu^{\mathrm{sp}}_k(C)$. Similarly, $A \subseteq C$ is an $\mathcal{F}_k$-reduct if and only if (F1$(k)''$) $\mu^{\mathrm{cl}}_k(A) = \mu^{\mathrm{cl}}_k(C)$ and (F2$(k)''$) $\nexists a \in A$, $\mu^{\mathrm{cl}}_k(A - \{a\}) = \mu^{\mathrm{cl}}_k(C)$.

As shown above, we obtain intermediate set functions between μ^{Q} and μ^{sp} as well as between μ^{Q} and μ^{cl}. Moreover, by a linear combination of those set functions μ^{sp}_k and μ^{cl}_k, we may obtain different attribute importance and interaction indices as well as different Möbius transforms.

We introduce one more set function μ^{∂}_k related to μ^{cl}_k. Similar to μ^{cl}_k, we may define

$$\mu_k^{\partial}(A) = \frac{\displaystyle\sum_{i:|\partial_C(x_i)|\leq k} [\partial_C(x_i) = \partial_A(x_i)]}{|U|}, \tag{21}$$

where $[statement]$ takes 1 if $statement$ is true and 0 otherwise. we have

$$\mu_k^{\rm cl}(A) \leq \mu_k^{\partial}(A), \ \forall A \subseteq C. \tag{22}$$

Although $\partial_C(x_i) = \partial_A(x_i)$ for some $x_i \in U$, x_i is not always counted in the numerator of $\mu_k^{\rm cl}$. For example, when there exists $x_j \in U$ such that $\partial_C(x_j) \subset \partial_C(x_i)$ but $\partial_A(x_j) = \partial_A(x_i) = \partial_C(x_i)$, $x_i \in U$ satisfying $\partial_A(x_i) = \partial_C(x_i)$ is not counted in the numerator of $\mu_k^{\rm cl}$ because $A_*(E_k) = \emptyset$ for $E_k \in \mathcal{E}$ such that $x_i \in E_k$. This evaluation is same as that for $x_l \in U$ such that $\partial_C(x_l) \neq \partial_A(x_l)$. From the viewpoint of the preservation of generalized decision classes of $x \in U$, $\mu_k^{\rm cl}$ underestimates the ability of A. $\mu_k^{\rm cl}$ evaluates the degree of the preservation of partition $\mathcal{E}$ while μ_k^{∂} evaluates the degree of the preservation of generalized classes of $x \in U$. The preservation of partition $\mathcal{E}$ would be less important than the preservation of generalized decision classes $\partial_C(x)$, $x \in U$. From this reason, we think μ_k^{∂} is more significant than $\mu_k^{\rm cl}$.

Measure μ_k^{∂} also corresponds to $\mathcal{F}_k$-reducts. Namely, $A \subseteq C$ is an $\mathcal{F}_k$-reduct if and only if (F1$(k)'''$) $\mu_k^{\partial}(A) = \mu_k^{\partial}(C)$ and (F2$(k)'''$) $\nexists a \in A$, $\mu_k^{\partial}(A - \{a\}) = \mu_k^{\partial}(C)$. Moreover, we have $\mu_k^{\partial} = \mu^{\rm Q}$ when $k = 1$.

As μ_k^{∂} is defined by generalized decision classes $\partial_A(x)$, $x \in U$, we express $\mu_k^{\rm sp}$ also by using generalized decision classes $\partial_A(x)$, $x \in U$. We obtain

$$\mu_k^{\rm sp}(A) = \frac{\displaystyle\sum_{i:|\partial_A(x_i)|\leq k} \binom{p - |\partial_A(x_i)|}{k - |\partial_A(x_i)|}}{\binom{p-1}{k-1}|U|}. \tag{23}$$

As shown in (23), $\mu_k^{\rm sp}$ depends on the sizes of generalized decision classes, $|\partial_A(x)|$, $x \in U$. As $|\partial_A(x)|$, $x \in U$ are small, the decision classes of objects $x \in U$ are specifically determined. Therefore, we recognize that $\mu_k^{\rm sp}$ evaluates well the degree of specificity. Moreover, we note that $\mu_k^{\rm sp}$ does not be influenced by $\partial_C(x)$, $x \in U$ but only $\partial_A(x)$, $x \in U$.

While $\mu_k^{\rm sp}$ is influenced by the size of generalized decision classes of objects x_i such that $|\partial_A(x_i)| \leq k$, μ_k^{∂} is influenced only by the preservation of $\partial_C(x_i)$. Therefore, μ_k^{∂} does not reflect size $|\partial_A(x_i)|$ very much and neither $\mu_k^{\rm cl}$ due to (22). From this observation, the linear combinations would be meaningful for μ_k^{∂} $k = 1, 2, \ldots, p-1$. For example, set functions

$$\hat{\mu}_k^{\partial}(A) = \frac{\displaystyle\sum_{q=1}^{k} \mu_q^{\partial}(A)}{k}, \ A \subseteq C, \ k = 1, 2, \ldots, p-1. \tag{24}$$

would be meaningful. Because of the linearity of Shapley interaction index and Möbius transform, we have

$$I^S_{\hat{\mu}^\partial_k}(A) = \frac{\sum_{q=1}^{k} I^S_{\mu^\partial_q}(A)}{k}, \; A \subseteq C, \; k = 1, 2, \ldots, p-1, \tag{25}$$

$$m_{\hat{\mu}^\partial_k}(A) = \frac{\sum_{q=1}^{k} m_{\mu^\partial_k}(A)}{k}, \; A \subseteq C, \; k = 1, 2, \ldots, p-1. \tag{26}$$

We note that μ^{sp}_k, μ^{cl}_k, μ^∂_k and $\hat{\mu}^\partial_k$ described in this section satisfy the monotonicity, i.e., $\mu : 2^C \to \mathbf{R}$ satisfies the monotonicity if and only if $\mu(A) \geq \mu(C), \forall A, B \subseteq C$. Therefore, the normalized those set functions can be seen as fuzzy measures.

4 A Numerical Example

Consider the decision table shown in Table 1. The decision table is given in a profile-wise way. There are four decision attribute values 1, 2, 3 and 4. In column d of Table 1, a frequency distribution of objects sharing a common profile (condition attribute values) is given. For example, (1, 0, 1, 0) on row w_1 implies that there are two objects taking 1 for a_1, 1 for a_2, 1 for a_3 and 1 for a_4 and one of them takes 1 for d while the other takes 3 for d. Similarly, (2, 0, 0, 0) on row w_3 implies that there are two objects taking 2 for a_1, 2 for a_2, 3 for a_3, 1 for a_4 and 1 for d.

By the proposed decision matrix method, we obtain $\{a_2\}$ and $\{a_1, a_3\}$ as $\mathcal{F}_1$-reducts, $\{a_1, a_3\}$ and $\{a_1, a_2, a_4\}$ as $\mathcal{F}_2$-reducts, and $\{a_1, a_2, a_4\}$ and $\{a_1, a_3, a_4\}$ as $\mathcal{F}_3$-reducts. We have $\{a_1\}$ as the $\mathcal{F}_2$-core and $\{a_1, a_4\}$ as the $\mathcal{F}_3$-core. We have no

Table 1 A decision table

Profile	a_1	a_2	a_3	a_4	d
w_1	1	1	1	1	(1, 0, 1, 0)
w_2	1	1	2	2	(1, 0, 0, 1)
w_3	2	2	3	1	(2, 0, 0, 0)
w_4	3	3	4	2	(0, 1, 0, 0)
w_5	4	1	2	2	(0, 1, 1, 0)
w_6	2	5	3	2	(1, 0, 0, 0)
w_7	2	4	4	2	(0, 0, 2, 0)
w_8	4	1	5	5	(0, 1, 1, 1)
w_9	4	1	5	4	(1, 0, 1, 1)

Table 2 Measures, Shapley interaction indices and Möbius transforms

A	μ_1^{sp}	$I^S_{\mu_1^{\mathrm{sp}}}$	$m_{\mu_1^{\mathrm{sp}}}$	μ_2^{sp}	$I^S_{\mu_2^{\mathrm{sp}}}$	$m_{\mu_2^{\mathrm{sp}}}$	μ_3^{sp}	$I^S_{\mu_3^{\mathrm{sp}}}$	$m_{\mu_3^{\mathrm{sp}}}$
a_1	0.0556	0.0556	0.0556	0.1481	0.1204	0.1481	0.3148	0.2099	0.3148
a_2	0.3333	0.1759	0.3333	0.3333	0.1512	0.3333	0.3333	0.1265	0.3333
a_3	0.1667	0.0926	0.1667	0.2593	0.1265	0.2593	0.3519	0.1481	0.3519
a_4	0	0.0093	0	0.0741	0.0463	0.0741	0.2593	0.1821	0.2593
a_1a_2	0.3333	−0.1296	−0.0556	0.3333	−0.1358	−0.1481	0.4074	−0.1420	−0.2407
a_1a_3	0.3333	0.0370	0.1111	0.4444	0.0123	0.0370	0.5556	−0.0494	−0.1111
a_1a_4	0.1667	0.0370	0.1111	0.3333	0.0494	0.1111	0.6111	0.0247	0.0370
a_2a_3	0.3333	−0.1852	−0.1667	0.3704	−0.1728	−0.2222	0.4074	−0.1605	−0.2778
a_2a_4	0.3333	−0.0185	0.0000	0.3704	−0.0247	−0.0370	0.5185	−0.0309	−0.0741
a_3a_4	0.1667	−0.0185	0	0.2593	−0.0988	−0.0741	0.4630	−0.1420	−0.1481
$a_1a_2a_3$	0.3333	−0.0556	−0.1111	0.4444	0.0556	0.0370	0.5556	0.1667	0.1852
$a_1a_2a_4$	0.3333	−0.0556	−0.1111	0.4444	−0.0185	−0.0370	0.6667	0.0185	0.0370
$a_1a_3a_4$	0.3333	−0.0556	−0.1111	0.4444	−0.0926	−0.1111	0.6667	−0.0556	−0.0370
$a_2a_3a_4$	0.3333	0.0556	0	0.3704	0.0556	0.0370	0.5185	0.0556	0.0741
C	0.3333	0.1111	0.1111	0.4444	0.0370	0.0370	0.6667	−0.0370	−0.0370

A	μ_1^{∂}	$I^S_{\mu_1^{\partial}}$	$m_{\mu_1^{\partial}}$	μ_2^{∂}	$I^S_{\mu_2^{\partial}}$	$m_{\mu_2^{\partial}}$	μ_3^{∂}	$I^S_{\mu_3^{\partial}}$	$m_{\mu_3^{\partial}}$
a_1	0.0556	0.0556	0.0556	0.0556	0.2037	0.0556	0.0556	0.2037	0.0556
a_2	0.3333	0.1759	0.3333	0.3333	0.1759	0.3333	0.3333	0.1759	0.3333
a_3	0.1667	0.0926	0.1667	0.2778	0.1852	0.2778	0.2778	0.1852	0.2778
a_4	0	0.0093	0	0.1111	0.1019	0.1111	0.4444	0.4352	0.4444
a_1a_2	0.3333	−0.1296	−0.0556	0.3333	−0.1296	−0.0556	0.3333	−0.1296	−0.0556
a_1a_3	0.3333	0.0370	0.1111	0.6667	0.1481	0.3333	0.6667	0.1481	0.3333
a_1a_4	0.1667	0.0370	0.1111	0.5000	0.1481	0.3333	0.8333	0.1481	0.3333
a_2a_3	0.3333	−0.1852	−0.1667	0.4444	−0.1852	−0.1667	0.4444	−0.1852	−0.1667
a_2a_4	0.3333	−0.0185	0	0.4444	−0.0185	0	0.7778	−0.0185	0
a_3a_4	0.1667	−0.0185	0	0.2778	−0.2407	−0.1111	0.6111	−0.2407	−0.1111
$a_1a_2a_3$	0.3333	−0.0556	−0.1111	0.6667	−0.0556	−0.1111	0.6667	−0.0556	−0.1111
$a_1a_2a_4$	0.3333	−0.0556	−0.1111	0.6667	−0.0556	−0.1111	1	−0.0556	−0.1111
$a_1a_3a_4$	0.3333	−0.0556	−0.1111	0.6667	−0.2778	−0.3333	1	−0.2778	−0.3333
$a_2a_3a_4$	0.3333	0.0556	0	0.4444	0.0556	0	0.7778	0.0556	0
C	0.3333	0.1111	0.1111	0.6667	0.1111	0.1111	1	0.1111	0.1111

$\mathcal{F}_1$-core, i.e., the $\mathcal{F}_1$-core is the empty set. Shapley interaction indices and Möbius transform with respect to μ_k^{sp} and μ_k^{∂} are obtained as in Table 2. In the first column of Table 2, objects in A are shown except case $A = C$. As shown in Table 2, the attribute importance degrees $I^S_\mu(\{a_i\})$, $i = 1, 2, 3, 4$ are different by k and between μ_k^{sp} and μ_k^{∂}. Moreover, we observe signs of $I^S_\mu(\{a_i\})$ and m_μ are similar. When $k = 1$, a_2 takes the largest attribute importance degree. This is consistent with the fact that $\{a_2\}$ is a unique singleton $\mathcal{F}_1$-reduct (L-reduct). Attributes a_1 and a_3 compose the other $\mathcal{F}_1$-reduct. Between a_1 and a_3, a_3 takes the larger attribute importance degree than a_1. This implies that attribute a_3 determines the decision attribute values more than a_1.

Indeed, $mu_1^{\rm sp} = \mu_1^{\partial}$ takes larger or equal values for attribute set A including a_3 than for attribute set A including a_1. As shown in Table 2, $I^S_{\mu_1^{\rm sp}}(\{a_1, a_3\}) = I^S_{\mu_1^{\partial}}(\{a_1, a_3\})$ and $I^S_{\mu_1^{\rm sp}}(\{a_1, a_4\}) = I^S_{\mu_1^{\partial}}(\{a_1, a_4\})$ take same positive values. Namely attributes a_1 and a_3 are complementary, i.e., their simultaneous existence creates a synergy effect. $I^S_{\mu_1^{\rm sp}}(B) = I^S_{\mu_1^{\partial}}(B)$, for any $B \subseteq C$ such that $|B| = 2$ and $a_2 \in B$ take negative values. This fact implies any condition attribute have a cancel effect with a_2 because $\{a_2\}$ is an $\mathcal{F}_1$-reduct (L-reduct) so that only a_2 is sufficient. $I^S_{\mu_1^{\rm sp}}(\{a_3, a_4\}) = I^S_{\mu_1^{\partial}}(\{a_3, a_4\})$ take a negative value because additional information about attribute a_4 does not give any effect on $\mu_1^{\rm sp}$ and μ_1^{∂} when values of a_3 are known.

When $k = 2$, $\{a_1\}$ is the $\mathcal{F}_2$-core. Namely, a_1 is qualitatively important. Nevertheless, attribute importance degree $I^S_{\mu_2^{\rm sp}}(\{a_1\})$ is less than $I^S_{\mu_2^{\rm sp}}(\{a_2\})$ and $I^S_{\mu_2^{\rm sp}}(\{a_3\})$. This is because the improvement of the specificity $I^S_{\mu_2^{\rm sp}}$ by adding attribute a_1 is not very large. Therefore, the attributes in the $\mathcal{F}_k$-core would not always be important quantitatively. On the other hand, attribute importance degree $I^S_{\mu_2^{\partial}}(\{a_1\})$ is larger than $I^S_{\mu_2^{\partial}}(\{a_i\})$, $i = 2, 3, 4$. This result coincides well to the fact that a_1 is in the $\mathcal{F}_2$-core. By the addition of a_1 to $\{a_3\}$ and $\{a_4\}$, there are big positive improvements in μ_2^{∂}. This fact can also be observed from values of $m_{\mu_2^{\partial}}(\{a_1, a_3\})$ and $m_{\mu_2^{\partial}}(\{a_1, a_4\})$. In this case, the preservation degree of generalized decision classes expresses well the importance degrees of attributes in the $\mathcal{F}_2$-core. Because we do not think the sizes of generalized decision classes in μ_2^{∂}, the importance degrees with respect to μ_2^{∂} would reflect the membership to $\mathcal{F}_2$-core more than those with respect to $\mu_2^{\rm sp}$. $I^S_{\mu_2^{\rm sp}}$ and $m_{\mu_2^{\rm sp}}$ are rather different from $I^S_{\mu_2^{\partial}}$ and $m_{\mu_2^{\partial}}$, respectively. This difference comes from the difference between underlying measures $\mu_2^{\rm sp}$ and μ_2^{∂}: $\mu_2^{\rm sp}$ takes into consideration the sizes of generalized decision while μ_2^{∂} does not.

When $k = 3$, $\{a_1, a_4\}$ is the $\mathcal{F}_3$-core. Namely, a_1 and a_4 are qualitatively important. In this case even in $\mu_3^{\rm sp}$, we observe that the attribute importance degrees $I^S_{\mu_3^{\rm sp}}(\{a_1\})$ and $I^S_{\mu_3^{\rm sp}}(\{a_4\})$ are larger than $I^S_{\mu_3^{\rm sp}}(\{a_2\})$ and $I^S_{\mu_3^{\rm sp}}(\{a_3\})$. Similarly, the attribute importance degrees $I^S_{\mu_3^{\partial}}(\{a_1\})$ and $I^S_{\mu_3^{\partial}}(\{a_4\})$ are larger than $I^S_{\mu_3^{\partial}}(\{a_2\})$ and $I^S_{\mu_3^{\partial}}(\{a_3\})$. Then, in both cases, the attribute importance degrees correspond to the membership to the $\mathcal{F}_3$-core. However, $I^S_{\mu_3^{\rm sp}}(\{a_1\})$ is larger than $I^S_{\mu_3^{\rm sp}}(\{a_4\})$ while $I^S_{\mu_3^{\partial}}(\{a_1\})$ is smaller than $I^S_{\mu_3^{\partial}}(\{a_4\})$. As described before, $\mu_3^{\rm sp}$ takes into consideration the sizes of generalized decision while μ_3^{∂} does not. This implies that adding a_1 would keep generalized decision classes small-sized, while adding a_4 would preserve a lot of generalized decision classes. $I^S_{\mu_3^{\rm sp}}$ and $m_{\mu_3^{\rm sp}}$ are different from $I^S_{\mu_3^{\partial}}$ and $m_{\mu_3^{\partial}}$ to some extent. For example, when $A = \{a_3, a_4\}$, they are very different. These differences imply that the evaluations are significantly different by viewpoints: specificity of decision class versus preservation of generalized decision classes.

Because μ_k^{∂} does not consider the sizes of generalized decision classes $\partial_C(x)$, $x \in U$, we calculate Shapley interaction indices and Möbius transforms for $\hat{\mu}_k^{\partial}$, $k = 2, 3$. The smaller the preserved generalized decision classes $\partial_C(x)$ is, the larger weights

Table 3 Shapley interaction indices and Möbius transforms of $\hat{\mu}_k^{\partial}$

A	$\hat{\mu}_2^{\partial}$	$I^S_{\hat{\mu}_2^{\partial}}$	$m_{\hat{\mu}_2^{\partial}}$	$\hat{\mu}_3^{\partial}$	$I^S_{\hat{\mu}_3^{\partial}}$	$m_{\hat{\mu}_3^{\partial}}$
a_1	0.0556	0.1296	0.0556	0.0556	0.1543	0.0556
a_2	0.3333	0.1759	0.3333	0.3333	0.1759	0.3333
a_3	0.2222	0.1389	0.2222	0.2407	0.1543	0.2407
a_4	0.0556	0.0556	0.0556	0.1852	0.1821	0.1852
a_1a_2	0.3333	−0.1296	−0.0556	0.3333	−0.1296	−0.0556
a_1a_3	0.5	0.0926	0.2222	0.5556	0.1111	0.2593
a_1a_4	0.3333	0.0926	0.2222	0.5000	0.1111	0.2593
a_2a_3	0.3889	−0.1852	−0.1667	0.4074	−0.1852	−0.1667
a_2a_4	0.3889	−0.0185	0	0.5185	−0.0185	0
a_3a_4	0.2222	−0.1296	−0.0556	0.3519	−0.1667	−0.0741
$a_1a_2a_3$	0.5	−0.0556	−0.1111	0.5556	−0.0556	−0.1111
$a_1a_2a_4$	0.5	−0.0556	−0.1111	0.6667	−0.0556	−0.1111
$a_1a_3a_4$	0.5	−0.1667	−0.2222	0.6667	−0.2037	−0.2593
$a_2a_3a_4$	0.3889	0.0556	0	0.5185	0.0556	0
C	0.5	0.1111	0.1111	0.6667	0.1111	0.1111

$\hat{\mu}_k^{\partial}$ puts. The results are shown in Table 3. The obtained values are different from those obtained for μ_k^{sp} and μ_k^{∂}. We recognize again that the attribute importance degree strongly depends on the viewpoint. The major differences between μ_k and $\hat{\mu}_k^{\partial}$ is in the following fact. While $\mu_k(A)$ considers the size of $\partial_A(x)$ for all $x \in U$, $\hat{\mu}_k^{\partial}$ considers the size of $\partial_A(x)$ only for $x \in U$ such that $\partial_A(x) = \partial_C(x)$.

5 Application Guide for the Proposed Approach

We have proposed a few measures showing the performances of attribute set. The attribute importance degrees and Shapley interaction indices are significantly different depending on the measures. In this section, we describe an application guide for the proposed attribute importance degrees and Shapley interaction indices. The usage of those important degrees and interaction indices depends on the situation of the analysis. We should two kinds of selection: parameter k and measure we use.

The selection of parameter k depends on the acceptable size of generalized decision class $\partial_C(x)$ of an object $x \in U$. If we regard $\partial_C(x) \leq \bar{k}$ informative, in other words, we do not think that $\partial_C(x) > \bar{k}$ is valuable information, we select $k = \bar{k}$. Now, we describe the selection of the measure. When we evaluate an attribute set $A \subseteq C$ by its specificity, i.e., to what extent the decision classes of objects are known precisely with the attributes in A, the measure of specificity μ_k^{sp} should be selected. When we evaluate an attribute set A by its preservation of generalized decision class $\partial_C(x)$,

$x \in U$, measure μ_k^∂ should be used. When the preservation of small-sized generalized decision class is more important than the preservation of large-sized generalized decision class, measure $\hat{\mu}_k^\partial$ should be selected, where we note that $\hat{\mu}_1^\partial = \mu_1^\partial$.

In such a way, when we have a meaningful measure for the evaluation of an attribute set $A \subseteq C$, we apply Shapley interaction indices as well as Möbius transforms to the measure for the evaluation of attribute importance degrees and interaction indices.

6 Possible Application and Modifications

There are many reducts as shown in the previous sections. Then there may have a question which reduct we should use. The selection of reduct basically depends on the aim of application. In this section, we would suggest a possible application to group decision making and give a modification in the case when decision attribute is ordinal.

Assume that there are several people who classify same objects into p classes. Each person can classify objects based on their experiences and feelings. We assume also that objects are characterized by a certain number of condition attributes and that people's evaluations are totally based on those condition attribute values. Using the classification results by all people, we would like to assign a class agreeable among all people to an object. Avoiding a big argument, we do not need to make such assignments for all objects but as many objects as we can.

Even if people assign different classes to an object, we may decide the agreeable class by some negotiation. However, if people assign many different classes to an object, it would be difficult to have a consensus in the assignment of the class. Therefore, we may restrict the negotiation into objects having small difference. Considering such a situation, we may restrict the negotiation into objects having k different classes assigned by people. Using $\mathcal{F}_k$-reducts, μ_k^{sp}, μ_k^∂ and $\hat{\mu}_k^\partial$, we may know which attributes are important. Therefore, we may discuss the classification of the object considering from the most important condition attributes to the least important conditions attributes.

Considering quality evaluations rather than classification, we may face situations where decision attribute is ordinal. Namely, we may have an order $D_1 \prec D_2 \prec \cdots \prec D_p$ on decision classes. We again consider the case when several people evaluate the decision attribute values on same objects and assign unique decision attribute value agreeable among all people to an object. In this case, we obtain the strength of difference in object evaluation between decision classes, i.e., the difference in object evaluation between D_i and D_{i+2} is larger than that between D_i and D_{i+1}. Moreover, let us assume that the difference in object evaluation between D_i and D_{i+1} is a constant for any $i \in [1, p-1]$.

Under this situation, the application of $\mathcal{F}_k$-reducts suggested above does not work well because the preservation of lower approximation of $D_1 \cup D_p$ is qualitatively different from that of the lower approximation of $D_i \cup D_{i+1}$, i.e., while the latter preserves the closeness of the individual evaluations into D_i or D_{i+1}, the former does not always but individual evaluations into two extreme classes. In order to treat the situation, we should take care of the order of decision attribute values and consider reducts, importance degrees suitable for the situation.

Considering the order, we propose the following modifications. Let us introduce the following notations: $D_{[l,r]} = \{D_k \mid k \in [l, r]\}$ $(l \leq r)$, $d_{\min}(x_i) = \min\{v \mid D_v \in \partial_P(x_i)\}$ and $d_{\max}(x_i) = \max\{v \mid D_v \in \partial_P(x_i)\}$. First, corresponding to the general decisions $\partial_P(x_i)$, we define a decision attribute interval $\mathcal{I}_P(x_i)$ by

$$\mathcal{I}_P(x_i) = \{D_l \mid l \in [d_{\min}(x_i), d_{\max}(x_i)]\}\,. \tag{27}$$

Moreover, we define $\mathcal{I}_k = \{C_*(D_{[l,r]}) \mid r - l = k - 1\}$. Then we can define $\mathcal{I}_k$-reducts by replacing $\mathcal{F}_k$ with $\mathcal{I}_k$. Since $\mathcal{I}_k$ is a family of a decision class intervals, $\mathcal{I}_k$-reducts are more suitable to the case when the decision attribute is ordinal.

The calculations of $\mathcal{I}_k$-reducts can be made by a discernibility matrix (17) with replacement of $\partial_C(x_i)$ with $\mathcal{I}_C(x_i)$. The measures corresponding to $\mathcal{I}_k$-reducts are defined by

$$\nu_k^{\mathrm{sp}}(A) = \sum_{i=1}^{p-k+1} \frac{|A_*(D_{[i,i+k-1]})|}{(p-k+1)|U|}, \tag{28}$$

$$\nu_k^{\partial}(A) = \frac{\displaystyle\sum_{i:|\mathcal{I}_C(x_i)|\leq k} [\mathcal{I}_C(x_i) = \mathcal{I}_A(x_i)]}{|U|}, \tag{29}$$

$$\hat{\nu}_k^{\partial}(A) = \frac{\displaystyle\sum_{q=1}^{k} \nu_q^{\partial}(A)}{k}. \tag{30}$$

The alternations described above are suitable for the situation where the decision attribute is ordinal.

7 Concluding Remarks

We have shown that several kinds of attribute reduction exist between previous two reducts: L- and U-reducts. We have shown that they are calculated also by discernibility matrices. We investigate measures corresponding to the proposed several kinds of attribute reduction. We proposed three measures showing specificity, preservation of generalized decision classes and the weighted preservation of generalized decision classes. Attribute importance degrees and interaction indices associated with the measures have been studied. We showed that the attribute importance degrees and interaction indices are significantly different by the underlying measure. Moreover, a possible application of the proposed approach is described and a modification for a case when the decision attribute is ordinal is given.

The applications of the proposed attribute importance degrees and interaction indices as well as a further development of the proposed approach are topics for the future research.

References

1. Greco, S., Matarazzo, B., Słowiński, R.: Fuzzy measure technique for rough set analysis. Proceedings of EUFIT'98 (1998) 99–103.
2. Inuiguchi, M., Tsurumi, M.: Measures based on upper approximations of rough sets for analysis of attribute importance and interaction. International Journal of Innovative Computing, Information & Control **2**(1) (2006) 1–12.
3. Kusunoki, Y., Inuiguchi, M.: Structure-based attribute reduction: A rough set approach. in: U. Stańczyk, L. C. Jain (Eds.), Feature Selection for Data and Pattern Recognition, Springer-Verlag, Belrin Heidelberg (2014) 113–160.
4. Miao, D.Q., Zhao, Y., Yao, Y. Y., Li, H. X., Xu, F. F.: Relative reducts in consistent and inconsistent decision tables of Pawlak rough set model, Information Science, **179** (2009) 4140–4150.
5. Pawlak, Z.: Rough sets, International Journal of Information and Computer Science, **11**(5) (1982) 341–356.
6. Pawlak, Z.: Rough Sets: Theoretical Aspects of Reasoning about Data Kluwer Academic Publishers, Dordrecht (1991).
7. Skowron, A., Rauser, C. M.: The discernibility matrix and function in information systems. in: R. Słowiński (Ed.), Intelligent Decision Support: Handbook of Application and Advances of Rough Set Theory Kluwer Academic Publishers, Dordrecht (1992) 331–362.
8. Ślęzak, D.: Various approaches to reasoning with frequency based decision reducts: a survey. in: L. Polkowski, S. Tsumoto, T.Y. Lin (Eds.), Rough Set Methods and Applications, Physica-Verlag, Heidelberg (2000) 235–285.

A Review on Rough Set-Based Interrelationship Mining

Yasuo Kudo and Tetsuya Murai

Abstract Interrelationship mining, proposed by the authors, aims at extracting characteristics of objects based on interrelationships between attributes. Interrelationship mining is an extension of rough set-based data mining, which enables us to extract characteristics based on comparison of values of two different attributes such that "the value of attribute a is higher than the value of attribute b." In this paper, we mainly review theoretical aspects of rough set-based interrelationship mining.

1 Introduction

Rough set theory, originally proposed by Z. Pawlak [12, 13], provides a mathematical basis of set-based approximation of concepts and logical data analysis. Various extensions of rough set models have been proposed to apply the concept of rough set to various kinds of data: Probabilistic rough sets (e.g., variable precision rough set [18], Bayesian rough set [15], three-way decisions with probabilistic rough sets [16]), rough set for incomplete information [11], and dominance-based rough set approach [1]. These extended rough set models are based on discernibility of objects by comparing attribute values. However, in Pawlak's rough set theory and these extended rough set models, interrelationship between attributes have not been disused as far as the authors know.

The authors have formulated interrelationship between attributes in the framework of rough set theory [3] and proposed a concept of rough set-based interrelationship mining to extract characteristics based on interrelations between different attributes in a given decision table [4]. Interrelationship mining is an extension of rough set-based

Y. Kudo (✉)
College of Information and Systems, Muroran Institute of Technology, Mizumoto 27-1, Muroran 050-8585, Japan
e-mail: kudo@csse.muroran-it.ac.jp

T. Murai
Faculty of Science and Technology, Chitose Institute of Science and Technology, 758-65 Bibi, Chitose 066-8655, Japan
e-mail: t-murai@photon.chitose.ac.jp

V. Torra et al. (eds.), *Fuzzy Sets, Rough Sets, Multisets and Clustering*,
Studies in Computational Intelligence 671, DOI 10.1007/978-3-319-47557-8_15

data mining, which enables us to extract characteristics based on comparison of values of two different attributes such that "the value of attribute a is higher than the value of attribute b." To the best of our knowledge, there is no other research that formulates interrelationship between attributes in the framework of rough set theory. In this paper, we review theoretical aspects of rough set-based interrelationship mining based on the authors' previous works [3, 4, 7, 9, 10].

The reminder of this paper is organized as follows: In Sect. 2, we introduce basic concepts of rough set theory. Next, a formulation of rough set-based interrelationship mining and its basic property [3, 4, 10] are introduced in Sect. 3. In Sect. 4, we review theoretical aspects of interrelated condition attributes that are newly introduced attributes to explicitly represent interrelationships between attributes [7, 9]. Finally, Sect. 5 provides conclusion of this paper. Appendix A describes proofs of theoretical properties in this paper.

2 Rough Set

In this section, we review the rough set theory, in particular, decision tables and relative reducts. Note that the contents of this section are mainly based on [13, 14].

2.1 Decision Table and Lower and Upper Approximations

Generally, data analysis subjects by rough sets are described by decision tables. Formally, a decision table is the following structure:

$$S = (U, AT, V, \rho), \tag{1}$$

where U is a finite and nonempty set of objects, $AT \stackrel{\text{def}}{=} C \cup \{\mathsf{d}\}$ is a finite and nonempty set of attributes, where C is a set of condition attributes and $\mathsf{d} \notin C$ is a decision attribute, V is a set of values of attributes, and $\rho : U \times AT \to V$ is a function that assigns a value $\rho(x, \mathsf{a}) \in V$ for each object $x \in U$ at each attribute $\mathsf{a} \in AT$.

Indiscernibility relations based on subsets of attributes provide classifications of objects in decision tables. For any set of attributes $A \subseteq AT$, the indiscernibility relation $IND(A)$ is the following binary relation on U:

$$IND(A) = \{(x, y) \in U \times U \mid \rho(x, \mathsf{a}) = \rho(y, \mathsf{a}), \forall \mathsf{a} \in A\}. \tag{2}$$

If a pair (x, y) is in $IND(A)$, then two objects x and y are indiscernible with respect to all attributes in A. It is well-known that any indiscernibility relation is an equivalence relation and, for each object $x \in U$, an equivalence class $[x]_A$ by an indiscernibility relation $IND(A)$ is defined by

$$[x]_A = \{y \in U \mid (x, y) \in IND(A)\}. \quad (3)$$

The set of all equivalence classes by an equivalence relation $IND(A)$ consists of a partition on the domain U, denoted by $U/IND(A)$. In particular, the indiscernibility relation $IND(\{\mathsf{d}\})$ based on the decision attribute d provides a partition $\mathcal{D} = \{D_1, \ldots, D_k\}$, and each element $D_i \in \mathcal{D}$ is called a decision class.

Classifying objects with respect to condition attributes provides approximation of decision classes. Formally, for any set $B \subseteq C$ of condition attributes and any decision class $D_i \in \mathcal{D}$, we let:

$$\underline{B}(D_i) = \{x \in U \mid [x]_B \subseteq D_i\}, \quad (4)$$
$$\overline{B}(D_i) = \{x \in U \mid [x]_B \cap D_i \neq \emptyset\}. \quad (5)$$

The sets $\underline{B}(D_i)$ and $\overline{B}(D_i)$ are called lower approximation and upper approximation of the decision class D_i with respect to the set B of condition attributes, respectively. Particularly, the lower approximation $\underline{B}(D_i)$ is the set of objects that are correctly classified to the decision class D_i by checking all attributes in B.

2.2 Relative Reduct

All discernible objects of the given decision table are able to be correctly classified to the corresponding decision classes by evaluating values of all condition attributes. In this case, some attributes may be essential for correct classification and other attributes may be redundant. A minimal set of condition attributes to classify all discernible objects to correct decision classes is called a relative reduct of the decision table.

For any subset $X \subseteq C$ of condition attributes in a decision table S, we let:

$$POS_X(\mathcal{D}) = \bigcup_{D_i \in \mathcal{D}} \underline{X}(D_i). \quad (6)$$

The set $POS_X(\mathcal{D})$ is called the positive region of $\mathcal{D}$ by X. All objects $x \in POS_X(\mathcal{D})$ are classified to correct decision classes by checking all attributes in X. In particular, the set $POS_C(\mathcal{D})$ is the set of all discernible objects in S.

The concept of relative reduct is formally introduce as follows: A set $A \subseteq C$ is called a relative reduct of the decision table S if the set A satisfies the following conditions:

$$1.\ \mathrm{POS}_A(\mathcal{D}) = \mathrm{POS}_C(\mathcal{D}),\ \text{and} \quad (7)$$
$$2.\ \mathrm{POS}_B(\mathcal{D}) \neq \mathrm{POS}_C(\mathcal{D})\ \text{for any proper subset}\ B \subset A. \quad (8)$$

In general, there are plural relative reducts in a decision table. The common part of all relative reducts is called the core of the decision table.

The authors [2] proposed an evaluation criterion for relative reducts based on the roughness of partitions generated from the relative reducts. For each relative reduct $A \subseteq C$, the evaluation score $ACov(A)$ of A is defined by

$$ACov(A) = \frac{|\mathcal{D}|}{\sum_{[x]_A \in U/IND(A)} |\{D_i \in \mathcal{D} \mid [x]_A \cap D_i \neq \emptyset\}|}, \quad (9)$$

where $|X|$ is the cardinality of the set X.

The score $ACov(A)$ of each relative reduct A corresponds to the arithmetic mean of the coverage scores of decision rules generated from the relative reduct A [2]. A relative reduct A generates more useful decision rules than any other relative reducts B such that $ACov(B) < ACov(A)$, where the usefulness is evaluated by the coverage. Therefore, the relative reduct A with higher evaluation score $ACov(A)$ generates more useful decision rules with higher coverage scores rather than any relative reducts with lower evaluation scores.

Example 1 ([9])
Table 1 is an example of a decision table S used in [7, 9]. The decision table consists of a set of six users of sample products; $U = \{u_1, \ldots, u_6\}$ as the set of objects, the set of attributes AT that is divided into the set of condition attributes $C = \{\text{Member}, \text{Sex}, \text{Before}, \text{After}\}$ that represents users' membership, sex, evaluation before/after using a sample product, respectively, and the decision attribute Purchase that represents users' answer to the question about purchase.

In Table 1, there are the following three relative reducts: {Member, Sex, Before}, {Member, Before, After}, and {Sex, Before, After}.

Let $A = \{\text{Member}, \text{Before}, \text{After}\}$ be a relative reduct in Table 1 and we consider the evaluation score of the relative reduct A defined by the Eq. (9). The partition $U/IND(A)$ by the indiscernibility relation $IND(A)$ consists of six singletons of objects, i.e., each object can be distinguished from every other object by A:

$$U/IND(A) = \{\{u1\}, \{u2\}, \{u3\}, \{u4\}, \{u5\}, \{u6\}\}.$$

For each equivalence class $[x]_A \in U/IND(A)$, it is obvious that there exists just one decision class $D_i \in \mathcal{D}$ such that $[x]_A \cap D_i \neq \emptyset$, and therefore, the denominator of the Eq. (9) is equal to the number of equivalence classes in $U/IND(A)$. It concludes that the evaluation score $ACov(A)$ is $\frac{1}{3}$:

$$ACov(A) = \frac{|\mathcal{D}|}{|U/IND(A)|} = \frac{2}{6} = \frac{1}{3}.$$

The relative reduct A generates the following six decision rules:

- $(\text{M} = \text{yes}) \wedge (\text{B} = \text{normal}) \wedge (\text{A} = \text{normal}) \rightarrow (\text{P} = \text{yes})$,

Table 1 An example of decision table [7]

U	Member	Sex	Before	After	Purchase
$u1$	yes	female	normal	normal	yes
$u2$	no	female	normal	v.g.	yes
$u3$	no	male	good	v.g.	yes
$u4$	no	female	good	good	yes
$u5$	no	male	normal	normal	no
$u6$	yes	female	good	normal	no

- $(\mathsf{M} = \text{no}) \wedge (\mathsf{B} = \text{normal}) \wedge (\mathsf{A} = \text{v.g.}) \rightarrow (\mathsf{P} = \text{yes})$,
- $(\mathsf{M} = \text{no}) \wedge (\mathsf{B} = \text{good}) \wedge (\mathsf{A} = \text{v.g.}) \rightarrow (\mathsf{P} = \text{yes})$,
- $(\mathsf{M} = \text{no}) \wedge (\mathsf{B} = \text{good}) \wedge (\mathsf{A} = \text{good}) \rightarrow (\mathsf{P} = \text{yes})$,
- $(\mathsf{M} = \text{no}) \wedge (\mathsf{B} = \text{normal}) \wedge (\mathsf{A} = \text{normal}) \rightarrow (\mathsf{P} = \text{no})$,
- $(\mathsf{M} = \text{yes}) \wedge (\mathsf{B} = \text{good}) \wedge (\mathsf{A} = \text{normal}) \rightarrow (\mathsf{P} = \text{no})$,

where M, B, A, and P are abbreviations of the attributes Member, Before, After, and Purchase, respectively. The coverage score of each decision rule is calculated by the number of objects that satisfy both the antecedent and conclusion of the decision rule divided by the number of objects that satisfy the conclusion of the rule. We then have the coverage score $\frac{1}{4}$ for each decision rule with the conclusion $(\mathsf{P} = \text{yes})$, and the coverage score $\frac{1}{2}$ for each decision rule with the conclusion $(\mathsf{P} = \text{no})$. Therefore, the arithmetic mean of the coverage scores of the six decision rules is

$$\frac{1}{6}\left(4 \times \frac{1}{4} + 2 \times \frac{1}{2}\right) = \frac{1}{3},$$

and it is equal to the evaluation score $ACov(A)$.

3 Interrelationship Mining

In this section, we review our formulation of rough set-based interrelationship mining with respect to the authors' previous manuscripts [3, 4, 10].

3.1 Observations and Motivations

Indiscernibility of objects, a basis of rough set data analysis, is based on comparison of attribute values between objects. For example, an indiscernibility relation $IND(B)$ by a subset of attributes $B \subseteq AT$ defined by (2) is based on the equality of attribute values in V_{a} of each attribute $\mathsf{a} \in B$. In the dominance-based rough set approach [1], a dominance relationship between objects is based on comparison of attribute values of objects according to the total order relationship among attribute values in each attribute.

Comparison of attribute values between objects in rough set data analysis is, however, restricted to compare attribute values of the same attribute, i.e., the comparison between two values of each attribute $\mathsf{a} \in AT$ in the following two cases:

- Between a value of an object x, $\rho(x, \mathsf{a})$, and a value $v \in V_{\mathsf{a}}$, e.g., the definition of semantics of decision logic, or
- Between a value of an object x, $\rho(x, \mathsf{a})$, and a value of another object y, $\rho(y, \mathsf{a})$, e.g., indiscernibility relations.

This restriction indicates that the interrelationship between attributes are not considered when we discuss indiscernibility of objects by values of attributes, even though values of different attributes are actually comparable. However, this fact indicates that it is difficult to extract the following characteristics that are based on comparison of attribute values between different attributes by using rough set-based data analysis [5]:

- The answer of question A is identical to the answer of question B,
- A sample A is similar to a sample B,
- The design of car A is preferred by users than the design of car B.

Therefore, by extending the domain of comparison of attribute values from each value set V_{a} to Cartesian product $V_{\mathsf{a}} \times V_{\mathsf{b}}$ with another attribute b, values of different attributes become comparable. This extension enable us to describe interrelationships between attributes by comparison between attribute values of different attributes in the framework of rough set theory.

3.2 A General Expression of Decision Tables

Information tables describe connections between objects and attributes by table-style format. In this paper, similar to the authors' recent works [7, 9], we use a general expression of information tables that was used by Yao et al. [17] defined by

$$S = (U, AT, \{V_{\mathsf{a}} \mid \mathsf{a} \in AT\}, \mathcal{R}_{AT}, \rho), \tag{10}$$

where U is a finite and nonempty set of objects, AT is a finite and nonempty set of attributes, V_{a} is a nonempty set of values for $\mathsf{a} \in AT$, $\mathcal{R}_{AT} = \{\{R_{\mathsf{a}}\} \mid \mathsf{a} \in AT\}$ is a set of families $\{R_{\mathsf{a}}\}$ of binary relations defined on each V_{a}, ρ is an information function $\rho : U \times AT \to V$ that assigns a value $\rho(x, \mathsf{a}) \in V_{\mathsf{a}}$ of the attribute $\mathsf{a} \in AT$ to each object $x \in U$, where $V = \bigcup_{\mathsf{a} \in AT} V_{\mathsf{a}}$ is the set of values of all attributes in AT.

The family $\{R_{\mathsf{a}}\}$ of binary relations for each attribute $\mathsf{a} \in AT$ can contain various binary relations; similarity, dissimilarity, dominance relation on V_{a} and usual information tables are implicitly assumed that the family $\{R_{\mathsf{a}}\}$ consists of only the equality relation $=$ on V_{a} [17]. We also assume that the equality relation $=$ is included into the family $\{R_{\mathsf{a}}\}$ for every attribute $\mathsf{a} \in AT$.

An information table is called a decision table if the set of attributes AT is partitioned into two disjoint sets; i.e., a set C of condition attributes and a set D of decision attributes. In this paper, without losing generality, we assume that D is a singleton, i.e., $D = \{\mathsf{d}\}$, and the attribute d is called the decision attribute.

Example 2 We extend the decision table $S = (U, AT, V, \rho)$ represented by Table 1 by adding the following family of binary relations defined on the set V_{a} for attribute $\mathsf{a} \in AT$ in S [7]:

$$\begin{gathered}\mathsf{Member} : \{=\}, \quad \mathsf{Sex} : \{=\}, \\ \mathsf{Before} : \{=, \succ_{\mathsf{Before}}, \succeq_{\mathsf{Before}}\}, \mathsf{After} : \{=, \succ_{\mathsf{After}}, \succeq_{\mathsf{After}}\}, \\ \mathsf{Purchase} : \{=, \succ_{\mathsf{Purchase}}, \succeq_{\mathsf{Purchase}}\},\end{gathered}$$

where each relation $\succ_{\mathsf{a}}$ is a preference relation that is defined as follows:

$$\begin{aligned}&\succ_{\mathsf{Before}}: && \text{v.g.} \succ \text{good} \succ \text{normal} \succ \text{bad} \succ \text{v.b.}, \\ &\succ_{\mathsf{After}}: && \text{v.g.} \succ \text{good} \succ \text{normal} \succ \text{bad} \succ \text{v.b.}, \\ &\succ_{\mathsf{Purchase}}: && \text{yes} \succ \text{no}.\end{aligned}$$

This extension enables us to treat the preference relationship of objects by comparing attribute values of each object.

3.3 Interrelationships Between attributes

We can consider many kinds of interrelations between attributes by comparison of attribute values, e.g., the equality, equivalence, order relations, similarity, etc. According to the observations and motivations in the previous subsection, interrelationships between attributes in a given decision table by a binary relation are characterized as follows [3].

Definition 1 Let $\mathsf{a}, \mathsf{b} \in C$ be any condition attributes of a given decision table S, and $R \subseteq V_{\mathsf{a}} \times V_{\mathsf{b}}$ be any binary relation. We call that attributes a and b are interrelated by R if and only if there exists an object $x \in U$ such that $(\rho(x, \mathsf{a}), \rho(x, \mathsf{b})) \in R$ holds.

We denote the set of objects that those values of attributes a and b satisfy the relation R as follows:

$$R(\mathsf{a}, \mathsf{b}) \stackrel{\text{def}}{=} \{x \in U \mid (\rho(x, \mathsf{a}), \rho(x, \mathsf{b})) \in R\}, \tag{11}$$

and we call the set $R(\mathsf{a}, \mathsf{b})$ the support set of the interrelation between a and b by R.

An interrelationship between two attributes by a binary relation provides a formulation of comparison of attribute values between different attributes. However, to simplify the formulation, we allow the interrelationship between the same attribute.

Indiscernibility relations in a given decision table by interrelationships between attributes are introduced [3].

Definition 2 Let S be a decision table, and suppose that condition attributes $\mathsf{a}, \mathsf{b} \in C$ are interrelated by a binary relation $R \subseteq V_{\mathsf{a}} \times V_{\mathsf{b}}$, i.e., $R(\mathsf{a}, \mathsf{b}) \neq \emptyset$ holds. The indiscernibility relation on U based on the interrelationship between a and b by R is defined by

$$IND(\mathsf{a}R\mathsf{b}) = \{(x, y) \in U \times U \mid x \in R(\mathsf{a}, \mathsf{b}) \text{ iff } y \in R(\mathsf{a}, \mathsf{b})\}. \tag{12}$$

For any objects x and y, $(x, y) \in IND(\mathsf{a}R\mathsf{b})$ means that x is not discernible from y from the viewpoint of whether the interrelationship between the attributes a and b by the relation R holds. Any binary relation $IND(\mathsf{a}R\mathsf{b})$ on U defined by (12) is an equivalence relation, and we can construct equivalence classes from an indiscernibility relation $IND(\mathsf{a}R\mathsf{b})$.

3.4 Decision Tables for Interrelationship Mining

To explicitly treat interrelationships between attributes, we need to reformulate the information table S by (10) by using the given binary relations between values of different attributes. This reformulation is based on revising the set $\mathcal{R}_{AT}$ of families of binary relations for comparing attribute values and expression of interrelationships by new condition attributes.

Definition 3 ([10]) Let S be an information table by (10). The information table S_{int} for interrelationship mining with respect to S is defined as follows:

$$S_{int} = (U, AT_{int}, V \cup \{0, 1\}, \mathcal{R}_{int}, \rho_{int}), \tag{13}$$

where U and $V = \bigcup_{\mathsf{a} \in AT} V_{\mathsf{a}}$ are identical to S.

The set $\mathcal{R}_{int}$ of families of binary relations is defined by

$$\mathcal{R}_{int} = \mathcal{R}_{AT} \cup \left\{ \{R_{\mathsf{a}_i \times \mathsf{b}_i}\} \,\middle|\, \begin{array}{l} R_{\mathsf{a}_i \times \mathsf{b}_i} \subseteq V_{\mathsf{a}_i} \times V_{\mathsf{b}_i}, \\ \exists \mathsf{a}_i, \mathsf{b}_i \in C, \ \mathsf{a}_i \neq \mathsf{b}_i \end{array} \right\} \cup \{\{=\} \mid \text{For each } \mathsf{aRb}\}, \tag{14}$$

where each family $\{R_{\mathsf{a}_i \times \mathsf{b}_i}\} = \{R^1_{\mathsf{a}_i \times \mathsf{b}_i}, \ldots, R^{n_i}_{\mathsf{a}_i \times \mathsf{b}_i}\}$ consists of n_i $(n_i \geq 0)$ binary relation(s) defined on $V_{\mathsf{a}_i} \times V_{\mathsf{b}_i}$. The expression aRb is defined below.

The set AT_{int} is defined by

$$AT_{int} = AT \cup \{\mathsf{aRb} \mid \exists R \in \{R_{\mathsf{a} \times \mathsf{b}}\}, R(\mathsf{a}, \mathsf{b}) \neq \emptyset\}, \tag{15}$$

and each expression aRb is called an interrelated condition attribute. $AT = C \cup \{\mathsf{d}\}$ is identical to S.

The information function ρ_{int} is defined by

$$\rho_{int}(x, \mathsf{c}) = \begin{cases} \rho(x, \mathsf{c}), & \text{if } \mathsf{c} \in AT, \\ 1, & \mathsf{c} = \mathsf{aRb} \text{ and } x \in R(\mathsf{a}, \mathsf{b}), \\ 0, & \mathsf{c} = \mathsf{aRb} \text{ and } x \notin R(\mathsf{a}, \mathsf{b}). \end{cases} \tag{16}$$

The redefined information table S_{int} intends to treat not only the information about interrelationships between attributes but also combinations of attribute values as in the original information system S. Therefore, S_{int} needs to contain all binary relations used in S.

Note that not all pairs of two attributes must have some binary relations and $n_i = 0$, i.e., $\{R_{\mathsf{a}_i \times \mathsf{b}_i}\} = \emptyset$, means that we do not compare attribute values between a_i and b_i. Note also that the set $\mathcal{R}_{AT}$ used in (14) is also identical to the case of the original information table S and every family $\{R_{\mathsf{a}}\}$ for each attribute $\mathsf{a} \in AT$ is assumed to contain at least the equality on V_{a}.

Each interrelated condition attribute aRb represents whether each object $x \in U$ supports the interrelationship between the attributes $\mathsf{a}, \mathsf{b} \in C$ by the binary relation $R \subseteq V_{\mathsf{a}} \times V_{\mathsf{b}}$. Therefore, all interrelated condition attributes are binary attributes. For every interrelated condition attribute, we only treat the equality relation for comparing attribute values of different objects. This is because interrelated condition attributes are nominal attributes.

Indiscernibility of objects by an interrelationship between two attributes a and b by a binary relation R in the original decision table S is representable by an indiscernibility relation by the singleton $\{\mathsf{aRb}\}$ in S_{int} as follows [10].

Proposition 1 *Let S be a decision table, $\mathsf{a}, \mathsf{b} \in C$ be condition attributes in S, $R \subseteq V_{\mathsf{a}} \times V_{\mathsf{b}}$ be a binary relation, and S_{int} be an information table for interrelationship mining with respect to S such that $R \in \{R_{\mathsf{a}\times\mathsf{b}}\}$. The following equality holds:*

$$IND_S(\mathsf{a}R\mathsf{b}) = IND_{S_{int}}(\{\mathsf{aRb}\}), \tag{17}$$

where $IND_S(\mathsf{a}R\mathsf{b})$ is the indiscernibility relation for S defined by (12), and $IND_{S_{int}}(\{\mathsf{aRb}\})$ is the indiscernibility relation for S_{int} by a singleton $\{\mathsf{aRb}\}$ of an interrelated condition attribute defined by (2).

Proposition 2 indicates, however, that the addition of interrelated condition attributes to the original decision table does not improve the total classification ability of the original decision table; interrelated condition attributes affect the representation ability of decision rules as we will discuss in Sect. 4.2.

Proposition 2 *Let S be a decision table and S_{int} be an information table for interrelationship mining with respect to S. The following equality holds:*

$$IND_{S_{int}}(AT) = IND_{S_{int}}(AT_{int}), \tag{18}$$

where $IND_{S_{int}}(AT)$ is the indiscernibility relation for S_{int} by using all attributes in AT of the original decision table S, and $IND_{S_{int}}(AT_{int})$ is the indiscernibility relation for S_{int} by using all attributes of S_{int} including interrelated condition attributes.

The following corollary represents that the addition of interrelated condition attributes to the original decision table does not enlarge the lower approximations of decision classes with respect to all condition attributes.

Corollary 1 *Let S be a decision table and S_{int} be an information table for interrelationship mining with respect to S. For every decision class $D \in \mathcal{D}$ of S_{int}, The following equality holds:*

$$\underline{C}(D) = \underline{C_{int}}(D), \tag{19}$$

where C is the set of all condition attributes of S and $C_{int} = AT_{int} \setminus \{\mathsf{d}\}$ is the set of all condition attributes of S_{int} including interrelated condition attributes.

Example 3 ([7])

We introduce an interrelationship between two attributes After and Before in Table 1 by comparing the values of these attributes. Because the range of these two attributes are identical, we can regard the preference relation $\succeq_{\mathsf{After}}$ as a binary relation $\succeq_{\mathsf{A}\times\mathsf{B}}$ defined on $V_{\mathsf{After}} \times V_{\mathsf{Before}}$.

We then construct an interrelated condition attribute $\mathsf{A} \succeq \mathsf{B}$. The support set of the interrelationship between After and Before by the relation $\succeq_{\mathsf{A}\times\mathsf{B}}$ is

$$\succeq_{\mathsf{A}\times\mathsf{B}} (\mathsf{A}, \mathsf{B}) = \{u1, u2, u3, u4, u5\}.$$

The information function is updated as follows:

$$\rho_{int}(x, \mathsf{A} \succeq \mathsf{B}) = \begin{cases} 1, & x \in \succeq_{\mathsf{A}\times\mathsf{B}} (\mathsf{A}, \mathsf{B}), \\ 0, & x \notin \succeq_{\mathsf{A}\times\mathsf{B}} (\mathsf{A}, \mathsf{B}). \end{cases}$$

Table 2 is the decision table S_{int} for interrelationship mining with respect to Table 1, i.e., the original table S, and the interrelated condition attribute $\mathsf{A} \succeq \mathsf{B}$. The indiscernibility relation $IND_{S_{int}}(\{\mathsf{A} \succeq \mathsf{B}\})$ is:

$$IND_{S_{int}}(\{\mathsf{A} \succeq \mathsf{B}\}) = \{(ui, uj) \mid 1 \leq i, j \leq 5\} \cup \{(u6, u6)\}.$$

It is easily confirmed that the indiscernibility relation $IND_{S_{int}}(\{\mathsf{A} \succeq \mathsf{B}\})$ is identical to the indiscernibility relation $IND_S(\mathsf{A} \succeq \mathsf{B})$ using the support set $\succeq_{\mathsf{A}\times\mathsf{B}} (\mathsf{A}, \mathsf{B})$ and defined by (12).

Table 2 has the following six relative reducts:

{Member, Sex, Before}, {Member, Before, After}, {Sex, Before, After},
{Member, After, $\mathsf{A} \succeq \mathsf{B}$}, {Sex, Before, $\mathsf{A} \succeq \mathsf{B}$}, {Sex, After, $\mathsf{A} \succeq \mathsf{B}$}.

It is obvious that all relative reducts for Table 1 are also relative reducts for Table 2.

Table 2 Decision table for interrelationship mining [7]

U	Member	Sex	Before	After	$\mathsf{A} \succeq \mathsf{B}$	Purchase
$u1$	yes	female	normal	normal	1	yes
$u2$	no	female	normal	v.g.	1	yes
$u3$	no	male	good	v.g.	1	yes
$u4$	no	female	good	good	1	yes
$u5$	no	male	normal	normal	1	no
$u6$	yes	female	good	normal	0	no

4 Theoretical Aspects of Interrelated Condition Attributes

In this section, we review theoretical aspects of interrelated decision attributes. The contents of this section is based on [7, 9].

4.1 Properties of Interrelated Attributes in Relative Reducts

In this subsection, we discuss a few properties of interrelated condition attributes that appear in relative reducts that are extracted from decision tables for interrelationship mining [7].

As we reviewed in the previous section, the indiscernibility relation $IND(\{\mathsf{aRb}\})$ by an interrelated attribute aRb is based on whether the interrelationship between two attributes a and b by a binary relation R on $V_{\mathsf{a}} \times V_{\mathsf{b}}$. The basis of the interrelationship between a and b is comparison between values $\rho_{int}(x, \mathsf{a})$ and $\rho_{int}(x, \mathsf{b})$ for each object $x \in U$.

We then consider that the discernibility of elements by an interrelated attribute aRb is strongly connected to the discernibility of either a or b. Proposition 3 describes connection between interrelated attributes aRb and attributes a and b.

Proposition 3 ([7]) *Let S_{int} be a decision table for interrelationship mining. If there exists a relative reduct $A \subseteq AT_{int}$ of S_{int} that contains an interrelated attribute aRb, then there exists at least one relative reduct $B \subseteq AT_{int}$ such that either $\mathsf{a} \in B$ or $\mathsf{b} \in B$ holds.*

Proposition 3 is applicable to any two condition attributes a and b and any binary relation R on $V_{\mathsf{a}} \times V_{\mathsf{b}}$, and it does not depend on any specific property of the binary relation R. Note also that Proposition 3 does not depend on the number of decision classes and this property is available to multi-valued decision attribute.

As an example, the set $\{\mathsf{Member}, \mathsf{After}, \mathsf{A} \succeq \mathsf{B}\}$ is a relative reduct of Table 2 in Example 3 and it includes an interrelated attribute $\mathsf{A} \succeq \mathsf{B}$. Proposition 3 indicates that there exists at least one relative reduct B such that $\mathsf{After} \in B$ or $\mathsf{Before} \in B$, and actually, a set $\{\mathsf{Member}, \mathsf{Before}, \mathsf{After}\}$ is also a relative reduct of Table 2.

The following corollary is easily obtained from Proposition 3.

Corollary 2 *([7]) Let S be a decision table, and* a *and* b *be two condition attributes of S that do not appear in any relative reduct of S. Then, for any binary relation R on* $V_{\mathsf{a}} \times V_{\mathsf{b}}$ *and the interrelated attribute* aRb *by the binary relation R, the interrelated attribute* aRb *does not appear in any relative reduct of the decision table* S_{int} *that are induced from the table S.*

This corollary indicates that the interrelationship between condition attributes that do not appear in any relative reduct is useless from a viewpoint of correct classification of elements. Selection of useful pairs of attributes is an important issue of interrelationship mining, and therefore, this property may provide a guideline for selecting pairs of condition attributes to consider interrelationships.

Obviously, the inverse of Proposition 3, i.e., if there is no relative reduct that contains the interrelated attribute aRb, then there is also no relative reduct that contains either a or b, is not satisfied. A counterexample of the converse of Proposition 3 is shown in [7].

4.2 Representation Ability of Interrelated Attributes

In this subsection, we discuss the representation ability of interrelated attributes [9].

Let $\mathsf{a}, \mathsf{b} \in C$ are two condition attributes in a given decision table S, and $\mathsf{aRb} \in AT_{int}$ be an interrelated attribute by a binary relation $R \in \{R_{\mathsf{a}\times\mathsf{b}}\}$ in the information table S_{int} for interrelationship mining. The following property has an important role for the representation ability of interrelated attributes.

Proposition 4 ([9]) *Let* $B = \{\mathsf{a}, \mathsf{b}\}$ *be any set of two condition attributes in C,* $\mathsf{aRb} \in AT_{int}$ *be an interrelated attribute based on the attributes* $\mathsf{a}, \mathsf{b} \in C$ *and a binary relation* $R \subseteq V_{\mathsf{a}} \times V_{\mathsf{b}}$*, and* $B' \subseteq AT_{int}$ *be a set of attributes generated by replacing either the attribute* a *or the attribute* b *in B by the interrelated attribute* aRb. *The following equation holds:*

$$[x]_B \subseteq [x]_{B'}, \ \forall x \in U. \tag{20}$$

This proposition indicates that the partition $U/IND(B')$ is equal to or coarser than the partition $U/IND(B)$, which implies the following important property about the evaluation scores of relative reducts with interrelated attributes.

Proposition 5 ([9]) *Suppose that* $B \subseteq AT_{int}$ *is a subset of condition attributes in the information table* S_{int} *for interrelationship mining. For any two condition attributes* $\mathsf{a}, \mathsf{b} \in B$*, let a set* B' *be either* $B' = (B \setminus \{\mathsf{a}\}) \cup \{\mathsf{aRb}\}$ *or* $B' = (B \setminus \{\mathsf{b}\}) \cup \{\mathsf{aRb}\}$. *Then the following inequality holds:*

$$ACov(B) \leq ACov(B'). \tag{21}$$

Corollary 3 ([9]) *Let B and B' in Proposition 5 be both relative reducts in the information table S_{int} for interrelationship mining such that $ACov(B) \leq ACov(B')$. The number of decision rules generated from B' is smaller than (or at least equal to) the number of decision rules generated from B.*

Therefore, if we replace an attribute a (or b) in a relative reduct by an interrelated attribute aRb, and the resulted set of this replacement is also a relative reduct, this replacement improves the representation ability of decision rules.

Example 4 ([9]) From Examples 1 and 3, we know that a subset of condition attributes $A = \{\mathsf{Member}, \mathsf{Before}, \mathsf{After}\}$ and a subset $A' = \{\mathsf{Member}, \mathsf{After}, \mathsf{A} \succeq \mathsf{B}\}$ are both the relative reducts in Table 2. Note that the relative reduct A' corresponds to the result of replacement of the attribute Before in the relative reduct A by the interrelated attribute $\mathsf{A} \succeq \mathsf{B}$.

From the relative reduct A', we have the following partition $U/IND(A')$:

$$U/IND(A') = \{\{u1\}, \{u2, u3\}, \{u4\}, \{u5\}, \{u6\}\},$$

i.e., two users $u_2, u_3 \in U$ are not discernible each other by the relative reduct A'

Similar to the case of the partition $U/IND(A)$ in Example 3, for each equivalence class $[x]_{A'} \in U/IND(A')$, it is obvious that there exists just one decision class $D_i \in \mathcal{D}$ such that $[x]_{A'} \cap D_i \neq \emptyset$. It implies that the denominator of the Eq. (9) is also equal to the number of equivalence classes in $U/IND(A')$, and therefore, the evaluation score $ACov(A')$ is

$$ACov(A') = \frac{|\mathcal{D}|}{|U/IND(A')|} = \frac{2}{5}.$$

The evaluation score $ACov(A')$ exceeds the evaluation score $ACov(A) = \frac{1}{3}$ in Example 1.

Actually, the relative reduct A' generates the following five decision rules and it can represent the characteristics in Table 2 with smaller number of decision rules rather than using the relative reduct A as in Example 1:

- $(\mathsf{M} = \text{yes}) \wedge (\mathsf{A} = \text{normal}) \wedge (\mathsf{A} \succeq \mathsf{B} = 1) \rightarrow (\mathsf{P} = \text{yes})$,
- $(\mathsf{M} = \text{no}) \wedge (\mathsf{A} = \text{v.g.}) \wedge (\mathsf{A} \succeq \mathsf{B} = 1) \rightarrow (\mathsf{P} = \text{yes})$,
- $(\mathsf{M} = \text{no}) \wedge (\mathsf{A} = \text{good}) \wedge (\mathsf{A} \succeq \mathsf{B} = 1) \rightarrow (\mathsf{P} = \text{yes})$,
- $(\mathsf{M} = \text{no}) \wedge (\mathsf{A} = \text{normal}) \wedge (\mathsf{A} \succeq \mathsf{B} = 1) \rightarrow (\mathsf{P} = \text{no})$,
- $(\mathsf{M} = \text{yes}) \wedge (\mathsf{A} = \text{normal}) \wedge (\mathsf{A} \succeq \mathsf{B} = 0) \rightarrow (\mathsf{P} = \text{no})$.

5 Conclusion

In this paper, we reviewed theoretical aspects of rough set-based interrelationship mining proposed by the authors: A formulation of rough set-based interrelationship mining [3, 4, 10] and properties of interrelated condition attributes [7, 9]. Even though we did not describe in this paper, the authors have also discussed decision logics for interrelationship mining [5], interrelationships between attributes from a viewpoint of rough sets on two universes [6], and interrelationship mining for incomplete decision tables [8].

Rough set-based interrelationship mining extends the range of application of rough set theory by extracting not only the characteristics in decision table by comparing the values of same attributes between different objects, but also the characteristics by comparing the values of different attributes with the same object. Applications of rough set-based interrelationship mining to life-oriented data analysis and bigdata analysis are planed as important future issues.

Acknowledgements The authors would like to thank the anonymous reviewers for their valuable comments.This work was supported by JSPS KAKENHI Grant Number JP25330315.

Proofs of Theoretical Properties

Proposition 1 *Let S be a decision table,* $\mathsf{a}, \mathsf{b} \in C$ *be condition attributes in S,* $R \subseteq V_{\mathsf{a}} \times V_{\mathsf{b}}$ *be a binary relation, and* S_{int} *be an information table* S_{int} *for interrelationship mining with respect to S such that* $R \in \{R_{\mathsf{a}\times\mathsf{b}}\}$. *The following equality holds:*

$$IND_S(\mathsf{a}R\mathsf{b}) = IND_{S_{int}}(\{\mathsf{aRb}\}),$$

where $IND_S(\mathsf{a}R\mathsf{b})$ *is the indiscernibility relation for S defined by (12), and* $IND_{S_{int}}(\{\mathsf{aRb}\})$ *is the indiscernibility relation for* S_{int} *by a singleton* $\{\mathsf{aRb}\}$ *of an interrelated condition attribute defined by (2).*

Proof Suppose $(x, y) \in IND_S(\mathsf{a}R\mathsf{b})$ holds. By the definition of $IND_S(\mathsf{a}R\mathsf{b})$ by (12), $x \in R(\mathsf{a}, \mathsf{b})$ holds if and only if $y \in R(\mathsf{a}, \mathsf{b})$ holds. Therefore, for the interrelated condition attribute aRb, it implies that either $\rho(x, \mathsf{aRb}) = \rho(y, \mathsf{aRb}) = 1$ or $\rho(x, \mathsf{aRb}) = \rho(y, \mathsf{aRb}) = 0$ holds, which concludes $(x, y) \in IND_{S_{int}}(\{\mathsf{aRb}\})$. The converse is also proved similarly.

Proposition 2 *Let S be a decision table and* S_{int} *be an information table for interrelationship mining with respect to S. The following equality holds:*

$$IND_{S_{int}}(AT) = IND_{S_{int}}(AT_{int}),$$

where $IND_{S_{int}}(AT)$ *is the indiscernibility relation for* S_{int} *by using all attributes in AT of the original decision table S, and* $IND_{S_{int}}(AT_{int})$ *is the indiscernibility relation for* S_{int} *by using all attributes of* S_{int} *including interrelated condition attributes.*

Proof Because $AT \subseteq AT_{int}$ by the definition of the information table S_{int} for interrelationship mining, $IND_{S_{int}}(AT) \supseteq IND_{S_{int}}(AT_{int})$ holds trivially. We show the converse set inclusion. Suppose $(x, y) \in IND_{S_{int}}(AT)$ holds. For any two attributes $\mathsf{a}, \mathsf{b} \in AT$ and any binary relation $R \in \{R_{\mathsf{a}\times\mathsf{b}}\}$ such that $R(\mathsf{a}, \mathsf{b}) \neq \emptyset$, the assumption $(x, y) \in IND_{S_{int}}(AT)$ implies that $\rho(x, \mathsf{a}) = \rho(y, \mathsf{a})$ and $\rho(x, \mathsf{b}) = \rho(y, \mathsf{b})$ hold.

This implies that $(\rho(x, \mathsf{a}), \rho(x, \mathsf{b})) \in R$ holds if and only if $(\rho(y, \mathsf{a}), \rho(y, \mathsf{b})) \in R$ holds, i.e., $x \in R(\mathsf{a}, \mathsf{b})$ holds if and only if $y \in R(\mathsf{a}, \mathsf{b})$ holds. By the definition of attribute value assignment for interrelated attributes by (16), it concludes that $\rho(x, \mathsf{aRb}) = \rho(y, \mathsf{aRb})$ holds. Therefore, x is indiscernible from y by any interrelated attribute aRb and $(x, y) \in IND_{S_{int}}(AT_{int})$ holds. This concludes the equality $IND_{S_{int}}(AT) = IND_{S_{int}}(AT_{int})$.

Corollary 1 *Let S be a decision table and S_{int} be an information table for interrelationship mining with respect to S. For every decision class $D \in \mathcal{D}$ of S_{int}, The following equality holds:*

$$\underline{C}(D) = \underline{C_{int}}(D),$$

where C is the set of all condition attributes of S and $C_{int} = AT_{int} \setminus \{\mathsf{d}\}$ is the set of all condition attributes of S_{int} including interrelated condition attributes.

Proof From the definitions of the decision table S, $C = AT \setminus \{\mathsf{d}\}$ holds. By Proposition 2, it is easily confirmed that $IND_{S_{int}}(C) = IND_{S_{int}}(C_{int})$ holds, which concludes $\underline{C}(D) = \underline{C_{int}}(D)$ for every decision class $D \in \mathcal{D}$ of S_{int}.

Proposition 3 ([7]) *Let S_{int} be a decision table for interrelationship mining. If there exists a relative reduct $A \subseteq AT_{int}$ of S_{int} that contains an interrelated attribute aRb, then there exists at least one relative reduct $B \subseteq AT_{int}$ such that either $\mathsf{a} \in B$ or $\mathsf{b} \in B$ holds.*

Proof ([7]) Suppose that the interrelated attribute aRb by a binary relation R on $V_{\mathsf{a}} \times V_{\mathsf{b}}$ appears in a relative reduct A. From the definition of relative reducts, there exist two elements $x, y \in U$ such that x and y are discernible each other by A, however, x and y are not discernible by $A' \stackrel{\text{def}}{=} A \setminus \{\mathsf{aRb}\}$. This means that the values of x and y at aRb are different each other, and without losing generality, we assume that $\rho_{int}(x, \mathsf{aRb}) = 1$ and $\rho_{int}(y, \mathsf{aRb}) = 0$. This assumption means $(\rho_{int}(x, \mathsf{a}), \rho_{int}(x, \mathsf{b})) \in R$ and $(\rho_{int}(y, \mathsf{a}), \rho_{int}(y, \mathsf{b})) \notin R$ hold, respectively, which implies that either the values of x and y at a are different or the values of x and y at b are different. Again, without loosing generality, we assume that the values of x and y at a are different. By this assumption, the subset $A' \cup \{\mathsf{a}\}$ satisfies the condition 1) of relative reducts; the attribute a can discern x from y and the subset A' can discern other elements that belong to different decision class each other. Moreover, the set $B \subseteq A' \cup \{\mathsf{a}\}$ with no redundant attributes is a relative reduct and it is easily confirmed that $\mathsf{a} \in B$. It concludes the proof.

Corollary 2 ([7]) *Let S be a decision table, and a and b be two condition attributes of S that do not appear in any relative reduct of S. Then, for any binary relation R on $V_{\mathsf{a}} \times V_{\mathsf{b}}$ and the interrelated attribute aRb by the binary relation R, the interrelated attribute aRb does not appear in any relative reduct of the decision table S_{int} that are induced from the table S.*

Proof It is obvious from Proposition 3.

Proposition 4 ([9]) *Let* $B = \{\mathsf{a}, \mathsf{b}\}$ *be any set of two condition attributes in* C, $\mathsf{aRb} \in AT_{int}$ *be an interrelated attribute based on the attributes* $\mathsf{a}, \mathsf{b} \in C$ *and a binary relation* $R \subseteq V_{\mathsf{a}} \times V_{\mathsf{b}}$, *and* $B' \subseteq AT_{int}$ *be a set of attributes generated by replacing either the attribute* a *or the attribute* b *in* B *by the interrelated attribute* aRb. *The following equation holds:*

$$[x]_B \subseteq [x]_{B'}, \ \forall x \in U. \tag{22}$$

Proof Suppose that $y \in [x]_B$ holds and let $\rho(x, \mathsf{a}) = v$ and $\rho(x, \mathsf{b}) = w$ be the values of x at the attributes a and b, respectively. Because $y \in [x]_B$, $\rho(y, \mathsf{a}) = v$ and $\rho(y, \mathsf{b}) = w$ also hold. In the binary relation $R \subseteq V_{\mathsf{a}} \times V_{\mathsf{b}}$ that is used for constructing the interrelated attribute $\mathsf{aRb} \in AT_{int}$, if $(v, w) \in R$ holds, it implies $\rho(x, \mathsf{aRb}) = \rho(y, \mathsf{aRb}) = 1$; otherwise, if $(v, w) \notin R$ holds, it implies $\rho(x, \mathsf{aRb}) = \rho(y, \mathsf{aRb}) = 0$. Therefore, the object y is still indiscernible from x by replacing either a or b by aRb, which concludes $y \in [x]_{B'}$.

Proposition 5 ([9]) *Suppose that* $B \subseteq AT_{int}$ *is a subset of condition attributes in the information table* S_{int} *for interrelationship mining. For any two condition attributes* $\mathsf{a}, \mathsf{b} \in B$, *let a set* B' *be either* $B' = (B \setminus \{\mathsf{a}\}) \cup \{\mathsf{aRb}\}$ *or* $B' = (B \setminus \{\mathsf{b}\}) \cup \{\mathsf{aRb}\}$. *then the following inequality holds:*

$$ACov(B) \leq ACov(B').$$

Proof Let N_B and $N_{B'}$ be the denominators of $ACov(B)$ and $ACov(B')$, respectively. By the definition of $ACov(\cdot)$ by (9), it is sufficient to show $N_{B'} \leq N_B$. Let $U/IND(B)$ and $U/IND(B')$ be the sets of equivalence classes by the indiscernibility relations $IND(B)$ and $IND(B')$, respectively, and suppose that there are m equivalence classes in $U/IND(B')$, i.e, there are m objects $x_1, \ldots, x_m \in U$ and $U/IND(B') = \{[x_1]_{B'}, \ldots, [x_m]_{B'}\}$. By Proposition 4, for each equivalence class $[x_i]_{B'} \in U/IND(B')$, there exist $k_i (\geq 1)$ equivalence classes $[y_{i_1}]_B, \ldots, [y_{i_{k_i}}]_B \in U/IND(B)$ such that $[x_i]_{B'} = \bigcup_{j=i_1}^{i_{k_i}} [y_j]_B$ holds. Therefore, $U/IND(B) = \{[y_{1_1}]_B, \ldots, [y_{1_{k_1}}]_B, \ldots, [y_{m_1}]_B, \ldots, [y_{m_{k_m}}]_B\}$. For each decision class $D \in \mathcal{D}$, it is clear that $D \cap [x_i]_{B'} \neq \emptyset$ holds if and only if there is at least one equivalence class $[y_{i_l}]_B$ such that $D \cap [y_{i_l}]_B \neq \emptyset$ holds. This fact implies that $|\{D \in \mathcal{D} \mid D \cap [x_i]_{B'} \neq \emptyset\}| \leq \sum_{j=i_1}^{i_{k_i}} |\{D \in \mathcal{D} \mid D \cap [y_j]_B \neq \emptyset\}|$ holds, which concludes $N_{B'} \leq N_B$.

Corollary 3 ([9]) *Let* B *and* B' *in Proposition 5 be both relative reducts in the information table* S_{int} *for interrelationship mining such that* $ACov(B) \leq ACov(B')$. *The number of decision rules generated from* B' *is smaller than (or at least equal to) the number of decision rules generated from* B.

Proof It is obvious from Proposition 5.

References

1. S. Greco, B. Matarazzo, and R. Słowiński, Rough set theory for multicriteria decision analysis, *European Journal of Operational Research*, Vol. 129, pp. 1–47, 2002.
2. Y. Kudo and T. Murai, An Evaluation method of Relative Reducts based on Roughness of partitions, *International Journal of Cognitive Informatics and Natural Intelligence*, Vol. 4, No. 2, pp. 50–62, 2010.
3. Y. Kudo and T. Murai, Indiscernibility Relations by Interrelationships between Attributes in Rough Set Data Analysis, *Proc. of IEEE GrC2012*, pp. 264–269, 2012.
4. Y. Kudo and T. Murai, A Plan of Interrelationship Mining Using Rough Sets, *Proc. of the 29th Fuzzy System Symposium*, pp. 33–36, 2013 (in Japanese).
5. Y. Kudo and T. Murai, Decision Logic for Rough Set-based Interrelationship Mining, *Proc. of IEEE GrC2013*, 2013.
6. Y. Kudo and T. Murai, Interrelationship mining from a viewpoint of rough sets on two universes, *Proc. of IEEE GrC2014*, pp. 137–140, 2014.
7. Y. Kudo and T. Murai, Some Properties of Interrelated Attributes in Relative Reducts for Interrelationship Mining, *Proc. of SCIS&ISIS 2014*, pp. 998–1001, 2014.
8. Y. Kudo and T. Murai, A Note on Application of Interrelationship Mining to Incomplete Information Systems, *Proc. of the 31st Fuzzy System Symposium*, pp. 777–778, 2015 (in Japanese).
9. Y. Kudo and T. Murai, On Representation Ability of Interrelated Attributes in Rough Set-based Interrelationship Mining *Proc. of ISIS 2015*, pp. 1229–1237, 2015.
10. Y. Kudo, Y. Okada, and T. Murai, On a Possibility of Applying Interrelationship Mining to Gene Expression Data Analysis, *Brain and Health Informatics*, LNAI 8211, pp. 379–388, 2013.
11. M. Kryszkiewicz, Rough set approach to incomplete information systems, *Information Sciences*, Vol. 112, pp. 39–49, 1998.
12. Z. Pawlak, Rough Sets, *International Journal of Computer and Information Science*, Vol. 11, pp. 341–356, 1982.
13. Z. Pawlak, *Rough Sets: Theoretical Aspects of Reasoning about Data*, Kluwer Academic Publishers, 1991.
14. L. Polkowski, *Rough Sets: Mathematical Foundations*, Advances in Soft Computing, Physica-Verlag, 2002.
15. D. Ślęzak and W. Ziarko,: The investigation of the Bayesian rough set model, *International Journal of Approximate Reasoning*, Vol. 40, pp. 81–91, 2005.
16. Y.Y. Yao, Three-way decisions with probabilistic rough sets, *Information Sciences*, Vol. 180, No. 3, pp. 341–353, 2010.
17. Y.Y. Yao, B. Zhou, and Y. Chen, Interpreting Low and High Order Rules: A Granular Computing Approach, *Proc. of RSEISP 2007*, LNCS 4585, pp. 371–380, 2007.
18. W. Ziarko,: Variable Precision Rough Set Model, *Journal of Computer and System Science*, Vol. 46, pp. 39–59, 1993.

Part IV
Fuzzy Sets and Decision Making

OWA Aggregation of Probability Distributions Using the Probabilistic Exceedance Method

Ronald R. Yager

Abstract We note the use of the OWA Operator in multi-criteria decision problems for aggregating the individual criteria satisfactions. We consider the situation where the criteria satisfactions have some uncertainty, are finite probability distributions. We note the requirement of needing to order these probability distributions. We note it is often not possible to obtain the required ordering over probability distributions. To circumvent this problem we introduce an approach called the Probabilistic Exceedance Method, PEM, which allows us to provide a surrogate for the OWA aggregation of probability distributions that doesn't require a linear ordering over the probability distributions.

1 Introduction

Multi-criteria decision-making often requires the aggregation of an alternative's satisfactions to the individual criteria to obtain the alternatives overall satisfaction to the task of interest. One commonly used method for performing this aggregation is the Ordered Weighted Averaging (OWA) Operator [1, 2]. Here the arguments in the OWA aggregation are the individual criteria satisfactions. In this work we look at the case in which there is some probabilistic uncertainty associated with the criteria satisfactions. The criteria satisfactions are probability distributions over a finite domain. Here we must perform the OWA aggregation where the arguments are discrete probability distributions. A central feature of the OWA operator is an ordering of the arguments being aggregated based upon their values, the bigger value the higher in the ordering. Here then we must an ordering over these probability distributions, a not necessarily easy task. One method for obtaining a linear ordering over a collection of finite probability distributions is to use stochastic dominance [3–6]. This requires a pairwise comparison of individual probability distributions to determine whether one stochastically dominates another. Using these pairwise comparisons we can get a binary relationship over the probability distributions from which we can generate the

R.R. Yager (✉)
Machine Intelligence Institute, Iona College, New Rochelle, NY 10801, USA
e-mail: yager@panix.com

V. Torra et al. (eds.), *Fuzzy Sets, Rough Sets, Multisets and Clustering*,
Studies in Computational Intelligence 671, DOI 10.1007/978-3-319-47557-8_16

desired linear ordering. A problem that arises here is that often the relationship is not complete, there are pairs of probability distributions for which neither stochastically dominates the other; as a result we can not obtain the required linear ordering. One way the address this problem is to provide a surrogate for the OWA aggregation of probability distributions, one that can be used even if we don't have the required linear ordering and is compatible with the OWA aggregation in the case in which there exists the required ordering. Here we introduce the Probabilistic Exceedance Method, PEM, as a surrogate.

2 OWA Operators in Multi-criteria Decision Problems

The OWA operator has found considerable use in multi-criteria decision-making. Assume $\boldsymbol{C} = \{C_1, \ldots, C_q\}$ are a collection of criteria of interest to a decision maker. Let X be a set of alternatives. For any alternative $x \in X$ let $C_k(x) \in [0, 1]$ indicate the degree to which x satisfies the alternative C_k. A commonly used decision procedure is to use a point wise decision function $M(C_1(x), \ldots, C_q(x)] = D(x)$ to evaluate each alternative and then select the alternative with the largest value for D. Here the value of D(x) does depend on any other alternative and satisfies Arrow's requirement that the decision procedure is independent of irrelevant alternatives [7].

In this approach the decision function captures the relationship between the criteria, the so-called decision imperative. One popular formulation for the decision function is to use the Ordered Weighted Averaging (OWA) operator. Here $D(x) = OWA(C_1(x), \ldots, C_q(x)]$. In the following for notational convenience we denote $a_k = C_k(x)$ then $D(x) = OWA(a_1, \ldots, a_q)$. The OWA operator is defined in the following.

Definition An OWA operator of dimension q is mapping OWA: $R^q \rightarrow R$ that has an associated collection of q weights $W = [w_1, \ldots, w_q]$such that all $w_j \in [0, 1]$ and $\Sigma_j w_j = 1$ where $OWA(a_1, \ldots, a_q) = \Sigma_j w_j b_j$ with b_j being the jth largest of the i.

We refer W to as the OWA weighting vector. We note that we can express this formulation as $OWA(a_1, \ldots, a_q) = \sum_{j=1}^{q} w_j a_{\rho(j)},$ where ρ is an index function so that $\rho(j)$ is the index of the jth largest of the arguments.

A fundamental aspect of this operator is the ordering step, the arguments are ordered in descending order with respect to their values. Thus here the argument a_i is not associated with a particular weight w_i but with the weight associated with a particular ordered position. This ordering step introduces a non-linearity into the aggregation process.

The OWA operator is known to be a mean (averaging) type operator in that it has the following properties: Boundedness, Monotonicity, Symmetry and Idempotency.

The OWA operator can implement many different types of aggregation depending on the choice of the weighting vector W. The different implementations of the OWA operator is related to the form of we weighting vector. For example weighting vector W_* such that $w_j = 0$ for $j = 1$ to $q - 1$ and $w_q = 1$ results in the formulation $OWA(a_1, \ldots, a_q) = Min_i[a_i]$. On the other hand the weighting vector W^* such that $w_i = 1$ and $w_j = 0$ for $j = 2$ to q results in the formulation $OWA(a_1, \ldots, a_q) = Max_j[a_i]$. The weighting vector W_{AVE} where $w_j = 1/q$ gives the simple average, $OWA(a_1, \ldots, a_q) = \frac{1}{q}\sum_{i=1}^{q} a_i$.

In the framework of multi-criteria decision making the choice of weighting vector W is a reflection of the decision imperative, the relationship between the criteria. For example of weighting vector W_*which results in the OWA aggregation $OWA(a_1, \ldots, a_q) = Min_l[a_i]$, can be seen as modeling the requirement to satisfy "all" the criteria. The case where $w_j = 1/q$ can be seen as implementing the imperative where we want to satisfy some of the criteria.

A crucial step in using the OWA operator to evaluate the overall satisfaction of a decision alternative x is the ordering of the alternative's satisfactions to the individual criteria, the $C_i(x)$. In the case in which the individual criteria satisfactions are scalar values this required ordering is a simple task. In situations in which the individual criteria satisfactions are more complex objects then scalar values the ordering of these satisfactions can become difficult. Here we look at one such situation.

3 Probabilistic Satisfactions to Criteria

We now want to consider the situation in which the satisfactions to the criteria by alternative x rather than being scalar values have some uncertainty that is expressed via a probability distribution.

Assume we have a collection of criteria, $\boldsymbol{C} = \{C_1, \ldots, C_k, \ldots, C_q\}$. Let our objective function for aggregating the criteria satisfaction be based on an OWA operator having weighting vector $W = [w_1, \ldots, w_i, \ldots, w_q]$. Let $Y = \{y_1, \ldots, y_j, \ldots, y_n\}$ be a finite set of numeric values from unit interval providing the possible degrees of satisfaction of an alternative to the criteria. In the following we will assume the indexing of the the elements in Y is such that $y_1 > y_2 > \cdots > y_j > y_{j+1} > \cdots > y_n$. The indexing of the y_j is in descending order. The set Y essentially reflects the decision maker's ability to distinguish different levels of satisfaction. The fact that Y is finite provides no real restriction it just takes advantage of the reality of the limited human ability to discriminate.

Here we shall assume the value of $C_k(x)$, the satisfaction of the kth criteria by x, is the probability distribution P_k where $P_k = [p_{k1}, \ldots, p_{kn}]$ is a probability distribution on the space Y such that p_{kj} is the probability that the satisfaction of the kth criteria by alternative x is equal to y_j. Here we see that all $p_{kj} \in [0, 1]$ and that for each criteria C_k we have $\sum_{j=1}^{n} p_{kj} = 1$.

To use the OWA aggregation to combine the $C_k(x)$ we must have an ordering of the $C_k(x)$, with respect to which are the bigger. Since these are probability distributions determining this ordering is not easy. Nevertheless, let us assume we have an ordering L over the criteria satisfactions by x. Here then L(i) is the index of criteria with the ith largest satisfaction by alternative x. Thus here $P_{L(i)}$ is the probability distribution with the ith best satisfaction.

Having this ordering using the OWA operator we can calculate the aggregated probability distribution $P = \sum_{i=1}^{q} w_i P_{L(i)}$. Here P is a n vector whose j^{th} component $p_j = \sum_{i=1}^{q} w_i p_{L(i)j}$ where $p_{L(i)j}$ is the probability of y_j for the ith probability distribution in the order L. Thus here $p = [p_1, \ldots, p_j, \ldots, p_n] = [\sum_{i=1}^{q} w_i p_{L(i)1}, \ldots, \sum_{i=1}^{q} w_i p_{L(i)j}, \ldots, \sum_{i=1}^{q} w_i p_{L(i)n}]$ where p_j is the aggregated probability associated with the satisfaction level y_j. This aggregated value is obtained using the OWA aggregation operator.

We see that P is a probability distribution on Y since each $p_j \in [0, 1]$ and

$$\sum_{i=1}^{n} P_j = \sum_{j=1}^{n}\left(\sum_{i=1}^{q} w_i p_{L(i)j}\right) = \sum_{i=1}^{q}\left(\sum_{j=1}^{n} w_i p_{L(i)j}\right) = \sum_{i=1}^{q} w_i \left(\sum_{j=1}^{n} p_{L(i)j}\right) = \sum_{j=1}^{q} w_i = 1$$

Thus here the aggregated value of the $C_k(x)$ is itself a probability distribution on the space Y.

Let us look at this for some special cases of W. Consider the case where $w_1 = 1$ and $w_i = 0$ for all i = 2 to q. Here $p_j = \sum_{i=1}^{q} w_i p_{L(i)j} = p_{L(1)j}$, it is the probability of y_j for the criteria with the largest criteria satisfaction. $p_{L(1)}$. Thus here our aggregated probability distribution, P, is simply the probability distribution of the most satisfied criteria. If the OWA weighting vector W is such that $w_q = 1$ and $w_i = 0$ for $i \neq q$ then $p_j = \sum_{i=1}^{q} w_i p_{L(i)j} = p_{L(q)j}$, it is the probability of y_j for the criteria with the smallest satisfaction. Thus here P is simple the probability distribution of the least satisfied criteria. Finally we see in the special case where the OWA weights are such that the $w_i = 1/q$ for all i then $P = \sum_{i=1}^{q} \frac{1}{q} p_{L(i)j} = \frac{1}{q}\sum_{k=1}^{q} p_{kj}$. It is simply the average of the $C_k(x)$ and no ordering is necessary.

Once have obtained the vector $P = [p_1, \ldots, p_j, \ldots, p_n]$ by aggregation of the $C_k(x)$ using the OWA operator we can use this to obtain an expected value for alternative x

$$\mathbf{EV}(\mathrm{OWA}(C_1(x), \ldots, C_q(x))) = \sum_{i=1}^{n} p_j y_j \tag{1}$$

We can use these expected values to compare the different alternatives.

4 Stochastic Dominance

In the proceeding, to be able to perform the OWA aggregation we assumed the availability of the ordering L over the criteria with respect to their satisfaction by alternate x. However, in the current situation, where the $C_k(x)$ are probability distributions over the space Y, it is not obvious how to determine that $C_{k_1}(x) > C_{k_2}(x)$. That is we need to obtain an ordering over probability distributions. One established way to obtain an ordering over probability distributions, like the $C_k(x)$, is via the idea of stochastic dominance [6]. Here for each $C_k(x)$, probability distribution p_k, we obtain an n vector $t_k = [t_k(1), \ldots, t_k(j), \ldots, t_k(n)]$ where $t_k(j) = \sum_{r=1}^{j} p_{kr}$. Since $y_j > y_{j+1}$ then $t_k(j) = \mathrm{Prob}(C_k(x) \geq y_j)$, thus $t_k(j)$ is the probability that x's satisfaction to criteria C_k is at least y_j. For given P_k, $C_k(x)$, t_k is known as its Exceedance Distribution Function [8]. Here we shall use EDF_k and $EDF_k(j)$ synonymously for t_k and $t_k(j)$. We note that for each criteria C_k, we can view $t_k(j)$ as a function of j, with j going from 1 to n, having the properties: **1.** $t_k(1) = p_{k1}$, **2.** $t_k(n) = 1$ and **3.** $t_k(j+1) \geq t_k(j)$ (monotonicity in j).

Using the exceedance distribution function we say that $C_{k_1}(x)$ ***stochastically dominates*** $C_{k_2}(x)$, denoted $C_{k_1}(x) >_{SD} C_{k_2}(x)$ if $t_{k_1} \geq t_{k_2}$ for all j = 1 to n and for at least one j we have $t_{k_1} > t_{k_2}$. Essentially we see that $C_{k_1}(x)$ stochastically dominates $C_{k_2}(x)$ if for each value $y_j \in Y$, $C_{k_1}(x)$ has at least as big a probability of having a satisfaction larger or equal y_j then does $C_{k_2}(x)$.

We observe that stochastic dominance provides a pairwise comparison between the probability distributions corresponding to the criteria satisfactions by x. Formally, it is providing a binary relationship $\boldsymbol{R}$ over the space $\boldsymbol{C} = \{C_1, \ldots, C_q\}$ so that C_{k_1} $\boldsymbol{R}$ C_{k_2} if $C_{k_1}(x) >_{SD} C_{k_2}(x)$. It is well known that in some cases the relationship $\boldsymbol{R}$ can generate a linear order L over the space $\boldsymbol{C}$ [9, 10]. However to be able to obtain a linear ordering L we need that the relationship $\boldsymbol{R}$ be transitive and complete, that is we need

(1) **Transitivity**: if C_{k_1} $\boldsymbol{R}$ C_{k_2} and C_{k_2} $\boldsymbol{R}$ C_{k_3} then C_{k_1} $\boldsymbol{R}$ C_{k_3}
(2) **Completion**: for all pairs C_{k_1}, C_{k_2} we have either C_{k_1} $\boldsymbol{R}$ C_{k_2} or C_{k_2} $\boldsymbol{R}$ C_{k_1}
Condition two, completion. requires that we must be able to establish for each pair P_{k_1} and P_{k_2} whether $P_{k_1} >_{SD} P_{k_2}$ or $P_{k_2} >_{SD} P_{k_1}$.

In order for the probability distributions, $p_1, \ldots, p_k, \ldots, p_q$, associated with the satisfactions of the criteria in $\boldsymbol{C}$ with a alternative x to induce a linear order L using stochastic dominance we observe that the following fundamental relationship must exist between the collection of associated exceedance distribution functions, the $t_k = [t_k(1), \ldots, t_k(j), \ldots, t_k(n)]$. If L(i) is the index of the ith criteria is $\boldsymbol{C}$ in the ordering L then at each j we must have

$$t_{L(1)}(j) \geq t_{L(2)}(j) \geq \cdots \geq t_{L(q)}(j) \quad (2)$$

Thus the ordering of the $t_k(j)$ with respect to k must be so that it is the same at each j. We shall refer to this condition as homogeneity among the exceedance distribution functions. Thus the assumption of a linear ordering among the P_k implies the satisfaction of the homogeneity condition.

We note in real applications it is frequently not possible to satisfy over the collection of P_k's the conditions of transitivity and completeness needed to generate a linear order L over the collection of criteria **C** needed implement to OWA aggregation, OWA($P_1, \ldots, P_k, \ldots, P_q$). In order to work around this real world difficulty we must look for other methods to calculate an aggregation of the P_k's that do not require the linear ordering but are "consistent" with stochastic dominance. We shall refer to these methods as surrogates [6].

We shall say that a method is a valid surrogate method for the calculation OWA($P_1, \ldots, P_k, \ldots, P_q$) if it is such that if there exists a stochastic dominance (SD) induced linear ordering L then the surrogate approach always gives the same result as using OWA($P_1, \ldots, P_k, \ldots, P_q$), it is consistent with the approach that uses the order L, furthermore however, we require that the surrogate method can be used in situations when there is not an available an SD induced linear ordering L.

Implicit in this consistency condition is the fact that for any surrogate method, if the P_k's are such that their associated exceedance distribution functions satisfy the homogeneity condition and hence induce an SD based linear ordering L then the calculation of the aggregation of the $P_1, \ldots, P_q$ using this surrogate method must be equal to the OWA aggregation OWA($P_1, \ldots, P_q$) calculated using the linear ordering L that is obtained to from the homogenous distribution.

We now turn to the task of obtaining a surrogate method for calculating the OWA aggregation of a collection of arguments that are probability distributions.

5 PEM the Probabilistic Exceedance Method

Let $\boldsymbol{C} = \{C_1, \ldots, C_k, \ldots, C_q\}$ be a collection of criteria. Assume $Y = \{y_1, \ldots, y_j, \ldots, y_n\}$ are the set of values used to indicate the degree of satisfaction to the criteria by an alternative x. Here we assume that the set Y has an ordering and the indexing of the elements in Y such that $y_j > y_{j+1}$. Furthermore, the overall satisfaction by alternative x to the collection of criteria is determined by an OWA aggregation OWA($C_1(x), \ldots, C_k(x), \ldots C_q(x)$) where $C_k(x)$ is the satisfaction of the criterion C_k by x and the OWA operator has a weighting vector $W = [w_1, \ldots, w_i, \ldots, w_q]$.

In the situation of interest here we assume there exists some uncertainty in our knowledge of an alternative's satisfactions to the criteria. Specifically in this case each $C_k(x)$ is a probability distribution, P_k, over the space Y. In particular $P_k = [p_{k1}, \ldots, p_{kj}, \ldots, p_{kn}]$ where p_{kj} is the probability that the satisfaction of C_k by x is equal to y_j. Here each $p_{kj}, \in [0, 1]$ and $\sum_{j=1}^{n} p_{kj} = 1$. Our objective her is to find the OWA aggregation of these probability distributions

$$P = OWA(C_1(x), \ldots, C_k(x), \ldots, C_q(x)) = OWA(P_1, \ldots, P_k, \ldots, P_q]$$

where P is a probability distribution over Y.

Here we shall suggest a surrogate method for accomplishing this aggregation one that can be used without having to provide an ordering over the probability distributions. We shall refer to this as the Probabilistic Exceedance Method, PEM, approach. Subsequently we shall show that this is a valid surrogate for the approach that uses the ordering. We note that one benefit of the PEM method is that it will allow us to perform the aggregation whether an order over the P_k exists or not.

The following is the PEM algorithm, a surrogate, for calculating an OWA aggregation of probability distributions. Here we denote $\tilde{P} = PEM(p_1, \ldots, p_k, \ldots, p_q)$

(1) For each probability distribution P_k calculate its associated exceedance distribution function. $EDF_k = [EDF_k(1), \ldots, EDF_k(j), \ldots, EDF_k(n)]$ where $EDF_k(j) = \sum_{t=1}^{j} p_{kt}$.

We note that $EDF_k(j)$ is the probability that the alternative's satisfaction to C_k is at least y_j.

(2) Let $g_j(i)$ be an index function so that $g_j(i)$ is the index of the ith largest of the EDF_k at j. That is $EDF_{g_j(i)}$ is the ith largest value of $EDF_k(j)$.

(3) Let $B_j = OWA(EDF_1(j), \ldots, EDF_k(j), \ldots, EDF_q(j))$ using the weighting vector W, that is $B_j = \sum_{i=1}^{q} w_i EDF_{g_j(i)}$. We note we can perform this OWA aggregation since each $EDF_k(j)$ is a scalar and hence we can obtain ordering g_j.

(4) Here our aggregated value $\tilde{P} = [\tilde{p}_1, \ldots, \tilde{p}_j, \ldots, \tilde{p}_n]$ is defined such that for j = 1 to n we have $\tilde{p}_j = B_j - B_{j-1}$. (Note by convention we define $B_0 = 0$)

Let us assure ourselves that $\tilde{P}$ is a probability distribution. We first note that since all $EDF_k(j) \in [0, 1]$ then all $B_j \in [0, 1]$. Furthermore since for each k, $EDF_k(j) \geq EDF_k(j-1)$ then $B_j \geq B_{j-1}$, hence each $\tilde{p}_j = B_j - B_{j-1} \geq 0$. Further since $B_j \in [0, 1]$ then each $\tilde{p}_j \in [0, 1]$. We further observe that since for all k, $EDF_k(n) = \sum_{j=1}^{n} p_{kj} = 1$ then $B_n = OWA[1, \ldots, 1] = 1$. Finally we see that $\sum_{j=1}^{n} \tilde{p}_j = \sum_{j=1}^{n} (B_j - B_{j-1}) = B_n - B_0 = 1$.

We can provide an intuitive understanding of what is happening in this PEM method. We now recall that $B_j = OWA(EDF_1(j), \ldots, EDF_q(j))$, since the OWA

operator is a type of averaging operator then essentially B_j is a kind of average of the $EDF_k(j)$ at j over k, here then B_j = Average of $EDF_k(j) = \overline{EDF}_k(j)$. Similarly B_{j-1} is a kind of average over k of the EDF_k's at $j - 1$, B_{j-1} = Average $EDF_k(j-1) = \overline{EDF}_k(j-1)$. Finally see that
$\tilde{p}_j = B_j - B_{j-1} = \overline{EDF}_k(j) - \overline{EDF}_k(j-1)$
it is a kind of difference of these averages. Now recall that for each k, $EDF_k(j) - EDF_k(j-1) = p_{kj}$ it is the probability of the element y_j in probability distribution p_k. Thus we see that $\tilde{p}_j$ can be viewed as a kind average of the p_{kj} over k.

Thus we see that the PEM provides a probability distribution $\tilde{P}$ over the space Y where $\tilde{p}_j$, the aggregated probability associated with outcome is y_j, can be viewed as a kind of average of the p_{kj} over k.

We provide the following example illustrating this PEM algorithm

Example: Let $Y = \{y_1, y_2, y_3, y_4, y_5\}$ with $y_j > y_{j+1}$. Assume we have four criteria whose respective satisfactions by alternative x are the probability distributions P_1, P_2, P_3, P_4, shown in Table 1. In Table 1 the entry at the intersection of row P_k and column j is the probability p_{kj}.

In Table 2 we provide the respective exceedance distribution functions. In Table 2 the entry at the intersection of row EDF_k and column j is $EDF_k(j) = \sum_{t=1}^{j} p_{kt}$.

In Table 3 we provide the ordering of the EDF_k's for each j. In this table the entry at the intersection of row i and column j is $EDF_{g_j(i)}$, the ith largest of the values in the column j.

Assume our desired aggregation is based on an OWA operator with weighting vector W having components w_i for i = 1–4. Using this OWA operator we have

$$B_j = OWA(EDF_{g_j(1)}(j), EDF_{g_j(2)}(j), EDF_{g_j(3)}(j), EDF_{g_j(4)}(j)).$$

Table 1 Probability Distributions

	j = 1	j = 2	j = 3	j = 4	j = 5
P_1	0.4	0.0	0.0	0.5	0.1
P_2	0.2	0.2	0.2	0.2	0.2
P_3	0.5	0.2	0.2	0.1	0
P_4	0	0	0.3	0.2	0.5

Table 2 Exceedance Distribution Functions

	j = 1	j = 2	j = 3	j = 4	j = 5
EDF_1	0.4	0.4	0.4	0.9	1
EDF_2	0.2	0.4	0.6	0.8	1
EDF_3	0.5	0.7	0.9	1	1
EDF_4	0	0	0.3	0.5	1

Table 3 Ordered EDF's

	j = 1	j = 2	j = 3	j = 4	j = 5
i = 1	0.5	0.7	0.9	1	1
i = 2	0.4	0.4	0.6	0.9	1
i = 3	0.2	0.4	0.4	0.8	1
i = 4	0	0	0.3	0.5	1

Inserting the values in Table 3 we get

$$B_1 = 0.5w_1 + 0.4w_2 + 0.2w_3 + 0w_4$$
$$B_2 = 0.7w_1 + 0.4w_2 + 0.4w_3 + 0w_4$$
$$B_3 = 0.9w_1 + 0.6w_2 + 0.4w_3 + 0.3w_4$$
$$B_4 = 1w_1 + 0.9w_2 + 0.8w_3 + 0.3w_4$$
$$B_5 = 1w_1 + 1w_2 + 1w_3 + 1w_4 = 1$$

and $B_0 = 0$. From this we obtain the aggregated vector $\tilde{p}$ with 5 components

$$\tilde{p}_j = B_j - B_{j-1} \text{ for } j = 1 \text{ to } 5.$$

The actual values of the B_j's and in turn the $\tilde{p}_j$'s will depend on the OWA weighting vector W. Let us look at this for some notable examples of W.

Case 1. $w_1 = 1, w_2 = 0, w_3 = 0, w_4 = 0$. Here

$$B_0 = 0, B_1 = 0.5, B_2 = 0.7, B_3 = 0.9, B_4 = 1 \text{ and } B_5 = 1$$

and hence

$$\tilde{p}_1 = B_1 - 0 = B_1 = 0.5, \tilde{p}_2 = B_2 - B_1 = 0.2, \tilde{p}_3 = B_3 - B_2 = 0.2$$
$$\tilde{p}_4 = B_4 - B_3 = 0.1 \text{ and } \tilde{p}_5 = B_5 - B_4 = 0$$

Case 2. $w_1 = w_2 = w_3 = 0$ and $w_4 = 1$. Here

$$B_1 = 0, B_2 = 0, B_3 = 0.3, B_4 = 0.5, B_5 = 1$$

and hence

$$\tilde{p}_1 = B_1 - 0 = B_1 = 0, \tilde{p}_2 = B_2 - B_1 = 0, \tilde{p}_3 = B_3 - B_2 = 0.3$$
$$\tilde{p}_4 = B_4 - B_3 = 0.2 \text{ and } \tilde{p}_5 = B_5 - B_4 = 0.5$$

Case 3. All $w_i = \frac{1}{4}$. In this case

$$B_1 = \frac{1}{4}(0.5 + 0.4 + 0.2) = \frac{1.1}{4}$$
$$B_2 = \frac{1}{4}(0.7 + 0.4 + 0.4) = \frac{1.5}{4}$$
$$B_3 = \frac{1}{4}(0.9 + 0.6 + 0.4 + 0.3) = \frac{2.2}{4}$$
$$B_4 = \frac{1}{4}(1 + 0.9 + 0.8 + 0.5) = \frac{3.2}{4}$$
$$B_5 = \frac{1}{4}(1 + 1 + 1 + 1) = \frac{4}{4} = 1$$

In this case

$$\tilde{p}_1 = \frac{1.1}{4}, \tilde{p}_2 = \frac{1}{4}(1.5 - 1.1) = \frac{0.4}{4}, \tilde{p}_3 = \frac{1}{4}(2.2 - 1.5) = \frac{0.7}{4}$$
$$\tilde{p}_4 = \frac{1}{4}(3.2 - 2.2) = \frac{1}{4}, \tilde{p}_5 = \frac{1}{4}(4 - 3.2) = \frac{0.8}{4}$$

6 Showing that PEM Algorithm Is a Surrogate

Earlier we indicated that for a procedure such as the PEM algorithm, $PEM(P_1, \ldots, P_k, \ldots, P_q)$, to be a valid surrogate for the calculation $OWA(P_1, \ldots P_k, \ldots, P_q)$ it must have the same value as $OWA(P_1, \ldots P_k, \ldots, P_q)$ in the case where the constituent probability distributions, the P_k, are such that they can induce a linear ordering with respect to the stochastic dominance relationship.

In the case where there exists a linear ordering L over the P_k, with $P_{L(i)}$ being the ith element in this ordering, we indicated that

$$OWA(P_1, \ldots P_k, \ldots, P_q) = \hat{P} = \sum_{i=1}^{q} w_i P_{L(i)} = [\hat{p}_1, \ldots, \hat{p}_j, \ldots \hat{p}_n].$$

Here $\hat{p}$ is a probability distribution on Y such that $\hat{p}_j = \sum_{i=1}^{q} w_i p_{L(i)j}$ where $p_{L(i)j}$ is the probability of y_j in the ith distribution in the ordering of the p_k.

If the PEM algorithm is to be a valid surrogate then in the case where there exists a stochastic dominance based linear ordering L over the p_k and if $\tilde{P} =$

$\text{PEM}(P_1, \ldots, P_k, \ldots, P_q)$ with $\tilde{P} = [\tilde{p}_1, \ldots, \tilde{p}_j, \ldots \tilde{p}_n]$ we must have $\tilde{p}_j = \hat{p}_j$ for $j = 1$ to q.

In order to provide the proof we recall an observation we made earlier; if the P_k induce a linear ordering L with respect to stochastic dominance then their associated EDF_k's have homogeneity property, the ordering of the EDF_k's is the same for j, at each j the ordering is L. We now show for the case where there is a linear ordering on the p_k, the case where the ordering of the EDF_k's is the same at each j, then the using the PEM algorithm leads to the same result as the $\text{OWA}(P_1, \ldots, P_q)$, $\tilde{P} = \hat{P}$.

Consider the special case of the PEM algorithm, $\text{PEM}(P_1, \ldots, P_q)$ in which the ordering of the $\text{EDF}_k(j)$ is the same at each j. In this case $g_j(i) = g(i) = L(i)$. Here when we implement the PEM algorithm we have to calculate

$$B_j = \text{OWA}(\text{EDF}_1(j), \text{EDF}_2(j), \ldots, \text{EDF}_q(j)) = \sum_{i=1}^{q} w_i \text{EDF}_{g_j(i)}(j),$$

however in this special case we have for all j that $g_j(i) = L(i)$ and hence

$$B_j = \sum_{i=1}^{q} w_i \text{EDF}_{L(i)}(j).$$

In this special case of linearly ordered probability distributions for all j

$$\tilde{p}_j = B_j - B_{j-1} = \sum_{i=1}^{q} w_i \text{EDF}_{L(i)}(j) - \sum_{i=1}^{q} w_i \text{EDF}_{L(i)}(j-1) = \sum_{i=1}^{q} w_i (\text{EDF}_{L(i)}(j) - \text{EDF}_{L(i)}(j-1)).$$

Since $\text{EDF}_{L(i)}(j) - \text{EDF}_{L(i)}(j-1) = p_{L(i)j}$ then $\tilde{p}_j = \sum_{i=1}^{q} w_i p_{L(i)}(j)$. We see this is the same as the value $\hat{p}_j$ obtained using the $\text{OWA}(P_1, \ldots, P_q)$ with the ordering L.

Thus the PEM algorithm approach provides a surrogate for calculating the OWA aggregation of probability distributions; it can be used when we don't have a linear ordering on the probability distributions being aggregated and if we do have a linear ordering it yields the same result as we would get using the OWA with the linear order.

In the following we shall look at the preceding example where we used the PEM algorithm and show that we actually used it in its surrogate mode. For our subsequent discussion we shall find it useful to provide a richer version of the Table 3. Here each entry is of the form $\text{EDF}_{g_j(i)}/K = g_j(i)$. Here in addition to providing i^{th} largest value of EDF for column j we also provide K the index of the criteria that provides this value (Table 4).

Here we see that ordering is not the same for all j. In particular columns 3 and 4 we have

Table 4 Enhanced EDF's

	j = 1	j = 2	j = 3	j = 4	j = 5
i = 1	0.5/K = 3	0.7/K = 3	0.9/K = 3	1/K = 3	1/K = 3
i = 2	0.4/K = 1	0.7/K = 1	0.6/K = 2	0.9/K = 1	1/K = 1
i = 3	0.2/K = 2	0.4/K = 2	0.4/K = 1	0.4/K = 2	1/K = 2
i = 4	0/K = 4	0/K = 4	0.3/K = 4	0.3/K = 4	1/K = 4

$j = 3$	$j = 4$
$g_3(1)$ occurs at K = 3	$g_4(1)$ occurs at K = 3
$g_3(2)$ occurs at K = 2	$g_4(2)$ occurs at K = 1
$g_3(3)$ occurs at K = 1	$g_4(3)$ occurs at K = 2
$g_3(4)$ occurs at K = 4	$g_4(4)$ occurs at K = 4

Thus the ordering is not the same and hence there is not linear ordering L over probability distributions, however the use of the PEM method allowed us to obtain an aggregated probability distribution.

In the preceding we showed how to find the OWA aggregation of $P_1, \ldots, P_q$ using the surrogate PEM algorithm. This aggregation resulted in a probability distribution $\tilde{P} = [\tilde{p}_1, \ldots, \tilde{p}_j, \ldots \tilde{p}_n]$ over the space $Y = \{y_1, \ldots, y_n\}$ where $\tilde{p}_j$ is the probability of y_j. It is important to emphasize that the elements y_j in Y need not be numeric values, all that is required is that we have an ordering on the elements which we assumed to be $y_j > y_{j+1}$.

Once having this aggregated probability distribution we can apply it in many of the kinds of operations required in decision-making. The operations we can perform, with the aggregated probability distribution, depend on whether the set Y is non-numeric or numeric. If it is numeric we have more structure and hence more available operations then if it is non-numeric. Actually any operation that can be performed in the case of a non-numeric set Y can be performed in a numeric space. Because it is richer we shall look at the situation when Y is a numeric space.

7 Conclusion

We discussed the use of the Ordered Weighted Averaging (OWA) Operator in multi-criteria decision problems as a means of aggregating the individual criteria satisfactions. We looked at the situation where the criteria satisfactions are probability distributions and commented on the need of getting linear order over these probability distributions. We introduced the idea of using pairwise stochastic dominance to provide the necessary ordering relationship over the probability distributions. We indicated that while this approach is appropriate, it is often not possible, since the presence of a stochastic dominance relationship between all pairs of probability distributions is not always the case, the relationship is not complete. To circumvent this problem we introduced an approach called the Probabilistic Exceedance Method, PEM, which allowed us to provide a surrogate for the OWA aggregation of probability distributions that doesn't require a linear ordering over the probability distributions.

References

1. Yager, R. R., On ordered weighted averaging aggregation operators in multi-criteria decision making, IEEE Transactions on Systems, Man and Cybernetics 18, pp. 183–190, 1988.
2. Yager, R. R., Kacprzyk J., and Beliakov G., Recent Developments in the Ordered Weighted Averaging Operators: Theory and Practice. Berlin: Springer, 2011.
3. Levy, H., Stochastic Dominance: Investment Decision Making Under Uncertainty. New York: Springer, 2006.
4. Sriboonchita, S., Wong K. S., Dhompongs S., and Nguyen H. T., Stochastic Dominance and Applications to Finance, Risk and Economics: Chapman and Hall/CRC, 2009.
5. Whitmore, G. A. and Findlay M. C., Stochastic Dominance: An Approach to Decision Making Under Risk. Lexington, Mass: Heath, 1978.
6. Yager, R. R. and Alajlan N., Probability weighted means as surrogates for stochastic dominance in decision making, Knowledge-Based Systems 66, pp. 92–98, 2014.
7. Arrow, K. J., Social Choice and Individual Values. New York: John Wiley & Sons, 1951.
8. Weik, M., Computer Science and Communications Dictionary. Heidelberg: Springer, 2001.
9. Roberts, F. S., Measurement Theory. Reading, MA: Addison-Wesley, 1979.
10. Roubens, M. and Vincke P., Preference Modeling. Berlin: Springer-Verlag, 1989.

A Dynamic Average Value-at-Risk Portfolio Model with Fuzzy Random Variables

Yuji Yoshida

Abstract A perception-based portfolio model under uncertainty is presented. In the proposed model, randomness and fuzziness are evaluated respectively by probabilistic expectation and the mean values with evaluation weights and λ-mean functions. Introducing average value-at-risks under conditions, this paper formulates average value-at-risks in dynamic stochastic environment. By dynamic programming approach, an optimality condition of the optimal portfolios for dynamic average value-at-risks is derived. It is shown that the optimal average value-at-risk is a solution of the optimality equation under a reasonable assumption, and an optimal portfolio weight is obtained from the equation.

1 Introduction

Fuzzy random variables, which were introduced by Kwakernaak [5], are applied to decision-making under uncertainty with fuzziness like linguistic data in statistics, engineering and economics. This paper deals with a financial portfolio model with fuzzy random variables. Soft computing like fuzzy logic works effectively for financial models in uncertain environment. To represent uncertainty in this paper, we use *fuzzy random variables* which have two kinds of uncertainties, i.e. randomness and fuzziness. In this model, randomness is used to represent the uncertainty regarding the belief degree of frequency, and fuzziness is applied to linguistic imprecision of data because of a lack of knowledge regarding the current stock market. At the financial crisis in October 2008 and January 2016, we have observed the serious distrust of the market that the risky information regarding banks and security companies, and it is surely a kind of risks occurring from the imprecision of information. The fuzziness comes from the imprecision of data because of a lack of knowledge, and such serious distrust in the stock market will be represented by the fuzziness of information in finance models.

Y. Yoshida (✉)
Faculty of Economics and Business Administration, University of Kitakyushu, 4-2-1 Kitagata, Kokuraminami, Kitakyushu 802-8577, Japan
e-mail: yoshida@kitakyu-u.ac.jp

V. Torra et al. (eds.), *Fuzzy Sets, Rough Sets, Multisets and Clustering*,
Studies in Computational Intelligence 671, DOI 10.1007/978-3-319-47557-8_17

In financial market, the portfolio is one of the most useful risk allocation techniques for stable asset management. The minimization of the financial risk as well as the maximization of the return are important themes in the asset management. In a classical portfolio theory, *Markowitz's mean-variance model* is studied by many researchers and fruitful results have been achieved, and then the variance is investigated as the risk for portfolios [6, 7]. In this paper we focus on the drastic decline of asset prices. Recently, *value-at-risk (VaR)* is used widely to estimate the risk that asset prices decline based on worst scenarios. VaR is a risk-sensitive criterion based on percentiles, and it is one of the standard criteria in asset management. *Average value-at-risk (AVaR)* is a kind of coherent risk-level indexes of the asset at a specified probability of decline and it is used to select portfolios after due consideration of worst scenarios in investment. We extend the AVaR for real-valued random variables to one regarding fuzzy random variables from the viewpoint of perception-based approach in Yoshida [10]. We formulate a portfolio problem with fuzzy random variables, and we discuss the fundamental properties of the extended AVaR. Estimation of uncertain quantities is important in decision making. Yoshida [9] introduced the mean, the variance and the covariances of fuzzy random variables, using *evaluation weights* and λ-mean functions. This paper estimates fuzzy numbers and fuzzy random variables by probabilistic expectation and these criteria, which are characterized by *possibility and necessity criteria* for subjective estimation and a *pessimistic-optimistic index* for subjective decision. In this paper, using the results in Yoshida [9–15], we discuss dynamic average value-at-risk portfolio optimization with fuzzy random variables, which is one of extended models from Yoshida [14].

2 Perception-Based Estimations for Fuzzy Random Variables

Let $\mathbb{R} = (-\infty, \infty)$ be the set of all real numbers. A *fuzzy number* is denoted by its membership function $\tilde{a} : \mathbb{R} \to [0, 1]$ which is normal, upper-semicontinuous, fuzzy convex and has a compact support (Zadeh [18]). Denote by $\mathcal{R}$ the set of all fuzzy numbers. For a fuzzy number $\tilde{a}$, its α-cuts are given by $\tilde{a}_\alpha = \{x \in \mathbb{R} \mid \tilde{a}(x) \geq \alpha\}$ $(\alpha \in (0, 1])$ and $\tilde{a}_0 = \mathrm{cl}\{x \in \mathbb{R} \mid \tilde{a}(x) > 0\}$, where cl denotes the closure of an interval. The α-cut is also written by closed intervals $\tilde{a}_\alpha = [\tilde{a}^-_\alpha, \tilde{a}^+_\alpha]$ $(\alpha \in [0, 1])$. Hence we introduce a partial order $\succeq$, so called the *fuzzy max order*, on fuzzy numbers $\mathcal{R}$: Let $\tilde{a}, \tilde{b} \in \mathcal{R}$ be fuzzy numbers. $\tilde{a} \succeq \tilde{b}$ means that $\tilde{a}^-_\alpha \geq \tilde{b}^-_\alpha$ and $\tilde{a}^+_\alpha \geq \tilde{b}^+_\alpha$ for all $\alpha \in [0, 1]$. An addition and a scalar multiplication for fuzzy numbers are defined as follows: For $\tilde{a}, \tilde{b} \in \mathcal{R}$ and $\lambda \in \mathbb{R}$ satisfying $\lambda \geq 0$, the addition $\tilde{a} + \tilde{b}$ of $\tilde{a}$ and $\tilde{b}$ and the nonnegative scalar multiplication $\lambda\tilde{a}$ of λ and $\tilde{a}$ are fuzzy numbers given by their α-cuts $(\tilde{a} + \tilde{b})_\alpha = [\tilde{a}^-_\alpha + \tilde{b}^-_\alpha, \tilde{a}^+_\alpha + \tilde{b}^+_\alpha]$ and $(\lambda\tilde{a})_\alpha = [\lambda\tilde{a}^-_\alpha, \lambda\tilde{a}^+_\alpha]$, where $\tilde{a}_\alpha = [\tilde{a}^-_\alpha, \tilde{a}^+_\alpha]$ and $\tilde{b}_\alpha = [\tilde{b}^-_\alpha, \tilde{b}^+_\alpha]$ $(\alpha \in [0, 1])$. We also use the following *metric* d_H on $\mathcal{R}$ induced from Hausdorff metric: $d_H(\tilde{a}, \tilde{b}) = \sup_{\alpha \in [0,1]} \max\{|\tilde{a}^-_\alpha - \tilde{b}^-_\alpha|, |\tilde{a}^+_\alpha - \tilde{b}^+_\alpha|\}$ for $\tilde{a}, \tilde{b} \in \mathcal{R}$.

Let Ω be a sample space and let P be a non-atomic probability on Ω. Let $\mathcal{X}$ be a set of real-valued random variables on Ω. A fuzzy-number-valued map $\tilde{X} : \Omega \to \mathcal{R}$ is called a *fuzzy random variable* if $\omega \mapsto \tilde{X}_\alpha^-(\omega)$ and $\omega \mapsto \tilde{X}_\alpha^+(\omega)$ are measurable for all $\alpha \in [0, 1]$, where $\tilde{X}_\alpha(\omega) = [\tilde{X}_\alpha^-(\omega), \tilde{X}_\alpha^+(\omega)] = \{x \in \mathbb{R} \mid \tilde{X}(\omega)(x) \geq \alpha\}$ is the α-cut (Kwakernaak [5]). Kruse and Meyer [4] gave a *perception-based definition regarding the expectation* $\tilde{E}(\tilde{X})$ *of a fuzzy random variable* $\tilde{X}$, which is induced from Zadeh's extension principle, as follows.

$$\tilde{E}(\tilde{X})(x) = \sup_{X \in \mathcal{X} : E(X) = x} \inf_{\omega \in \Omega} \tilde{X}(\omega)(X(\omega)) \tag{1}$$

for $x \in \mathbb{R}$, where $\mathcal{X}$ is the set of all integrable real-valued random variables and $E(X) = \int X dP$. Then, the expectation $\tilde{E}(\tilde{X})$ is a fuzzy number whose α-cut is given by

$$\tilde{E}(\tilde{X})_\alpha = [E(\tilde{X}_\alpha^-), E(\tilde{X}_\alpha^+)] \tag{2}$$

for $\alpha \in (0, 1]$. The α-cut of fuzzy number (1) can be generally given by the following *Aumann integral*: $\tilde{E}(\tilde{X})_\alpha = \{E(X) | X \in \mathcal{X}$ and $X(\omega) \in \tilde{X}_\alpha(\omega)$ for all $\omega \in \Omega\}$. Puri and Ralescu [8] discussed the conditional expectation of fuzzy random variables by Aumann integral, and López-Díaz et al. [2] studied it for statistics with fuzzy data. For a functional $\varphi : \mathcal{X} \to \mathcal{R}$ as a general estimation of real-valued random variables, by the perception-based approach we can discuss *fuzzy extensions* $\tilde{\varphi}$ of the estimation φ which is defined as follows:

$$\tilde{\varphi}(\tilde{X})(x) = \sup_{X \in \mathcal{X} : \varphi(X) = x} \inf_{\omega \in \Omega} \tilde{X}(\omega)(X(\omega)), \qquad x \in \mathbb{R} \tag{3}$$

for a fuzzy random variable $\tilde{X} \in \tilde{\mathcal{X}}$ (Fig. 1, Yoshida [11]).

Denote by Φ a family of functionals $\varphi : \mathcal{X} \to \mathbb{R}$ which satisfy the following conditions (a) and (b):

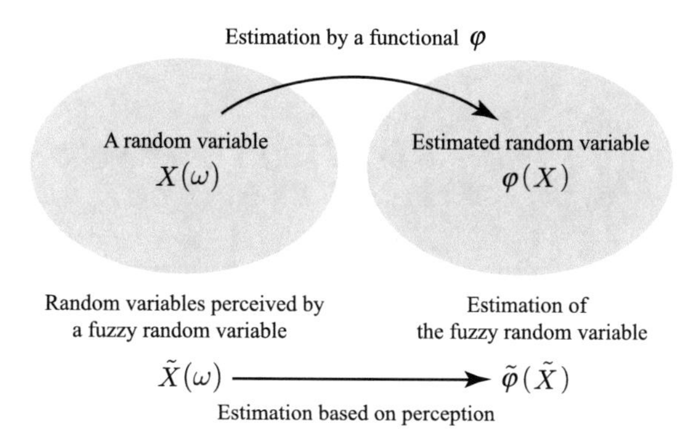

Fig. 1 Perception-based estimation of a fuzzy random variable

(a) $\varphi(X) \leq \varphi(Y)$ holds for all real-valued random variables $X, Y \in \mathcal{X}$ satisfying $X(\omega) \leq Y(\omega)$ for all $\omega \in \Omega$. (non-decreasing property)
(b) Let $\{X_n\}_n \subset \mathcal{X}$ and $X \in \mathcal{X}$ be a sequence and its limit, i.e. $\lim_{n\to\infty} X_n(\omega) = X(\omega)$ for all $\omega \in \Omega$. Then it holds that $\lim_{n\to\infty} \varphi(X_n) = \varphi(X)$. (continuity)

Then the family Φ has the following properties (i)–(ii):

(i) If $\varphi^1, \varphi^2 \in \Phi$, then it holds that $\varphi^1 + \varphi^2 \in \Phi$.
(ii) If $\varphi \in \Phi$ and $\lambda \in [0, \infty)$, then it holds that $\lambda\varphi \in \Phi$.

The following lemma gives fundamental properties for the perception-based estimations (Yoshida [11]).

Lemma 1 *Let $\varphi \in \Phi$. Define a functional $\tilde{\varphi} : \tilde{\mathcal{X}} \to \mathcal{R}$ by*

$$\tilde{\varphi}(\tilde{X})(x) = \sup_{X \in \mathcal{X} : \varphi(X) = x} \inf_{\omega \in \Omega} \tilde{X}(\omega)(X(\omega)), \qquad x \in \mathbb{R} \tag{4}$$

for a fuzzy random variable $\tilde{X} \in \tilde{\mathcal{X}}$. Then the following (i)–(v) hold.

(i) The α-cut of $\tilde{\varphi}(\tilde{X})$ is given by a closed interval

$$\tilde{\varphi}(\tilde{X})_\alpha = [\varphi(\tilde{X}_\alpha^-), \varphi(\tilde{X}_\alpha^+)] \tag{5}$$

for $\alpha \in (0, 1]$.
(ii) It holds that $\tilde{\varphi}(\tilde{X}) \preceq \tilde{\varphi}(\tilde{Y})$ for all fuzzy random variables $X, Y \in \tilde{\mathcal{X}}$ satisfying $\tilde{X}(\omega) \preceq \tilde{Y}(\omega)$ for all $\omega \in \Omega$, where $\preceq$ is the fuzzy max order on $\mathcal{R}$.
(iii) Let $\{\tilde{X}_n\}_n \subset \tilde{\mathcal{X}}$ and $\tilde{X} \in \tilde{\mathcal{X}}$ be a sequence and its limit, i.e. $\lim_{n\to\infty} \tilde{X}_n(\omega) = \tilde{X}(\omega)$ for all $\omega \in \Omega$. Then it holds that $\lim_{n\to\infty} \tilde{\varphi}(\tilde{X}_n) = \tilde{\varphi}(\tilde{X})$, where we use the metric d_H on $\mathcal{R}$.
(iv) If φ satisfies $\varphi(X + Y) \leq \varphi(X) + \varphi(Y)$ for all $X, Y \in \mathcal{X}$, then it holds that $\tilde{\varphi}(\tilde{X} + \tilde{Y}) \preceq \tilde{\varphi}(\tilde{X}) + \tilde{\varphi}(\tilde{Y})$ for all $\tilde{X}, \tilde{Y} \in \tilde{\mathcal{X}}$, where

$$(\tilde{X} + \tilde{Y})(\omega)(x) = \sup_{X, Y \in \mathcal{X} : X(\omega) + Y(\omega) = x} \min\{\tilde{X}(\omega)(X(\omega)), \tilde{Y}(\omega)(Y(\omega))\} \tag{6}$$

for $\omega \in \Omega$ and $x \in \mathbb{R}$.
(v) If φ satisfies $\varphi(\lambda X) = \lambda\varphi(X)$ for all $\lambda \in \mathbb{R}$ and $X \in \mathcal{X}$, then it holds that $\tilde{\varphi}(\lambda\tilde{X}) = \lambda\tilde{\varphi}(\tilde{X})$ for all $\lambda \in \mathbb{R}$ and $\tilde{X} \in \tilde{\mathcal{X}}$, where

$$(\lambda\tilde{X})(\omega)(x) = \sup_{X \in \mathcal{X} : \lambda X(\omega) = x} \tilde{X}(\omega)(X(\omega)) \tag{7}$$

for $\omega \in \Omega$ and $x \in \mathbb{R}$.

Example (*Perception-based estimation*).

(i) Let $\mathcal{M}$ be a σ-field on Ω and let $\mathcal{G}$ be a sub-σ-field of $\mathcal{M}$. The *conditional expectation* of an integrable fuzzy random variable $\tilde{X}$ is a fuzzy random variable

$$\tilde{E}(\tilde{X}|\mathcal{G})(\omega')(x) = \sup_{X \in \mathcal{X} : E(X|\mathcal{G})(\omega') = x} \inf_{\omega \in \Omega} \tilde{X}(\omega)(X(\omega)) \tag{8}$$

for $\omega' \in \Omega$ and $x \in \mathbb{R}$, where we take $\varphi = E(\cdot|\mathcal{G})(\omega') \in \Phi$ in (3). Its α-cut is given by $\tilde{E}(\tilde{X}|\mathcal{G})_\alpha = [E(\tilde{X}_\alpha^-|\mathcal{G}), E(\tilde{X}_\alpha^+|\mathcal{G})]$. Puri and Ralescu [8] discussed the properties of this extension.

(ii) For a real-valued random variable $X \in \mathcal{X}$ with a continuous cumulative distribution function, $x \mapsto F_X(x) = P(X < x)$ for which there exists a non-empty open interval I such that $F_X(\cdot) : I \to (0, 1)$ is a strictly increasing and onto. Then there exists a strictly increasing and continuous inverse function $F_X^{-1} : (0, 1) \to I$. We note that $F_X(\cdot) : I \to (0, 1)$ and $F_X^{-1} : (0, 1) \to I$ are one-to-one and onto, and we have $F_X(\inf I) = \lim_{x \downarrow \inf I} F_X(x) = 0$ and $F_X(\sup I) = \lim_{x \uparrow \sup I} F_X(x) = 1$. The *value-at-risk (VaR)* at a risk probability p is given by the percentile of the distribution function F_X.

$$\mathrm{VaR}_p(X) = \begin{cases} \inf I & \text{if } p = 0 \\ \sup\{x \in I \mid F_X(x) \le p\} & \text{if } p \in (0, 1) \\ \sup I & \text{if } p = 1. \end{cases} \tag{9}$$

Then we have $F_X(\mathrm{VaR}_p(X)) = p$ and $\mathrm{VaR}_p(X) = F_X^{-1}(p)$ for $p \in (0, 1)$. The *average value-at-risk (AVaR)* at a probability p is given by

$$\mathrm{AVaR}_p(X) = \frac{1}{p} \int_0^p \mathrm{VaR}_q(X)\, dq \tag{10}$$

if $p \in (0, 1]$ and $\mathrm{AVaR}_p(X) = \inf I$ if $p = 0$. Average value-at-risk of a fuzzy random variable $\tilde{X}$ is a fuzzy number such that

$$\widetilde{\mathrm{AVaR}}_p(\tilde{X})(x) = \sup_{X \in \mathcal{X} : \mathrm{AVaR}_p(X) = x} \inf_{\omega \in \Omega} \tilde{X}(\omega)(X(\omega)) \tag{11}$$

for $x \in \mathbb{R}$, where we take $\varphi = \mathrm{AVaR}_p(\cdot) \in \Phi$ in (3). Its α-cut is given by $\widetilde{\mathrm{AVaR}}_p(\tilde{X})_\alpha = [\mathrm{AVaR}_p(\tilde{X}_\alpha^-), \mathrm{AVaR}_p(\tilde{X}_\alpha^+)]$, and this estimation (11) also has the following properties [13].

Lemma 2 *Let $\tilde{X}, \tilde{Y} \in \tilde{\mathcal{X}}$ be fuzzy random variables. Let a probability $p \in (0, 1)$. Then the following (i)–(iii) hold.*

(i) If $\tilde{X} \preceq \tilde{Y}$, then $\widetilde{AVaR}_p(\tilde{X}) \preceq \widetilde{AVaR}_p(\tilde{Y})$.
(ii) $\widetilde{AVaR}_p(\tilde{a}\tilde{X}) = \tilde{a}\, \widetilde{AVaR}_p(\tilde{X})$ for fuzzy numbers $\tilde{a} \in \mathcal{R}$ satisfying $\tilde{a} \succeq 0$.
(iii) $\widetilde{AVaR}_p(\tilde{X} + \tilde{a}) = \widetilde{AVaR}_p(\tilde{X}) + \tilde{a}$ for fuzzy numbers $\tilde{a} \in \mathcal{R}$.

(iii) Let $\mathcal{G}$ be a sub-σ-field of $\mathcal{M}$. Let $F_X(\cdot \mid \mathcal{G})$ be a function on $\mathbb{R}$ given by $x(\in \mathbb{R}) \mapsto F_X(x \mid \mathcal{G}) = P(X < x \mid \mathcal{G}) = E(1_{\{X<x\}} \mid \mathcal{G})$. Then we define a *value-at-risk of* $X(\in \mathcal{X})$ *under a condition* $\mathcal{G}$ at a risk probability p by

$$\mathrm{VaR}_p(X \mid \mathcal{G}) = \begin{cases} \inf I & \text{if } p = 0 \\ \sup\{x \in I \mid F_X(x \mid \mathcal{G}) \le p\} & \text{if } p \in (0,1) \\ \sup I & \text{if } p = 1. \end{cases} \tag{12}$$

We note that $\mathrm{VaR}_p(X \mid \mathcal{G})$ is a random variable adapted to the σ-field $\mathcal{G}$, and we also have $P(X \le \mathrm{VaR}_p(X \mid \mathcal{G})) = p$ for $p \in (0,1)$. The *average value-at-risk (AVaR)* at a probability p is given by

$$\mathrm{AVaR}_p(X \mid \mathcal{G}) = \frac{1}{p}\int_0^p \mathrm{VaR}_q(X \mid \mathcal{G})\, dq \tag{13}$$

if $p \in (0,1]$ and $\mathrm{AVaR}_p(X \mid \mathcal{G}) = \inf I$ if $p = 0$. Average value-at-risk of a fuzzy random variable $\tilde{X}$ under a condition $\mathcal{G}$ at a risk probability p is a fuzzy random variable

$$\widetilde{\mathrm{AVaR}}_p(\tilde{X} \mid \mathcal{G})(\omega')(x) = \sup_{X \in \mathcal{X}:\, \mathrm{AVaR}_p(X \mid \mathcal{G})(\omega') = x} \inf_{\omega \in \Omega} \tilde{X}(\omega)(X(\omega)) \tag{14}$$

for $\omega' \in \Omega$ and $x \in \mathbb{R}$. Then the α-cut of (14) is given by $\widetilde{\mathrm{AVaR}}_p(\tilde{X} \mid \mathcal{G})_\alpha = [\widetilde{\mathrm{AVaR}}_p(\tilde{X}_\alpha^- | \mathcal{G}), \widetilde{\mathrm{AVaR}}_p(\tilde{X}_\alpha^+ | \mathcal{G})]$. This estimation (14) also has the following similar properties, which are easily checked by Lemma 2 and Yoshida [14].

Lemma 3 *Let* $\tilde{X}, \tilde{Y}, \tilde{Z} \in \tilde{\mathcal{X}}$ *be fuzzy random variables and let* $\mathcal{G}$ *be a sub-*σ*-field of* $\mathcal{M}$ *such that* $\mathcal{G}$ *is the* σ*-field generated by the fuzzy random variable* $\tilde{Z}$. *Let* $\tilde{Y}$ *and* $\mathcal{G}$ *be independent and let a probability* $p \in (0,1)$. *Then (i)–(v) hold.*

(i) If $\tilde{X} \preceq \tilde{Y}$, *then* $\widetilde{AVaR}_p(\tilde{X} \mid \mathcal{G}) \preceq \widetilde{AVaR}_p(\tilde{Y} \mid \mathcal{G})$.
(ii) $\widetilde{AVaR}_p(\tilde{Y} \mid \mathcal{G}) = \widetilde{AVaR}_p(\tilde{Y})$.
(ii) $\widetilde{AVaR}_p(\tilde{Z} \mid \mathcal{G}) = \tilde{Z}$.
(iv) $\widetilde{AVaR}_p(\tilde{Z}\tilde{X} \mid \mathcal{G}) = \tilde{Z}\,\widetilde{AVaR}_p(\tilde{X} \mid \mathcal{G})$ *if* $\tilde{Z} \succeq 0$.
(v) $\widetilde{AVaR}_p(\tilde{X} + \tilde{Z} \mid \mathcal{G}) = \widetilde{AVaR}_p(\tilde{X} \mid \mathcal{G}) + \tilde{Z}$.

3 Estimation of Fuzzy Numbers with Evaluation Weights

Yoshida [9] has studied an evaluation of fuzzy numbers by *evaluation weights* which are induced from fuzzy measures to evaluate a confidence degree that a fuzzy number takes values in an interval. With respect to fuzzy random variables, the randomness is evaluated by probabilistic expectation and the fuzziness is estimated by the evaluation weights and the following function. Let $\lambda \in [0, 1]$ and let $g^\lambda : \mathcal{I} \to \mathbb{R}$ be a map such that

$$g^\lambda([x, y]) = \lambda x + (1 - \lambda) y \tag{15}$$

for $[x, y] \in \mathcal{I}$, where $\mathcal{I}$ denotes the set of all bounded closed intervals. This scalarization is used for the estimation of fuzzy numbers to give a mean value of the interval $[x, y]$ with a weight λ, and g^λ is called a *λ-mean function* and λ is called a *pessimistic-optimistic index* which indicates the pessimistic degree of attitude in decision making [3]. Let a fuzzy number $\tilde{a} \in \mathcal{R}$. A mean value of the fuzzy number $\tilde{a}$ with respect to λ-mean functions g^λ and an evaluation weight $w(\alpha)$, which depends only on $\tilde{a}$ and α, is given as follows [11]:

$$E^\lambda(\tilde{a}) = \frac{\int_0^1 g^\lambda(\tilde{a}_\alpha)\, w(\alpha)\, d\alpha}{\int_0^1 w(\alpha)\, d\alpha}, \tag{16}$$

where $\tilde{a}_\alpha = [\tilde{a}_\alpha^-, \tilde{a}_\alpha^+]$ is the α-cut of the fuzzy number $\tilde{a}$. In (16), $w(\alpha)$ indicates a confidence degree that the fuzzy number $\tilde{a}$ takes values in the interval $\tilde{a}_\alpha$ at each level α. Hence, an evaluation weight $w(\alpha)$ is called the *possibility evaluation weight* $w^P(\alpha)$ if $w^P(\alpha) = 1$ for $\alpha \in [0, 1]$, and $w(\alpha)$ is called the *necessity evaluation weight* $w^N(\alpha)$ if $w^N(\alpha) = 1 - \alpha$ for $\alpha \in [0, 1]$. Especially, for a fuzzy number $\tilde{a} \in \mathcal{R}$, the means in the possibility and necessity cases are represented respectively by $E^P(\tilde{a})$ and $E^N(\tilde{a})$, and we can consider their combination $\nu E^P(\tilde{a}) + (1 - \nu) E^N(\tilde{a})$ with a parameter $\nu \in [0, 1]$ (Yoshida [9, 10]). The mean E^λ has the following natural properties of the addition and scalar multiplication and the monotonicity regarding the fuzzy max order $\succeq$.

Lemma 4 ([9]). *Let $\lambda \in [0, 1]$. For fuzzy numbers $\tilde{a}, \tilde{b} \in \mathcal{R}$ and real numbers θ, ζ, the following (i)–(iv) hold.*

(i) $E^\lambda(\tilde{a} + 1_{\{\theta\}}) = E^\lambda(\tilde{a}) + \theta$.
(ii) $E^\lambda(\zeta \tilde{a}) = \zeta E^\lambda(\tilde{a})$ *if* $\zeta \geq 0$.
(iii) $E^\lambda(\tilde{a} + \tilde{b}) = E^\lambda(\tilde{a}) + E^\lambda(\tilde{b})$.
(iv) *If* $\tilde{a} \succeq \tilde{b}$, *then* $E^\lambda(\tilde{a}) \geq E^\lambda(\tilde{b})$ *holds.*

For a fuzzy random variable $\tilde{X} (\in \tilde{\mathcal{X}})$, the mean of the expectation $E(E^\lambda(\tilde{X}))$ is a real number

$$E(E^\lambda(\tilde{X})) = E\left(\frac{\int_0^1 g^\lambda(\tilde{X}_\alpha)\, w(\alpha)\, d\alpha}{\int_0^1 w(\alpha)\, d\alpha} \right). \tag{17}$$

Then, from Lemma 4, we obtain the following results.

Lemma 5 ([9, 11]). *Let $\lambda \in [0, 1]$. For a fuzzy number $\tilde{a} \in \mathcal{R}$, integrable fuzzy random variables $\tilde{X}, \tilde{Y}$, an integrable real-valued random variable Z and a nonnegative real number ζ, the following (i)–(v) hold.*

(i) $E(E^\lambda(\tilde{X})) = E^\lambda(\tilde{E}(\tilde{X}))$.

(ii) $E(E^{\lambda}(\tilde{a})) = E^{\lambda}(\tilde{a})$ *and* $E(E^{\lambda}(Z)) = E(Z)$.
(iii) $E(E^{\lambda}(\zeta\tilde{X})) = \zeta E(E^{\lambda}(\tilde{X}))$.
(iv) $E(E^{\lambda}(\tilde{X}+\tilde{Y})) = E(E^{\lambda}(\tilde{X})) + E(E^{\lambda}(\tilde{Y}))$.
(v) *If* $\tilde{X} \succeq \tilde{Y}$, *then* $E(E^{\lambda}(\tilde{X})) \geq E(E^{\lambda}(\tilde{Y}))$ *holds.*

Let $\tilde{\mathcal{X}}_c$ be a family of fuzzy random variables $\tilde{X} \in \tilde{\mathcal{X}}$ such that $\{\tilde{X}_{\alpha}^{\pm} \mid \alpha \in [0,1]\}$ are *comonotonic*, i.e., there exists a real-valued random variable $X \in \mathcal{X}$ such that for $\tilde{X}_{\alpha}^{\pm}$ $(\alpha \in [0,1])$ there exists a non-decreasing function $h_{\alpha}^{\pm} : R(X) \to \mathbb{R}$ satisfying $\tilde{X}_{\alpha}^{\pm}(\omega) = h_{\alpha}^{\pm}(X(\omega))$ for all $\omega \in \Omega$, where $R(X)$ is the range of X.

Example Let $X \in \mathcal{X}$ be an integrable real-valued random variable and let $\tilde{a} \in \mathcal{R}$ be a fuzzy number $\tilde{a}(x) = \max\{1 - |x|/c, 0\}$ for $x \in \mathbb{R}$, where c is a positive number. Let $\tilde{X} \in \tilde{\mathcal{X}}$ be a triangle-type fuzzy random variable such that

$$\tilde{X}(\omega)(\cdot) = 1_{\{X(\omega)\}}(\cdot) + \tilde{a}(\cdot) \tag{18}$$

for $\omega \in \Omega$, where $1_{\{\cdot\}}$ denotes the characteristic function of a singleton. Then $\tilde{X}_{\alpha}^{\pm}(\omega) = X(\omega) \pm (1-\alpha)c = h_{\alpha}^{\pm}(X(\omega))$ for $\omega \in \Omega$, where $h_{\alpha}^{\pm}(x) = x \pm (1-\alpha)c$ $\alpha \in [0,1]$. Therefore $\tilde{X}_{\alpha}^{\pm}$ $(\alpha \in [0,1])$ are comonotnic [10, 16], and we obtain $\tilde{X} \in \tilde{\mathcal{X}}_c$.

Because AVaR_p is *comonotonically additive* from [10, Proposition 3(iii)], we can easily check the following proposition by (15), (16), [13, Lemma 1(ii)] and [17, Lemma 2.1].

Proposition 1 *Let* $\lambda \in [0,1]$ *and* $p \in (0,1)$. *Then it holds that*

$$E^{\lambda}(\widetilde{AVaR_p}(\tilde{X})) = AVaR_p(E^{\lambda}(\tilde{X})) \tag{19}$$

for integrable fuzzy random variables $\tilde{X} \in \tilde{\mathcal{X}}_c$.

Finally we introduce variances and covariances of fuzzy random variables from the viewpoint of λ-mean functions and evaluation weights. From (19), for fuzzy random variables $\tilde{X}$ and $\tilde{Y}$, we define variances and covariances as follows:

$$V(E^{\lambda}(\tilde{X})) = E\left((E^{\lambda}(\tilde{X}) - E(E^{\lambda}(\tilde{X})))^2\right), \tag{20}$$

$$Cov(E^{\lambda}(\tilde{X}), E^{\lambda}(\tilde{Y})) = E\left((E^{\lambda}(\tilde{X}) - E(E^{\lambda}(\tilde{X})))(E^{\lambda}(\tilde{Y}) - E(E^{\lambda}(\tilde{Y})))\right) \tag{21}$$

for $\lambda \in [0,1]$, where $V(\cdot)$ and $Cov(\cdot,\cdot)$ denote the variance and the covariance of real-valued random variables. Then we can easily check the following lemma.

Lemma 6 *Let* $\lambda \in [0,1]$. *For fuzzy numbers* $\tilde{a}, \tilde{b} \in \mathcal{R}$, *integrable fuzzy random variables* $\tilde{X}, \tilde{Y}$ *and a nonnegative real number* ζ, *the following (i)–(v) hold.*

(i) $V(E^{\lambda}(\tilde{a})) = 0$.
(ii) $V(E^{\lambda}(\tilde{X} + \tilde{a})) = V(E^{\lambda}(\tilde{X}))$.
(iii) $V(E^{\lambda}(\zeta\tilde{X})) = \zeta^2 V(E^{\lambda}(\tilde{X}))$.
(iv) $Cov(E^{\lambda}(\tilde{X}), E^{\lambda}(\tilde{a})) = Cov(E^{\lambda}(\tilde{a}), E^{\lambda}(\tilde{X})) = 0$.
(v) $Cov(E^{\lambda}(\tilde{X} + \tilde{a}), E^{\lambda}(\tilde{Y} + \tilde{b})) = Cov(E^{\lambda}(\tilde{X}), E^{\lambda}(\tilde{Y}))$.

4 A Dynamic Portfolio Model with Fuzzy Random Variables

First we explain a portfolio model with n stocks, where n is a positive integer. Let $\{0, 1, 2, \ldots, T\}$ be the time space with an expiration date T, where T is a positive integer. For an asset $i = 1, 2, \ldots, n$, a *stock price process* $\{S_t^i\}_{t=0}^T$ is given generally by *rates of return* R_t^i as follows. Let $S_t^i = S_{t-1}^i(1 + R_t^i)$ for $t = 1, 2, \ldots, T$, where $\{R_t^i\}_{t=1}^T$ is an integrable sequence of independent real-valued random variables satisfying $1 + R_t^i \geq 0$ for $t = 1, 2, \ldots, T$.

In this paper, we deal with rates of return have fuzzy element, i.e., they are given by independent fuzzy random variables $\{\tilde{R}_t^i\}_{t=1}^T \subset \tilde{\mathcal{X}}$. Hence we assume that

$$1 + \tilde{R}_t^i \succeq 0 \tag{22}$$

for $i = 1, 2, \ldots, n$ and $t = 1, 2, \ldots, T$. Then $w_t = (w_t^1, w_t^2, \ldots, w_t^n)$ is called a *portfolio weight vector* if it satisfies $w_t^1 + w_t^2 + \cdots + w_t^n = 1$, and further a portfolio $(w_t^1, w_t^2, \ldots, w_t^n)$ is said to *allow for short selling* if $w_t^i \geq 0$ for all $i = 1, 2, \ldots, n$. The rate of return with a portfolio $(w_t^1, w_t^2, \ldots, w_t^n)$ is given by

$$\tilde{R}_t = w_t^1\tilde{R}_t^1 + w_t^2\tilde{R}_t^2 + \cdots + w_t^n\tilde{R}_t^n, \tag{23}$$

and the reward at time $t (= 1, 2, \ldots, T)$ follows

$$\tilde{S}_t = \tilde{S}_{t-1}\sum_{i=1}^n w_t^i(1 + \tilde{R}_t^i) = \tilde{S}_{t-1}(1 + \tilde{R}_t), \tag{24}$$

and we take an initial stock price by a real number $\tilde{S}_0 = 1$ for simplicity. In (23), portfolio weights $w_t = (w_t^1, w_t^2, \ldots, w_t^n)$ are decided sequentially and predictably. The risk of $\tilde{S}_t$ is related to the information $\mathcal{M}_{t-1}$ up to time $t - 1$, and the average value-at-risk of $\tilde{S}_t$ under information $\mathcal{M}_{t-1}$ at a probability p is

$$\begin{aligned}\widetilde{\mathrm{AVaR}}_p(\tilde{S}_t \mid \mathcal{M}_{t-1}) &= \widetilde{\mathrm{AVaR}}_p\left(\tilde{S}_{t-1}\sum_{i=1}^n w_t^i(1 + \tilde{R}_t^i) \mid \mathcal{M}_{t-1}\right) \\ &= \widetilde{\mathrm{AVaR}}_p\left(\tilde{S}_{t-1}(1 + \tilde{R}_t) \mid \mathcal{M}_{t-1}\right).\end{aligned} \tag{25}$$

The term (25) means the risk of worst scenarios which occur on the transition from time $t-1$ to time t. Therefore, taking the sum of the risks which occur at each time, this paper deals with the following dynamic portfolio problem regarding the total of average value-at-risks (25) under information $\{\mathcal{M}_{t-1}\}_{t=1}^{T}$. Let a discount rate β be a positive constant and let $\lambda \in [0, 1]$.

Dynamic Portfolio Problem 1 (D1): Maximize the total average value-at-risk

$$E\left(E^{\lambda}\left(\sum_{t=1}^{T}\beta^{t-1}\widetilde{\mathrm{AVaR}}_p(\tilde{S}_t \mid \mathcal{M}_{t-1})\right)\right) \tag{26}$$

with portfolio weights $w_t = (w_t^1, w_t^2, \ldots, w_t^n)$ satisfying $w_t^1 + w_t^2 + \cdots + w_t^n = 1$ and $w_t^i \geq 0$ $(i = 1, 2, \ldots, n; t = 1, 2, \ldots, T)$.

By Lemmas 2 and 3, Dynamic Portfolio Problem 1 (D1) is reduced to the following problem.

Dynamic Portfolio Problem 2 (D2): Maximize the total average value-at-risk

$$\begin{aligned} &E^{\lambda}\left(\sum_{t=1}^{T}\beta^{t-1}\prod_{s=1}^{t-1}(1+\tilde{E}(\tilde{R}_s))\cdot(1+\widetilde{\mathrm{AVaR}}_p(\tilde{R}_t))\right) \\ &= E^{\lambda}\left(\sum_{t=1}^{T}\beta^{t-1}\prod_{s=1}^{t-1}\left(1+\tilde{E}\left(\sum_{i=1}^{n}w_s^i\tilde{R}_s^i\right)\right)\cdot\left(1+\widetilde{\mathrm{AVaR}}_p\left(\sum_{i=1}^{n}w_t^i\tilde{R}_t^i\right)\right)\right) \end{aligned} \tag{27}$$

with portfolios $w_t = (w_t^1, w_t^2, \ldots, w_t^n)$ satisfying $w_t^1 + w_t^2 + \cdots + w_t^n = 1$ and $w_t^i \geq 0$ $(i = 1, 2, \ldots, n; t = 1, 2, \ldots, T)$.

Define the set of portfolios by $\mathcal{W} = \{(w^1, w^2, \ldots, w^n) \in \mathbb{R}^n \mid w^1 + w^2 + \cdots + w^n = 1 \text{ and } w^i \geq 0\ (i = 1, 2, \ldots, n)\}$. By Lemmas 4 and 5, we can easily check the following dynamic programming [14].

Theorem 1 *The optimal average value-at-risk for (27) in Dynamic Portfolio Problem 2 (D2) is given by v_1 which is defined inductively by the sequence $\{v_t\}$ of sub-total-sum average value-at-risks after time $t-1$ satisfying the following backward optimality equations:*

$$\begin{aligned} v_{t-1} = \max_{(w^1,w^2,\ldots,w^n)\in\mathcal{W}} \Bigg\{ & 1 + E^{\lambda}\left(\widetilde{AVaR}_p\left(\sum_{i=1}^{n}w^i\tilde{R}_{t-1}^i\right)\right) \\ & + \beta\left(1+\sum_{i=1}^{n}w^iE^{\lambda}(\tilde{E}(\tilde{R}_{t-1}^i))\right)v_t\Bigg\} \end{aligned} \tag{28}$$

for $t = 2, 3, \ldots, T$, and

$$v_T = \max_{(w^1,w^2,\ldots,w^n)\in\mathcal{W}} \left\{ 1 + E^\lambda \left(\widetilde{AVaR}_p \left(\sum_{i=1}^{n} w^i \tilde{R}_T^i \right) \right) \right\}. \tag{29}$$

In the next section we focus on the average value-at-risks at each time in (28) and (29).

5 Portfolio Optimization for Average Value-at-Risks

First we estimate the rate of return for a portfolio [12]. Let $t = 1, 2, \ldots, T$ and let $\lambda \in [0, 1]$. Let the mean, the variance and the covariance of (23) respectively by

$$\begin{aligned}
\mu_t^i &= E(E^\lambda(\tilde{R}_t^i)),\\
\sigma_t^{ii} &= V(E^\lambda(\tilde{R}_t^i)),\\
\sigma_t^{ij} &= Cov(E^\lambda(\tilde{R}_t^i), E^\lambda(\tilde{R}_t^j)) \ (i \neq j)
\end{aligned}$$

for $i, j = 1, 2, \ldots, n$, where $V(\cdot)$ and $Cov(\cdot, \cdot)$ denote the variance and the covariance of real-valued random variables. In this section, we assume the rate of return $\tilde{R}_t^i \in \tilde{\mathcal{X}}_c$ for $i = 1, 2, \ldots, n$. Further we assume that the determinant of the variance-covariance matrix $\Sigma_t = [\sigma_t^{ij}]$ is not zero and there exists its inverse matrix Σ_t^{-1}. This assumption is natural and it can be realized easily by taking care of the combinations of assets. For a portfolio $w = (w^1, w^2, \ldots, w^n)$ satisfying $w^1 + w^2 + \cdots + w^n = 1$ and $w^i \geq 0$ $(i = 1, 2, \ldots, n)$, we calculate the expectation and the variance regarding $\tilde{R}_t = w^1\tilde{R}_t^1 + w^2\tilde{R}_t^2 + \cdots + w^n\tilde{R}_t^n$. The expectation μ_t of the rate of return $\tilde{R}_t$ with the portfolio w is

$$\mu_t = E(E^\lambda(\tilde{R}_t)) = \sum_{i=1}^{n} w^i E(E^\lambda(\tilde{R}_t^i)) = \sum_{i=1}^{n} w^i \mu_t^i. \tag{30}$$

On the other hand, the variance $(\sigma_t)^2$ of the rate of return $\tilde{R}_t$ with the portfolio w is

$$(\sigma_t)^2 = V(E^\lambda(\tilde{R}_t)) = \sum_{i=1}^{n}\sum_{j=1}^{n} w^i w^j \sigma_t^{ij} \tag{31}$$

for $i = 1, 2, \ldots, n$. In this paper we deal with a portfolio model which has the following representation:

$$\text{(VaR)} = \text{(the mean)} - \text{(a positive constant } \kappa(p)) \times \text{(the standard deviation)}, \tag{32}$$

where the positive constant $\kappa(p)$ is given corresponding to probability p (Fig. 2). One of the most popular sufficient condition for (32) is what the distribution of the rate

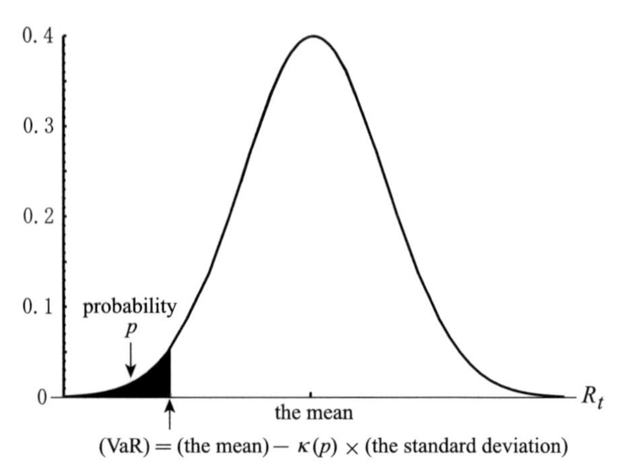

Fig. 2 Value-at-risk at a probability p

of return is Gaussian [1]. From Proposition 1, for a given positive probability p the mean of the average value-at-risk $E^\lambda(\widetilde{\mathrm{AVaR}_p(\tilde{R}_t)}) = \mathrm{AVaR}_p(E^\lambda(\tilde{R}_t))$ is evaluated as

$$E^\lambda(\widetilde{\mathrm{AVaR}_p(\tilde{R}_t)}) = \sum_{i=1}^{n} w^i \mu_t^i - \kappa \sqrt{\sum_{i=1}^{n}\sum_{j=1}^{n} w^i w^j \sigma_t^{ij}} \tag{33}$$

with a positive constant $\kappa = \frac{1}{p}\int_0^p \kappa(q)\,dq$.

Let

$$\mu_t = \begin{bmatrix} \mu_t^1 \\ \mu_t^2 \\ \vdots \\ \mu_t^n \end{bmatrix}, \quad \Sigma_t = \begin{bmatrix} \sigma_t^{11} & \sigma_t^{12} & \cdots & \sigma_t^{1n} \\ \sigma_t^{21} & \sigma_t^{22} & \cdots & \sigma_t^{2n} \\ \vdots & \vdots & \ddots & \vdots \\ \sigma_t^{n1} & \sigma_t^{n2} & \cdots & \sigma_t^{nn} \end{bmatrix}, \quad \mathbf{1} = \begin{bmatrix} 1 \\ 1 \\ \vdots \\ 1 \end{bmatrix},$$

$$A_t = \mathbf{1}^{\mathrm{T}}\Sigma_t^{-1}\mathbf{1}, \qquad B_t = \mathbf{1}^{\mathrm{T}}\Sigma_t^{-1}\mu_t, \qquad C_t = \mu_t^{\mathrm{T}}\Sigma_t^{-1}\mu_t, \qquad \Delta_t = A_t C_t - B_t^2,$$

where T denotes the transpose of a vector. Now we discuss the following AVaR portfolio problem without allowance for short selling. The following form comes from the average value-at-risk $\widetilde{\mathrm{AVaR}_p(\tilde{R}_t)}$ given in (33).

AVaR-portfolio problem (AP): Let $t = 1, 2, \ldots, T$. Maximize the average value-at-risk

$$E^\lambda(\widetilde{\mathrm{AVaR}_p(\tilde{R}_t)}) = \sum_{i=1}^{n} w^i \mu_t^i - \kappa \sqrt{\sum_{i=1}^{n}\sum_{j=1}^{n} w^i w^j \sigma_t^{ij}} \tag{34}$$

with respect to portfolios $w = (w^1, w^2, \ldots, w^n)$ satisfying $w^1 + w^2 + \cdots + w^n = 1$ and $w^i \geq 0$ for $i = 1, 2, \ldots, n$.

We have the analytical solutions regarding AVaR-portfolio problem (AP) under the following condition.

Condition 1: $1 + \beta v_t > 0$ for all $t = 1, 2, \ldots, T$.

If Condition 1 is not satisfied at some time t, it means that the portfolio is bankrupt at the time t. Under Condition 1, Theorem 1 is written as following theorem.

Theorem 2 *Suppose Condition 1 is satisfied. The optimal average value-at-risk v_1 in Theorem 1 is given by the sequence $\{v_t\}$ of sub-total-sum average value-at-risks after time $t-1$ satisfying the following backward optimality equations:*

$$v_{t-1} = \max_{(w^1, w^2, \ldots, w^n) \in \mathcal{W}} (1 + \beta v_t) \left(1 + \sum_{i=1}^n w^i \mu_{t-1}^i - \frac{\kappa}{1 + \beta v_t} \sqrt{\sum_{i=1}^n \sum_{j=1}^n w^i w^j \sigma_{t-1}^{ij}} \right) \tag{35}$$

for $t = 2, 3, \ldots, T$, and

$$v_T = \max_{(w^1, w^2, \ldots, w^n) \in \mathcal{W}} \left(1 + \sum_{i=1}^n w^i \mu_T^i - \kappa \sqrt{\sum_{i=1}^n \sum_{j=1}^n w^i w^j \sigma_T^{ij}} \right). \tag{36}$$

Lemma 7 ([12]). *Let $t = 1, 2, \ldots, T$. Let A_t and Δ_t be positive. Let the constant κ satisfy $\kappa^2 > C_t$. Then the following (i) and (ii) hold.*

(i) The solution of AVaR-portfolio problem (AP) is given by

$$w^* = \xi_t \Sigma_t^{-1} \mathbf{1} + \eta_t \Sigma_t^{-1} \mu_t, \tag{37}$$

and then the corresponding average value-at-risk is

$$E^\lambda \left(\widetilde{AVaR}_p \left(\sum_{i=1}^n w^{*i} \tilde{R}_t^i \right) \right) = \frac{B_t - \sqrt{A_t \kappa^2 - \Delta_t}}{A_t}, \tag{38}$$

where $w^* = (w^{*1}, w^{*2}, \ldots, w^{*n})$, $\gamma_t = \frac{B_t}{A_t} + \frac{\Delta_t}{A_t \sqrt{A_t \kappa^2 - \Delta_t}}$, $\xi_t = \frac{C_t - B_t \gamma_t}{\Delta_t}$ and $\eta_t = \frac{A_t \gamma_t - B_t}{\Delta_t}$.

(ii) Further, if $\Sigma_t^{-1} \mathbf{1} \geq \mathbf{0}$ and $\Sigma_t^{-1} \mu_t \geq \mathbf{0}$, then the portfolio (37) satisfies $w^* \geq \mathbf{0}$, i.e. w^* is a portfolio without allowance for short selling. Here, $\mathbf{0}$ denotes the zero vector.

Assume the condition (32) holds. Then, for a risk probability p, we take a constant κ in (33) by

$$\kappa = -\frac{1}{p} \int_0^p \Phi^{-1}(q)\, dq, \tag{39}$$

where Φ^{-1} is the inverse function of the cumulative normal distribution function

$$\Phi(z) = \frac{1}{\sqrt{2\pi}} \int_{-\infty}^{z} e^{-\frac{t^2}{2}} \, dt \tag{40}$$

for $z \in \mathbb{R}$. Applying Lemma 7 to Theorem 2, we obtain the following results.

Theorem 3 *Suppose Condition 1 is satisfied. Assume $E^{\lambda}(R_t^i)$ $(t = 1, 2, \ldots, T; i = 1, 2, \ldots, n)$ have normal distributions, and let $\kappa = -\frac{1}{p}\int_0^p \Phi^{-1}(q)\,dq$ in (33). Then the optimal average value-at-risk v_1 in Theorem 2 is calculated by the sequence $\{v_t\}$ of sub-total-sum average value-at-risks after time $t-1$ satisfying the following backward optimality equations:*

$$v_{t-1} = \frac{(A_{t-1} + B_{t-1})(1 + \beta v_t) - \sqrt{A_{t-1}\kappa^2 - \Delta_{t-1}(1 + \beta v_t)^2}}{A_{t-1}} \tag{41}$$

for $t = 2, 3, \ldots, T$ and

$$v_T = \frac{A_T + B_T - \sqrt{A_T \kappa^2 - \Delta_T}}{A_T}. \tag{42}$$

On the other hand, the optimal portfolios in Theorem 2 and Lemma 7 are obtain as follows [15].

Corollary 1 *Suppose Condition 1 is satisfied. Let A_t and Δ_t be positive for $t = 1, 2, \ldots, T$. Let $\kappa = -\frac{1}{p}\int_0^p \Phi^{-1}(q)\,dq$, and let $\kappa_{t-1} = \frac{\kappa}{1+\beta v_t}$ $(t = 2, 3, \ldots, T)$ and $\kappa_T = \kappa$. Assume κ_{t-1} satisfies $\kappa_{t-1}^2 > C_{t-1} (t = 2, 3, \ldots, T)$. Then the following (i) and (ii) hold.*

(i) The optimal portfolios (33) in Theorem 2 are given by

$$w_t^* = \xi_t \Sigma_t^{-1} \mathbf{1} + \eta_t \Sigma_t^{-1} \mu_t \tag{43}$$

for $t = 1, 2, \ldots, T$, where $\gamma_t = \frac{B_t}{A_t} + \frac{\Delta_t}{A_t\sqrt{A_t\kappa^2 - \Delta_t}}$, $\xi_t = \frac{C_t - B_t\gamma_t}{\Delta_t}$ and $\eta_t = \frac{A_t\gamma_t - B_t}{\Delta_t}$.

(ii) Further, if $\Sigma_t^{-1}\mathbf{1} \geq \mathbf{0}$ and $\Sigma_t^{-1}\mu_t \geq \mathbf{0}$ for $t = 1, 2, \ldots, T$, then the portfolio (43) satisfies $w_t \geq \mathbf{0}$, i.e. w_t^* is a portfolio without allowance for short selling.

In the rest of this section we demonstrate the *falling of asset prices* in AVaR-portfolio problem (AP) at each time t. Let $\lambda \in [0, 1]$. Regarding asset price $\tilde{S}_t$ with a portfolio w_t, the theoretical *bankruptcy* at time t occurs on a scenario $\omega(\in \Omega)$ satisfying $\tilde{S}_t(\omega) \preceq 0$, i.e. it follows $1 + \tilde{R}_t(\omega) \preceq 0$ from (24). By similar idea, for a constant $\bar{\rho}$ satisfying $\bar{\rho} \in [0, 1]$, a set of sample paths

$$\{\omega \in \Omega \mid 1 + E^{\lambda}(\tilde{R}_t(\omega)) \leq 1 - \bar{\rho}\} = \{\omega \in \Omega \mid E^{\lambda}(\tilde{R}_t(\omega)) \leq -\bar{\rho}\} \tag{44}$$

is the event of scenarios where the asset price $\tilde{S}_t$ will fall from the current price $\tilde{S}_{t-1}$ to a lower level than $100(1 - \bar{\rho})$ % of the current price $\tilde{S}_{t-1}$, i.e. the rate of falling is 100 $\bar{\rho}$ %. Then, the parameter $\bar{\rho}$ is called *the rate of falling*, and the *probability of falling* is also given by

$$\bar{p} = P(E^{\lambda}(\tilde{R}_t) \leq -\bar{\rho}). \tag{45}$$

For example, $\bar{p}$ denotes the probability of the falling below par value if '$\bar{\rho} = 0$' and it indicates the probability of the bankruptcy if '$\bar{\rho} = 1$'. From this observation, we obtain the following results.

Let p be a positive probability. A value-at-risk regarding the rate of return $\tilde{R}_t$ at probability p is given by a real number $v(p)$ satisfying

$$P(E^{\lambda}(\tilde{R}_t) \leq v(p)) = p. \tag{46}$$

The value-at-risk $v(p)$ is the upper bound of the rate of return $\tilde{R}_t$ at the worst scenarios under a given risk probability p, and then $v(p) = \mathrm{VaR}_p(E^{\lambda}(\tilde{R}_t))$. Then, from Proposition 1, it holds that

$$E^{\lambda}(\widetilde{\mathrm{AVaR}_p(\tilde{R}_t)}) = \mathrm{AVaR}_p(E^{\lambda}(\tilde{R}_t)) = \frac{1}{p}\int_0^p \mathrm{VaR}_q(E^{\lambda}(\tilde{R}_t))\,dq = \frac{1}{p}\int_0^p v(q)\,dq. \tag{47}$$

On the other hand, from (45) and (46), the *average rate of falling* is given by $\rho = -\frac{1}{p}\int_0^p v(q)\,dq$. Therefore the average rate of falling follows $\rho = -E^{\lambda}(\widetilde{\mathrm{AVaR}_p(\tilde{R}_t)})$. Let the *expected rate of return* by $\gamma = E(E^{\lambda}(\tilde{R}_t))$. Then for the optimal portfolios (43) at each time t we can find the following corresponding relation holds among the expected rate of return γ, the average rate of falling ρ and the risk probability of falling p [15]:

Expected rate of return γ

$$\Updownarrow\ \gamma = \frac{B_t\rho + C_t}{A_t\rho + B_t}$$

Average rate of falling ρ

$$\Updownarrow\ p^2 \exp\left(\Phi^{-1}(p)^2\right) = \frac{1}{2\pi(A_t\rho^2 + 2B_t\rho + C_t)}$$

Risk probability of falling p

This relation will be useful not only for theoretical analysis but also for actual management in finance.

Acknowledgements This research is supported from JSPS KAKENHI Grant Number JP 16K05282.

References

1. L.El Chaoui, M.Oks and F.Oustry, Worst-case value at risk and robust portfolio optimization: A conic programming approach, *Operations Research*, 51:543–556, 2003.
2. López-Díaz, M., Gil, M.A., Ralescu, D.A.: Overview on the Development of Fuzzy Random Variables. Fuzzy Sets and Systems **147** (2006) 2546–2557.
3. P. Fortemps and M. Roubens, Ranking and defuzzification methods based on area compensation, *Fuzzy Sets and Systems*, 82:319–330, 1996.
4. Kruse, R., Meyer, K.D.: Statistics with Vague Data. Riedel Publ. Co., Dortrecht (1987).
5. H.Kwakernaak, Fuzzy random variables – I. Definitions and theorem, *Inform. Sci.*, 15:1–29, 1978.
6. H.Markowitz, *Mean-Variance Analysis in Portfolio Choice and Capital Markets*, Blackwell, Oxford, 1990.
7. S.R.Pliska, *Introduction to Mathematical Finance: Discrete-Time Models*, Blackwell Publ., New York, 1997.
8. M.L.Puri and D.A.Ralescu, Fuzzy random variables, *J. Math. Anal. Appl.*, 114:409–422, 1986.
9. Y.Yoshida, Mean values, measurement of fuzziness and variance of fuzzy random variables for fuzzy optimization, *Proceedings of SCIS & ISIS 2006*, 2277–2282, Sept. 2006.
10. Y.Yoshida, A risk-minimizing model under uncertainty in portfolio, in: V.Torra et al. eds., *MDAI 2007*, LNAI 4529, 295–306, Springer, August, 2007.
11. Y.Yoshida, Fuzzy extension of estimations with randomness: The perception-based approach, in: P.Melin et al. eds., *IFSA2007*, LNAI 4617, 381–391, Springer, Sept., 2007.
12. Y.Yoshida, An estimation model of value-at-risk portfolio under uncertainty, *Fuzzy Sets and Systems*, 160:3250–3262, 2009.
13. Y.Yoshida, A perception-based portfolio under uncertainty: Minimization of average rates of falling, in: V.Torra, Y.Narukawa and M.Inuiguchi, eds., *MDAI 2009*, LNAI 5861, 149–160, Springer, Nov., 2009.
14. Y.Yoshida, A dynamic value-at-risk portfolio model, in: V.Torra, Y.Narukawa, J.Yin and J.Long eds., *MDAI 2011*, LNAI 6820, 43–24, Springer, Jul., 2011.
15. Y.Yoshida, An average value-at-risk portfolio model under uncertainty: A perception-based approach by fuzzy random variables, *Journal of Advanced Computational Intelligence and Intelligent Informatics*, 15:56–62, 2011.
16. Y.Yoshida, An ordered weighted average with a truncation weight on intervals, in: V.Torra, Y.Narukawa B.López and M.Villaret eds., *MDAI 2012*, LNAI 7647, 45–55, Springer, Nov., 2012.
17. Y.Yoshida, Aggregation of Dynamic Risk Measures in Financial Management, in: V.Torra and Y.Narukawa eds., *MDAI 2012*, LNAI 8825, 38–49, Springer, Oct., 2014.
18. L.A.Zadeh, Fuzzy sets, *Inform. and Control*, 8:338–353, 1965.

Group Decision Making: Consensus Approaches Based on Soft Consensus Measures

Francisco Javier Cabrerizo, Ignacio Javier Pérez, Francisco Chiclana and Enrique Herrera-Viedma

Abstract A group decision making situation involves multiple decision makers communicating with others to reach a decision. In such a situation, the most important issue is to obtain a decision that is best acceptable by the decision makers, and, therefore, consensus has attained a great attention and it is a major goal of group decision making situations. To measure the closeness among the opinions given by the decision makers, different approaches have been proposed. At the beginning, consensus was meant to be a unanimous and full agreement. However, because this situation is often not reachable in practice, the use of a softer consensus, which assesses the level of agreement in a more flexible way and reflects the large spectrum of possible partial agreements, is a more reasonable approach. Soft consensus approaches better reflects a real human perception of the essence of consensus and, therefore, they have been widely used. The purpose of this contribution is to review the different consensus approaches based on soft consensus measures that have been proposed.

Keywords Group decision making · Consensus · Fuzzy set theory

---

F.J. Cabrerizo (✉) · E. Herrera-Viedma
Department of Computer Sciences and Artificial Intelligence, University of Granada,
C/ Periodista Daniel Saucedo Aranda S/n, 18071 Granada, Spain
e-mail: cabrerizo@decsai.ugr.es

E. Herrera-Viedma
e-mail: viedma@decsai.ugr.es

I.J. Pérez
Department of Computer Sciences and Engineering, University of Cádiz,
Avenida Universidad de Cádiz 10, 11519 Puerto Real, Spain
e-mail: ignaciojavier.perez@uca.es

F. Chiclana
Faculty of Technology, Centre for Computational Intelligence, De Montfort University,
Leicester LE1 9BH, UK
e-mail: chiclana@dmu.ac.uk

V. Torra et al. (eds.), *Fuzzy Sets, Rough Sets, Multisets and Clustering*,
Studies in Computational Intelligence 671, DOI 10.1007/978-3-319-47557-8_18

1 Introduction

One of the most crucial human activities is decision making, which consists of finding the best alternative, variant, opinion, and so on, from among some possible ones. This task generally involves multiple decision makers to make the decision [10, 34, 47] and, then, it is called a multiperson or group decision making (GDM) situation [34].

In a GDM situation there is a group of decision makers expressing their preferences or opinions concerning a set of different alternatives. In such a context, the question is to find a solution which is best acceptable by the whole group of decision makers. Here, the process arriving at an agreed-upon opinion, perhaps consensus, by using in a democratic way knowledge of the different decision makers, leads to better decision [5].

The term consensus has been used for centuries in different areas and contexts. When it is used in GDM contexts, an important issue is the very meaning of consensus and the problems related to its essence.

First, consensus makes reference to the state of accordance with a group of decision makers in the sense that they show similar preferences or opinions related to the alternatives in question. In this sense, consensus was initially meant as a complete agreement. In such a way, some authors proposed consensus measures assuming values in-between 0, meaning no consensus or partial consensus, and 1, meaning full consensus [3, 51]. The above situation has however been considered impractical in most real world situations as decision makers on rare occasions arrive at that complete agreement. Therefore, the very essence of consensus was reconsidered, and it was admitted that the decision makers are not willing to fully change their preferences or opinions so that the consensus will not be a complete agreement. A milestone was here a special issue of the Synthese journal [40]. In particular, the paper written by Loewer and Laddaga [41] is the most relevant for this purpose, in which, these authors clearly made the case for a soft concept of consensus stating that:

> … It can correctly be said that there is a *consensus* among biologists that Darwinian natural selection is an important cause of evolution though there is currently *no consensus* concerning Gould's hypothesis of speciation. This means that there is a *widespread agreement* among biologists concerning the first matter but *disagreement* concerning the second…

It was suggested that a fuzzy majority is suitable, and that it makes sense to speak about a degree of consensus, or a distance from ideal consensus. The linguistic quantifiers, exemplified, for example, by 'most", "almost all", "much more than a half", and so on, are a natural manifestation of this fuzzy majority. Linguistic quantifiers can be handled by a calculus of linguistically quantified propositions [56], and also by using aggregation operators or aggregation functions [20, 53], in particular, Yager's OWA (Ordered Weighted Average) operators [54], which offer a much needed generality and flexibility [57]. Janusz Kacprzyk introduced the concept of a fuzzy majority related to a fuzzy linguistic quantifier into GDM situations [32–34]. Since then, the concept of a fuzzy majority has been the key point for new definitions of soft consensus [6, 31, 36, 37], which assess the degree of agreement in a more flexible way, reflecting the large spectrum of possible partial agreements and guiding the discussion process until widespread agreement, not always full, is achieved among the decision makers.

Second, consensus refers as a process to reach agreement. This process involves an evolution of the preferences expressed by the decision makers towards agreement with respect to their preferences. In such a situation, the point of departure is a the set of preferences given by the particular decision makers concerning in general opinions as to the values of some quantities. At the beginning, the preferences expressed by the decision makers was equated with some utilities resulting from some courses of actions, the probabilities of them, and alike [12, 18, 22]. Nevertheless, since GDM situations are centered on decision makers, coming with inherent subjectivity, imprecision and vagueness in the articulation of preferences, the theory of fuzzy sets [55], has delivered new tools in this field for a long time, as it is a more adequate tool to represent often not clear-cut human preferences encountered in most practical cases. Fuzzy logic has played here a considerable role by providing means for the representation and processing of imprecise information and preferences [17].

Because it is important to obtain an approved solution by all the decision makers, the consensus is one of the major goals of GDM situations. Concretely, consensus approaches based on soft consensus measures have been widely proposed in the literature [6, 31], as it is more human-consistent and suitable for reflecting human perceptions of the meaning of consensus.

The objective of this contribution is to review the different consensus approaches based on soft consensus measures that have been proposed in the literature. To do so, the pioneering contributions are described and a comprehensive presentation of the state of the art of all kinds of consensus related problems is shown. After some decades of fruitful research in this field, it is a good time for looking backward an review what the research has been developed on this topic.

This contribution is outlined as follows. In Sect. 2, we introduce the GDM framework and the usual consensus process. Section 3 highlights the pioneering and most important contributions existing on consensus approaches based on soft consensus measures. In Sect. 4, we describe the main consensus approaches based on soft consensus measures. Finally, in Sect. 5, we present some conclusions and future work.

2 Preliminaries

This section is devoted to introduce the GDM framework to develop consensus processes. Concretely, the GDM situation is defined, the formats of preferences utilized by the decision makers to provide their opinions are presented, and the usual consensus process is described.

2.1 GDM Framework

There have been several efforts in the specialized literature to create different models to correctly address and solve GDM situations. Some of them make use of fuzzy set theory as it is a good tool to model and deal with vague or imprecise opinions [19, 23, 38, 48].

In a classical GDM situation [9, 19, 34], there is a problem to solve, a solution set of possible alternatives, $X = \{x_1, x_2, \ldots, x_n\}$ $(n \geq 2)$, and a group of two or more decision makers, $E = \{e_1, e_2, \ldots, e_m\}$ $(m \geq 2)$, characterized by their background and knowledge, who express their opinions about the alternatives to achieve a common solution. In a fuzzy context, the objective is to classify the alternatives from best to worst, associating with them some degrees of preference expressed in the [0, 1] interval.

Decision makers can use several preference representation structures to provide their preferences or opinions about the alternatives in a GDM situation. The most common ones that have been widely used in the literature are the following:

- *Preference orderings*. Using this preference representation structure, the opinions of a decision maker $e_l \in E$ about a set of feasible alternatives X are described as a preference ordering $O^l = \{o_1^l, \ldots, o_n^l\}$, where $o^l(\cdot)$ is a permutation function over the indexes set $\{1, \ldots, n\}$ [52]. Hence, a decision maker gives an ordered vector of alternatives from best to worst.
- *Utility values*. Using this preference representation structure, a decision maker $e_l \in E$ expresses his/her opinions about a set of feasible alternatives X by means of a set of n utility values $U^l = \{u_1^l, \ldots, u_n^l\}$, $u_i^l \in [0, 1]$. Here, the higher the value for an alternative, the better it satisfies decision maker's objective [28].
- *Preference relations*. In this case, the preferences given by the decision maker on X are described by a function $\mu_{P^l} : X \times X \rightarrow D$ where $\mu_{P^l}(x_i, x_j) = p_{ij}^l$ can be interpreted as the preference degree or intensity of the alternative x_i over x_j expressed in the information representation domain D. Different types of preference relations can be used according to the domain used to evaluate the intensity of the preference:

  1. *Fuzzy preference relations* [34]: If $D = [0, 1]$, every value p_{ij}^l in the matrix P^l represents the preference degree or intensity of preference of the alternative x_i over x_j: $p_{ij}^l = 1/2$ indicates indifference between x_i and x_j, $p_{ij}^l = 1$ indicates that x_i is absolutely preferred to x_j, and $p_{ij} > 1/2$ indicates that x_i is preferred to x_k. It is usual to assume the additive reciprocity property $p_{ij}^l + p_{ji}^l = 1 \; \forall i, j$.
  2. *Multiplicative preference relations* [49]: If $D = [1/9, 9]$, and then every value p_{ij}^l in the matrix P^l represents a ratio of the preference intensity of the alternative x_i to that of x_j, i.e., it is interpreted as x_i is p_{ij}^l times good as x_j: $p_{ij}^l = 1$ indicates indifference between x_i and x_j, $p_{ij}^l = 9$ indicates that x_i is unanimously preferred and $p_{ij}^l \in \{2, 3, \ldots, 8\}$ indicates intermediate evaluations. It is usual to assume the multiplicative reciprocity property $p_{ij}^l \cdot p_{ji}^l = 1 \; \forall i, j$ too.

3. *Linguistic preference relations* [23, 24]: If $D = S$, where S is a linguistic term set $S = \{s_0, \ldots, s_g\}$ with odd cardinality (g+1), $s_{g/2}$ being a neutral label (meaning "equally preferred") and the rest of labels distributed homogeneously around it, then every value p^l_{ij} in the matrix P^l represents the linguistic preference degree or linguistic intensity of preference of the alternative x_i over x_j.

Among the different representation formats that decision makers may use to provide their opinions, fuzzy preference relations [34, 44] are one of the most used because of their effectiveness as a tool for modelling decision processes. In particular, they are very useful when we want to aggregate decision makers' preferences into a collective one [34, 52], which is carried out by using aggregation functions or aggregation operators [20, 53].

Finally, according to the importance of each decision maker, GDM situations are usually classified into two groups [11, 46]:

- A GDM situation is heterogeneous when the opinions of the decision makers are not equally important.
- A GDM situation is homogeneous if every opinion is treated equally.

A way to implement this heterogeneity is to assign a weight to every decision maker. Weights are qualitative or quantitative values that can be assigned in several different ways [11]: (i) weights can be assigned directly, (ii) or they can be obtained automatically from the opinions provided by the decision makers. The weights can be interpreted as a fuzzy subset, I, with a membership function, $\mu_I : E \rightarrow [0, 1]$, in such a way that $\mu_I(e_l) \in [0, 1]$ denotes the importance degree of the decision maker within the group, or how relevant is the decision maker in relation with the problem to be solved [14, 15]. Finally, it should be pointed out that fuzzy measures and fuzzy integrals [21, 43] can also be used to implement the heterogeneity among the decision makers.

2.2 *Consensus Process*

A way of solving GDM situations is by carrying out a selection process to choose a solution set of alternatives from the opinions provided by the decision makers [19, 48], without taking into account the level of agreement. It involves two different steps [7, 48]: (i) aggregation of individual preferences, and (ii) exploitation of the collective preference.However, this process can lead sometimes solutions that are not well accepted by some decision makers in the group [5, 50], because they could consider that their opinions have not been considered properly to obtain the solution, and, hence, they might reject it. To avoid this situation, it is advisable that decision makers carry out a consensus process. For this reason, GDM problems are usually faced by applying a consensus process and a selection process before a final solution can be obtained [38].

Two approaches may be distinguished in the formulation of a consensus process. The traditional one, in which the process is modeled by using matrix calculus or Markov chains to model the time evolution of changes of opinions toward consensus [12, 18, 22]. The approaches exemplified by the above citations have contributed much to the understanding of the process and its dynamics. However, it has been considered much more promising to run the consensus process with the help of a special agent, called a moderator, whose task is to help the decision makers involved while changing their testimonies towards consensus, by rational argument, persuasion, and so on. This second approach, in which there is a moderator, is more promising in practice and the most used.

According to this second approach, a consensus process is an iterative process composed of several consensus rounds, where the decision makers accept to change their opinions following the advice given by a moderator, which knows the agreement in each moment of the consensus process by means of the computation of some consensus measures.

3 Pioneering Contributions

In this section, the innovative and prominent contributions in the field of consensus approaches based on soft consensus measures are revised. As we have aforementioned, people are generally willing to accept that consensus has been reached when most actors agree with the opinions associated with the most relevant alternatives. The milestone was a special issue published in the Synthese journal:

B. Loewer. *Special issue on consensus*. Synthese 62 (1), 1–122 (1985).

Among many papers therein, Loewer and Laddaga wrote the most important one for our purpose:

B. Loewer, R. Laddaga. *Destroying the consensus*. Synthese 62 (1), 79–96 (1985).

Here the first approach for a soft concept of consensus was clearly made, suggesting that a fuzzy majority is appropriate, and that it makes sense to speak about a degree of consensus, or a distance from (ideal) consensus.

According to Loewer and Laddaga, Kacprzyk and Fedrizzi introduced the concept of a fuzzy majority using Zadeh's fuzzy linguistic quantifier to compute soft consensus measures in the following prominent contributions:

J. Kacprzyk, M. Fedrizzi. *Soft consensus measure for monitoring real consensus reaching processes under fuzzy preferences*. Control and Cybernetics 15 (3–4), 309–323 (1986).

J. Kacprzyk, M. Fedrizzi. *A 'soft' measure of consensus in the setting of partial (fuzzy) preferences*. European Journal of Operational Research 34 (3), 316–325 (1988).

J. Kacprzyk, M. Fedrizzi. *A 'human-consistent' degree of consensus based on fuzzy logic with linguistic quantifiers*. Mathematical Social Sciences 18 (3), 275–290 (1989).

Then, the classical operational definition of consensus was expressed by a linguistically quantified proposition as:

$$\text{“}Most\ (Q1)\text{ of the }important\ (B)\text{ individuals }agree \\ \text{as to }almost\ all\ (Q2)\ relevant\ (I)\text{ alternatives”} \quad (1)$$

where: $Q1$ and $Q2$ are fuzzy linguistic quantifiers [56], e.g.,“most” and “almost all”, and B and I stand for fuzzy sets denoting the importance/relevance of the individuals and alternatives.

The above works constituted the basis of many consensus approaches based on soft consensus measures proposed later. In the following, some of the most prominent contributions are presented:

- Herrera, Herrera-Viedma and Verdegay defined the first soft consensus model in GDM problems in a fuzzy linguistic context:

  F. Herrera, E. Herrera-Viedma, J. L. Verdegay, *A model of consensus in group decision making under linguistic assessments*. Fuzzy Sets and Systems 78 (1), 73–87 (1996).

  This prominent contribution has shown a high impact in the fuzzy decision making community, and it is considered a highly cited paper according to the Essential Science Indicators (ESI) database, published by Thomson Reuters. Here, the authors present a new consensus model for GDM problems based on fuzzy linguistic preference relations defined in an ordinal fuzzy linguistic approach [26, 27]. As main novelty, two types of soft consensus measures to guide the consensus process are defined: (i) consensus degrees, and (ii) proximity measures. In addition, they are applied in three activity levels: (i) level of preference, (ii) level of alternative, and (iii) level of preference relation. The consensus degrees indicate how far a group of decision makers is from the maximum consensus, and the proximity measures indicate how far each decision maker is from current consensus labels over the preferences. In such a way, the moderator is provided with a complete consensus instrument to control the consensus process.
- Later, assuming also a fuzzy linguistic context, the same authors presented the first consensus model which is guided by both consensus and consistency measures:

  F. Herrera, E. Herrera-Viedma, J. L. Verdegay. *A rational consensus model in group decision making using linguistic assessments*. Fuzzy Sets and Systems 88 (1), 31–49 (1997).

  In this new approach, the moderator is provided with consistency measures to guide the consensus process too. This consensus approach offers the possibility of achieving more rational consensus solutions, i.e., less distorted consensus solutions due to inconsistencies in the decision makers' preferences.
- Other prominent contribution in soft consensus was proposed by Herrera-Viedma, Herrera, and Chiclana:

  E. Herrera-Viedma, F. Herrera, F. Chiclana. *A consensus model for multiperson decision making with different preference structures*. IEEE Transactions on Systems Man and Cybernetics-Part A: Systems and Humans 32 (3), 394–402 (2002).

In this consensus approach, the decision makers can provide their preferences with different preference representation structures. Two main novelties are also contained in this contribution. Firstly, soft consensus measures are computed by comparison between decision makers' solutions and not between decision makers' preferences, as it usually happens in previous consensus approaches. In such a way, the problem of computing consensus measures is overcome when we use different preference representation structures in GDM problems. And secondly, using these measures, a feedback mechanism based on simple and easy rules to help decision makers change their preferences is defined. Therefore, the consensus process could be guided automatically, without a moderator, avoiding the possible subjectivity that he/she could introduce into the process. We should point out that this consensus contribution is a highly cited paper according to the ESI database too.

- Herrera-Viedma, Martinez, Mata and Chiclana dealt with the consensus problem when the GDM problem is defined in a fuzzy multi-granular linguistic context, i.e., by assuming that decision makers could use different linguistic term sets to provide their preferences:

  E. Herrera-Viedma, L. Martinez, F. Mata, F. Chiclana. *A consensus support system model for group decision-making problems with multigranular linguistic preference relations.* IEEE Transactions on Fuzzy Systems 13 (5), 644–658 (2005).

  The main novelty of this contribution is to present an automatic control system to guide the consensus process that substitutes the moderator's actions. To do so, this approach uses the consensus degrees to decide when the consensus process should finish and the proximity measures to define a recommendation system that recommends decision makers about the preferences that they should change in the next consensus rounds. This contribution is also considered a highly cited paper according to the ESI database.
- Finally, other seminal consensus contribution was proposed by Herrera-Viedma, Alonso, Chiclana, and Herrera in:

  E. Herrera-Viedma, S. Alonso, F. Chiclana, and F. Herrera. *A consensus model for group decision making with incomplete fuzzy preference relations.* IEEE Transactions on Fuzzy Systems 15 (5), 863–877 (2007).

  The main novelty of this soft consensus approach is that it provides tools to support the consensus processes in the presence of missing values or incomplete information in GDM situations. Here, the authors define the first consensus approach based on soft consensus measures which is carried out automatically (without a moderator) by three kinds of measures: consensus measures, consistency measures and incompleteness measures, too. Similarly, this contribution is considered a highly cited paper in the ESI database.

Finally, the main novelties of the above prominent consensus approaches based on soft consensus measures are summarized in Table 1.

Table 1 Most prominent soft consensus approaches

Contribution	Novelties
F. Herrera, E. Herrera-Viedma, J. L. Verdegay, *A model of consensus in group decision making under linguistic assessments*. Fuzzy Sets and Systems 78 (1), 73–87 (1996)	• It defines the first soft consensus approach in a fuzzy linguistic context • It uses both consensus degrees and proximity measures to guide the consensus process
F. Herrera, E. Herrera-Viedma, J. L. Verdegay. *A rational consensus model in group decision making using linguistic assessments*. Fuzzy Sets and Systems 88 (1), 31–49 (1997).	• It is guided by both consensus and consistency measures
E. Herrera-Viedma, F. Herrera, F. Chiclana. *A consensus model for multiperson decision making with different preference structures*. IEEE Transactions on Systems Man and Cybernetics-Part A: Systems and Humans 32 (3), 394–402 (2002)	• Different preference representation structures can be used • Soft consensus measures are computed by comparison between decision makers' solutions • A feedback mechanism is incorporated
E. Herrera-Viedma, L. Martinez, F. Mata, F. Chiclana. *A consensus support system model for group decision-making problems with multigranular linguistic preference relations*. IEEE Transactions on Fuzzy Systems 13 (5), 644–658 (2005)	• It is defined in a fuzzy multi-granular linguistic context • It presents an automatic control system substituting the moderator's actions
E. Herrera-Viedma, S. Alonso, F. Chiclana, and F. Herrera. *A consensus model for group decision making with incomplete fuzzy preference relations*. IEEE Transactions on Fuzzy Systems 15 (5), 863–877 (2007)	• It supports consensus processes in the presence of incomplete information

4 Consensus Approaches Based on Soft Consensus Measures

Consensus approaches based on soft consensus measures have been a hot topic in recent years [31], and different approaches can be found in the literature according to different criteria: (i) reference domain used to compute soft consensus measures, (ii) coincidence concept used to compute the soft consensus measures, (iii) generation method of recommendations, and (iv) guiding measures.

In the following subsections, these different consensus approaches are described in more detail.

4.1 Consensus Approaches Based on the Reference Domain

Two different consensus approaches can be found according to the reference domain utilized to compute the consensus measures.

Firstly, consensus measures focused on the decision maker set have been presented in [16, 34, 36, 37], in which consensus measures are computed in three steps: (i) for each pair of decision makers, a degree of agreement as to their preferences between all the pair of alternatives are computed, (ii) these degrees are aggregated to obtain a degree of agreement of each pair of decision makers as to their preferences between $Q1$ pairs of alternatives, and (iii) these degrees are aggregated to obtain a degree of agreement of $Q2$ pairs of decision makers as to their preferences between $Q1$ pair of alternatives, which is the degree of consensus sought.

Secondly, consensus measures focused on the alternative set have been presented in [23, 24, 29, 30], in which the consensus measures are computed at the three different levels of representation of a preference relation: (i) level of preference, indicating the consensus degree existing among all the m preference values attributed by the m decision makers to a specific preference, (ii) level of alternative, which allows us to measure the consensus existing over all the alternative pairs where a given alternative is present, and (iii) level of preference relation, which evaluates the social consensus, that is, the current consensus existing among all the decision makers about all the preferences. It allows us, for example, to identify which decision makers are close to the consensus solutions, or in which alternatives the decision makers are having more trouble to reach consensus.

Comparing both approaches, the latter seems better to design consensus processes allowing us to guide the decision makers to modify their opinions during the discussion process.

4.2 Consensus Approaches Based on the Coincidence Method

In the literature, we can find soft consensus measures valued in [0, 1], where a value close to 1 indicates a high level of consensus and a value close to 0 indicates a low level of consensus [4, 30, 34, 36, 38]. On the other hand, instead of using numerical values in [0, 1], soft consensus measures based on linguistic labels have been proposed [24, 25] to evaluate the level of consensus. Anyway, to obtain the level of consensus achieved in each round of the consensus process, the similarity among the preferences provided by the decision makers on the alternatives is measured. Soft consensus measures are based on the coincidence concept [25], and we can identify three different methods for computing them [7]:

1. *Consensus measures based on strict coincidence among preferences* [24, 35]. Here, similarity criteria among preferences are used to compute the coincidence concept. In such a case, only two possible results are assumed: 1 if the opinions

are equal and, otherwise, a value of 0. The advantage of this approach is that the computation of the consensus degrees is simple and easy. The drawback of this approach is that the consensus degrees obtained do not reflect the real consensus situation.

2. *Consensus measures based on soft coincidence among preferences* [1, 29, 34, 36]. As above, similarity criteria among preferences are used to compute the coincidence concept, but, now, a major number of possible coincidence degrees is considered. It is assumed that the coincidence concept is a gradual concept assessed with different degrees defined in the unit interval [0, 1]. The advantage of this approach is that the consensus degrees obtained reflect better the real consensus situation. The drawback of this approach is that the computation of the consensus degrees is more difficult because we need to define similarity criteria to compute the consensus measures, and, sometimes it is not possible to define them directly.
3. *Consensus measures based on coincidence among solutions* [2, 28]. Here, similarity criteria among the solutions obtained from the decision makers' preferences are used to compute the coincidence concept and different degrees assessed in [0, 1] are assumed. The advantage of this approach is that the consensus degrees are obtained comparing not the opinions but the position of the alternatives in each solution, what allows us to reflect the real consensus situation in each moment of the consensus reaching process. The drawback of this approach is that the computation of the consensus degrees is more difficult than in the above approaches because we need to define similarity criteria and it is necessary to apply a selection process before obtaining the consensus degrees.

It should be pointed out that the second and third methods, which reflect the real consensus state within the group of decision makers [6], are the most useful approaches to design consensus processes allowing us to advice the decision makers during the consensus process [31]. In particular, the second method is applied in contexts under preference relations and the third one is applied in decision situations under different formats of preference representation.

4.3 Consensus Approaches Based on the Generation Method of Recommendations

The generation method of recommendations to the decision makers is very important in order to increase the consensus level. From this point of view, the first consensus approaches proposed in the literature [23, 24, 34, 36, 38] can be considered as basic approaches based on a moderator, who monitors the agreement in each moment of the consensus process and is in charge of supervising and addressing the consensus process towards success. However, the moderator can introduce some subjectivity in the process.

To overcome this drawback, consensus approaches have been proposed by substituting the moderator figure or providing moderator with better analysis tools, making more effective and efficient the decision making processes:

- Consensus approaches incorporating a feedback mechanism substituting the moderator's actions have been developed [28–30]. In these approaches, proximity measures are calculated to evaluate the proximity between the individual decision makers' preferences and the collective one. These proximity measures allow to identify the preference values provided by the decision makers that are contributing less to reach a high consensus state. In such a way, the feedback mechanism gives advice to those decision makers to find out the changes they need to make in their opinions to obtain a solution with a better consensus degree.
- Consensus approaches have been proposed using a novel data mining tool [39], the so called action rules [45], to stimulate and support the discussion in the group. The purpose of an action rule is to show how a subset of flexible attributes should be changed to obtain an expected change in the decision attribute for a subset of objects characterized by some values of the subset of stable attributes. According to it, these action rules are used to indicate and suggest to the moderator with which decision makers and with respect to which options it may be expedient to deal.

It should be pointed out that the current consensus trends are committed to develop automated feedback mechanisms replacing the moderator, in particular, when consensus processes are developed in crowded social environments [1]. In addition, new feedback mechanisms which implement strategies that adjust the number of changes required depending on the level of consensus among decision makers in each consensus round are being proposed [42].

4.4 Consensus Approaches Based on Guidance Measures

The pairwise comparison in preference relations helps the decision makers to provide their preferences by focusing only on two elements once at a time. It allows to reduce uncertainty and hesitation while leading to the higher of consistency. The problem is that the definition of a preference relation does not imply any kind of consistency property, and the decision makers' preferences can be inconsistent [13]. Luckily, the lack of consistency can be quantified and monitored, and it has been used as a parameter to validate the final solution obtained after a consensus process [11, 24]. In such a way, consensus approaches using both consistency and consensus measures to guide the consensus process have been presented in [8, 24, 30]. Here, a consensus/consistency level is usually calculated as a weighted aggregation of the consistency level and the consensus degree, and it is used as a control parameter to decide if the consensus process has to finish.

It should be pointed out that the incorporation of other additional criteria in the consensus process, as, for instance, consistency measures, contributes to enrich the consensus processes and to achieve more adequate solutions in the GDM. For example, the use of the consistency measures avoids misleading solutions, which cannot be detected by the consensus approaches using only consensus degrees.

5 Conclusions and Future Work

In this contribution, we have reviewed the different consensus approaches based on soft consensus measures that have been proposed in the literature in which the consensus process is guided by a moderator. To do so, some basic concepts to understand the topic have been introduced, and both the pioneering and most relevant contributions on consensus approaches have been highlighted. In addition, several approaches of consensus in GDM according to different criteria have been analyzed.

In the future, it is worth continuing this research by studying the current trends in the development of consensus approaches and by bringing out several issues that could represent challenges to be faced.

Acknowledgements The authors would like to acknowledge FEDER financial support from the Project TIN2013-40658-P, and also the financial support from the Andalusian Excellence Project TIC-5991.

References

1. Alonso, S., Pérez, I.J., Cabrerizo, F.J., Herrera-Viedma, E.: A linguistic consensus model for web 2.0 communities. Applied Soft Computing **13**(1), 149–157 (2013)
2. Ben-Arieh, D., Chen, Z.: Linguistic-labels aggregation and consensus measure for autocratic decision making using group recommendations. IEEE Transactions on Systems Man and Cybernetics - Part A: Systems and Humans **36**(3), 558–568 (2006)
3. Bezdek, J., Spillman, B., Spillman, R.: A fuzzy relation space for group decision theory. Fuzzy Sets and Systems **1**(4), 255–268 (1978)
4. Bryson, N.: Group decision-making and the analytic hierarchy process: Exploring the consensus-relevant information content. Computers & Operations Research **23**(1), 27–35 (1996)
5. Butler, C.T., Rothstein, A.: On conflict and consensus: A handbook on formal consensus decision making. Tahoma Park (2006)
6. Cabrerizo, F.J., Moreno, J.M., Pérez, I.J., Herrera-Viedma, E.: Analyzing consensus approaches in fuzzy group decision making: Advantages and drawbacks. Soft Computing **14**(5), 451–463 (2010)
7. Cabrerizo, F.J., Heradio, R., Pérez, I.J., Herrera-Viedma, E.: A selection process based on additive consistency to deal with incomplete fuzzy linguistic information. Journal of Universal Computer Science **16**(1), 62–81 (2010)
8. Cabrerizo, F.J., Pérez, I.J., Herrera-Viedma, E.: Managing the consensus in group decision making in an unbalanced fuzzy linguistic context with incomplete information. Knowledge-Based Systems **23**(2), 169–181 (2010)
9. Chen, S.J., Hwang, C.L.: Fuzzy multiple attributive decision making: Theory and its applications. Springer, Berlin (1992)

10. Chen, X., Zhang, H., Dong, Y.: The fusion process with heterogeneous preference structures in group decision making: A survey. Information Fusion **24**, 72–83 (2015)
11. Chiclana, F., Herrera-Viedma, E., Herrera, F., Alonso, S.: Some induced ordered weighted averaging operators and their use for solving group decision making problems based on fuzzy preference relations. European Journal of Operational Research **182**(1), 383–399 (2007)
12. Coch, L., French, J.R.P.: Overcoming resistance to change. Human Relations **1**(4), 512–532 (1948)
13. Cutello, V., Montero, J.: Fuzzy rationality measures. Fuzzy Sets and Systems **62**(1), 39–54 (1994)
14. Dubois, D., Prade, H., Testemale, C.: Weighted fuzzy pattern matching. Fuzzy Sets and Systems **28**(3), 313–331 (1988)
15. Dubois, D., Koning, J.L.: Social choice axioms for fuzzy set aggregation. Fuzzy Sets and Systems **43**(3), 257–274 (1991)
16. Fedrizzi, M., Kacprzyk, J., Zadrozny, S.: An interactive multi-user decision support system for consensus reaching processes using fuzzy logic with linguistic quantifiers. Decision Support Systems **4**(3), 313–327 (1988)
17. Fedrizzi, M., Pasi, G.: Fuzzy logic approaches to consensus modeling in group decision making. In: Ruan, D., Hardeman, F., Van Der Meer, K. (Eds) Intelligent Decision and Policy Making Support Systems, pp. 19–37. Springer-Verlag, Berlin-Heidelberg (2008)
18. French, J.R.P.: A formal theory of social power. Psychological Review **63**(3), 181–194 (1956)
19. Fodor, J., Roubens, M.: Fuzzy preference modeling and multicriteria decision support. Kluwer, Dordrecht (1994)
20. Grabisch, M., Marichal, J.-L., Mesiar, R., Endre, P.: Aggregation functions (Encyclopedia of Mathematics and its Applications). Cambridge University Press, New York (2009)
21. Grabisch, M., Labreuche, C.: Fuzzy measures and integrals in MCDA. In: Greco, S., Ehrgott, M., Figueira, J.R. (Eds) Multiple Criteria Decision Analysis, pp. 553–603. Springer, New York (2016)
22. Harary, F.: On the measurement of structural balance. Behavioral Science **4**(4), 316–323 (1959)
23. Herrera, F., Herrera-Viedma, E., Verdegay, J.L.: A model of consensus in group decision making under linguistic assessments. Fuzzy Sets and Systems **78**(1), 73–87 (1996)
24. Herrera, F., Herrera-Viedma, E., Verdegay, J.L.: A rational consensus model in group decision making using linguistic assessments. Fuzzy Sets and Systems **88**(1), 31–49 (1997)
25. Herrera, F., Herrera-Viedma, E., Verdegay, J.L.: Linguistic measures based on fuzzy coincidence for reaching consensus in group decision making. International Journal of Approximate Reasoning **16**(3–4), 309–334 (1997)
26. Herrera, F., Herrera-Viedma, E.: Linguistic decision analysis: Steps for solving decisions problems under linguistic information. Fuzzy Sets and Systems **115**(1), 67–82 (2000)
27. Herrera, F., Alonso, S., Chiclana, F., Herrera-Viedma, E.: Computing with words in decision making: Foundations, trends and prospects. Fuzzy Optimization and Decision Making **8**(4), 337–364 (2009)
28. Herrera-Viedma, E., Herrera, F., Chiclana, F.: A consensus model for multiperson decision making with different preference structures. IEEE Transactions on Systems, Man, and Cybernetics - Part A: Systems and Humans **32**(3), 394–402 (2002)
29. Herrera-Viedma, E., Martinez, L., Mata, F., Chiclana, F.: A consensus support system model for group decision-making problems with multigranular linguistic preference relations. IEEE Transactions on Fuzzy Systems **13**(5), 644–658 (2005)
30. Herrera-Viedma, E., Herrera, F., Alonso, S.: Group decision-making model with incomplete fuzzy preference relations based on additive consistency. IEEE Transactions on Systems, Man and Cybernetics - Part B: Cybernetics **37**(1), 176–189 (2007)
31. Herrera-Viedma, E., Cabrerizo, F.J., Kacprzyk, J., Pedrycz, W.: A review of soft consensus models in a fuzzy environment. Information Fusion **17**, 4–13 (2014)
32. Kacprzyk, J.: Group decision-making with a fuzzy majority via linguistic quantifiers. Part I: A consensory-like pooling. Cybernetics and Systems: An International Journal **16**(2–3), 119–129 (1985)

33. Kacprzyk, J.: Group decision-making with a fuzzy majority via linguistic quantifiers. Part I: A competitive-like pooling. Cybernetics and Systems: An International Journal **16**(2–3), 131–144 (1985)
34. Kacprzyk, J.: Group decision making with a fuzzy linguistic majority. Fuzzy Sets and Systems **18**(2), 105–118 (1986)
35. Kacprzyk, J.: On some fuzzy cores and 'soft' consensus measures in group decision making. In: Bezdek, J.C. (Ed) The Analysis of Fuzzy Information, pp. 119–130. CRC Press, Boca Raton (1987)
36. Kacprzyk, J., Fedrizzi, M.: A 'soft' measure of consensus in the setting of partial (fuzzy) preferences. European Journal of Operational Research **34**(3), 316–325 (1988)
37. Kacprzyk, J., Fedrizzi, M.: A 'human-consistent' degree of consensus based on fuzzy logic with linguistic quantifiers. Mathematical Social Sciences **18**(3), 275–290 (1989)
38. Kacprzyk, J., Fedrizzi, M., Nurmi, H.: Group decision making and consensus under fuzzy preferences and fuzzy majority. Fuzzy Sets and Systems **49**(1), 21–31 (1992)
39. Kacprzyk, J., Zadrozny, S., Ras, Z.W.: How to support consensus reaching using action rules: a novel approach. International Journal of Uncertainty, Fuzziness and Knowledge-Based Systems **18**(4), 451–470 (2010)
40. Loewer, B.: Special issue on consensus. Synthese **62**(1), 1–122 (1985)
41. Loewer, B., Laddaga, R.: Destroying the consensus. Synthese **62**(1), 79–96 (1985)
42. Mata, F., Martinez, L., Herrera-Viedma, E.: An adaptive consensus support model for group decision-making problems in a multigranular fuzzy linguistic context. IEEE Transactions on Fuzzy Systems **17**(2), 279–290 (2009)
43. Narukawa, Y., Torra, V.: Fuzzy measures and integrals in evaluation of strategies. Information Sciences **177**(21), 4686–4695 (2007)
44. Orlovski, S.A.: Decision-making with a fuzzy preference relation. Fuzzy Sets and Systems **1**(3), 155–167 (1978)
45. Pawlak, A.: Information systems theoretical foundations. Information Systems **6**(3), 205–218 (1981)
46. Pérez, I.J., Cabrerizo, F.J., Alonso, S., Herrera-Viedma, E.: A new consensus model for group decision making problems with non homogeneous experts. IEEE Transactions on Systems, Man, and Cybernetics: Systems **44**(4), 494–498 (2014)
47. Pérez, L.G., Mata, F., Chiclana, F., Kou, G., Herrera-Viedma, E.: Modeling influence in group decision making. Soft Computing **20**(4), 1653–1665 (2016)
48. Roubens, M.: Fuzzy sets and decision analysis. Fuzzy Sets and Systems **90**(2), 199–206 (1997)
49. Saaty, T.L.: The analytic hierarchy process: Planning, priority setting, resource allocation. McGraw-Hill, New York (1980)
50. Saint, S., Lawson, J.R.: Rules for reaching consensus: A modern approach to decision making. Jossey-Bass (1994)
51. Spillman, B., Bezdek, J., Spillman, R.: Coalition analysis with fuzzy sets. Kybernetes **8**(3), 203–211 (1979)
52. Tanino, T.: Fuzzy preference orderings in group decision making. Fuzzy Sets and Systems **12**(2), 117–131 (1984)
53. Torra, V., Narukawa, Y.: Modeling decisions: Information fusion and aggregation operators. Springer-Verlag (2007)
54. Yager, R.R.: On ordered weighted averaging aggregation operators in multicriteria decision making. IEEE Transactions on Systems Man and Cybernetics **18**(1), 183–190 (1988)
55. Zadeh, L.A.: Fuzzy sets. Information and Control **8**(3), 338–353 (1965)
56. Zadeh, L.A.: A computational approach to fuzzy quantifiers in natural languages. Computers & Mathematics with Applications **9**(1), 149–184 (1983)
57. Zadrozny, S., Kacprzyk, J.: Issues in the practical use of the OWA operators in fuzzy querying. Journal of Intelligent Information Systems **33**(3), 307–325 (2009)

Construction of Capacities from Overlap Indexes

José Antonio Sanz, Mikel Galar, Radko Mesiar, Anna Kolesárová, Humberto Bustince, Javier Fernandez and Javier Montero

Abstract In this chapter, we show how the concepts of overlap function and overlap index can be used to define fuzzy measures which depend on the specific data of each considered problem.

Keywords Overlap function · Capacity · Fuzzy measure

1 Introduction

In many problems, it is crucial to find a relation between groups of data. Such relation can be expressed, for instance, in terms of an appropriate fuzzy measure or capacity [10, 21] which reflects the way the different data are connected to each other [20].

In this chapter, taking into account this fact and following the developments in [8], we introduce a method to build capacities [20, 21] directly from the data (inputs) of a given problem. In order to do so, we make use of the notions of overlap function and overlap index [4, 5, 7, 12–14, 16] for constructing capacities which reflect how different data are related to each other.

J.A. Sanz · M. Galar · H. Bustince (✉) · J. Fernandez
Departamento of Automática y Computación and the Institute of Smart Cities, Universidad Publica de Navarra, 31006 Navarra, Spain
e-mail: bustince@unavarra.es

R. Mesiar
Slovak University of Technology, Radlinskeho 11, Bratislava, Slovakia

R. Mesiar
Institute of Information Theory and Automation, Academy of Sciences of the Czech Republic, 18208 Prague, Czech Republic

A. Kolesárová
Institute of Information Engineering, Automation and Mathematics, Slovak University of Technology, 810 05 Bratislava, Slovakia

J. Montero
Department of Statistics and Operations Research I, Faculty of Mathematics, Universidad Complutense de Madrid, Plaza de Ciencias 3, 28040 Madrid, Spain

V. Torra et al. (eds.), *Fuzzy Sets, Rough Sets, Multisets and Clustering*, Studies in Computational Intelligence 671, DOI 10.1007/978-3-319-47557-8_19

This paper is organized as follows: after providing some preliminaries, we analyse, in Sect. 3, some properties of overlap functions and indexes. In Sect. 4 we discuss a method for constructing capacities from overlap functions and overlap indexes. Finally, we present some conclusions and references.

2 Preliminaries

Given a referential set (or universe) U, a fuzzy set A over U is defined in terms of a mapping μ_A:

$$\mu_A : U \to [0, 1] .$$

For simplicity, we write $A(i)$ instead of $\mu_A(i)$. We denote by $FS(U)$ the space of all fuzzy sets defined over U. We will only deal with fuzzy sets defined over a finite referential set U, so we can consider $U = \{1, \ldots, n\}$. $card(U)$ stands for the cardinality of U.

We consider in $FS(U)$ the usual partial order defined by Zadeh. Union and intersection between fuzzy sets are defined by means of the max and min, respectively. We say that $(x_1, \ldots, x_n) \leq (y_1, \ldots, y_n)$ if and only if $x_i \leq y_i$ for every $i \in U$.

By abuse of notation, we denote by $\emptyset$ empty set (that is, the fuzzy set where the membership values of all the elements are equal to 0), and by U the fuzzy set with all its memberships equal to 1.

The support of a fuzzy set $A \in FS(U)$ is given by:

$$supp(A) = \{i \in U \mid A(i) \neq 0\}.$$

We say that A is a full fuzzy set if $supp(A) = U$. To distinguish fuzzy sets from the classical subsets of U, we will use the notation $\widetilde{A}$ for the latter.

Let $\widetilde{A} \subseteq U$ and $t \in [0, 1]$. By $t\widetilde{A}$ we denote the fuzzy set given by:

$$t\widetilde{A}(i) = \begin{cases} t & \text{if } i \in \widetilde{A} \text{ ;} \\ 0 & \text{otherwise.} \end{cases}$$

By abuse of notation, we write $1_{\widetilde{A}}$ to denote $1\widetilde{A}$ for every $\widetilde{A} \subseteq U$, since $1\widetilde{A}$ equals the characteristic function of the set $\widetilde{A}$. Note that this definition corresponds to the basic function $b(\widetilde{A}, t)$ introduced by Benvenuti et al. in 2002 [2].

Given a function $F : [0, 1]^k \to [0, 1]$ (with $k \in \mathbb{N}$) and k fuzzy sets $A_k \in FS(U)$, the symbol $F(A_1, \ldots, A_k)$ denotes the fuzzy set over U whose membership function is given by:

$$F(A_1, \ldots, A_k)(i) = F(A_1(i), \ldots, A_k(i)).$$

Definition 1 Let $n \geq 2$. An n-dimensional aggregation function [1, 6, 9, 11, 15, 17, 18] is a mapping $M : [0, 1]^n \to [0, 1]$ such that:

1. $M(0, \ldots, 0) = 0$ and $M(1, \ldots, 1) = 1$;
2. M is increasing.

2.1 Capacities

In the following, we recall some basic notions concerning capacities [21].

Definition 2 Let $U = \{1, 2, \ldots, n\}$. A capacity (or non-additive measure) over U is a mapping $m : 2^U \to [0, 1]$ such that

1. $m(\emptyset) = 0$ and $m(U) = 1$;
2. If $\widetilde{A} \subset \widetilde{B}$ then $m(\widetilde{A}) \leq m(\widetilde{B})$.

Example 1 1. Any probability measure yields an example of a capacity.

2. The bottom capacity is defined by

$$m_*(\widetilde{A}) = \begin{cases} 1 & \text{if } \widetilde{A} = U; \\ 0 & \text{otherwise.} \end{cases}$$

3. The top capacity is defined by

$$m^*(\widetilde{A}) = \begin{cases} 0 & \text{if } \widetilde{A} = \emptyset; \\ 1 & \text{otherwise.} \end{cases}$$

Definition 3 If m is a capacity over $U = \{1, \ldots, n\}$, then:

1. m is called additive if $m(\widetilde{A} \cup \widetilde{B}) = m(\widetilde{A}) + m(\widetilde{B})$ whenever $\widetilde{A} \cap \widetilde{B} = \emptyset$.
2. m is called symmetric if $m(\widetilde{A}) = m(\widetilde{B})$ whenever $card(\widetilde{A}) = card(\widetilde{B})$.
3. m is called supermodular (submodular) if $m(\widetilde{A} \cup \widetilde{B}) + m(\widetilde{A} \cap \widetilde{B}) \geq m(\widetilde{A}) + m(\widetilde{B})$ $(m(\widetilde{A} \cup \widetilde{B}) + m(\widetilde{A} \cap \widetilde{B}) \leq m(\widetilde{A}) + m(\widetilde{B}))$ for every $\widetilde{A}, \widetilde{B} \in 2^U$.
4. m is called modular if it is supermodular and submodular.

Remark 1 Since we have $m(\emptyset) = 0$ for every capacity m, additivity and modularity are equivalent properties of capacities.

Capacities can be obtained from aggregation functions as follows.

Proposition 1 ([3, 21]) *Let $m : 2^U \to [0, 1]$ be a set function. The following items are equivalent.*

1. *m is a capacity.*
2. *There exists an aggregation function $M : [0, 1]^n \to [0, 1]$ such that, for every $\widetilde{A} \in 2^U$*

$$m(\widetilde{A}) = M(1_{\widetilde{A}}) .$$

3 Overlap Functions and Overlap Indexes

The concept of overlap function was extensively studied in [7]. Here we make a revision of the most relevant definitions and results for the present work.

Definition 4 An overlap function is a mapping $G_O : [0, 1]^2 \to [0, 1]$ such that:

1. $G_O(x, y) = G_O(y, x)$ for every $x, y \in [0, 1]$;
2. $G_O(x, y) = 0$ if and only if $xy = 0$;
3. $G_O(x, y) = 1$ if and only if $xy = 1$;
4. G_O is increasing;
5. G_O is continuous.

Overlap function can be seen as a generalization of continuous t-norms without divisors of zero. The class of all overlap functions is convex.

Overlap functions can be used to build overlap indexes by aggregating them. We start by recalling some basic notions about the idea of an overlap index and we will formalize the construction method in Theorem 1.

Definition 5 An overlap index is a mapping $O : FS(U) \times FS(U) \to [0, 1]$ such that

(O1) $O(A, B) = 0$ if and only if A and B have disjoint supports; that is, $A(i)B(i) = 0$ for every $i \in U$;
(O3) $O(A, B) = O(B, A)$;
(O4) If $B \subseteqq C$, then $O(A, B) \leq O(A, C)$.
An overlap index such that
(O2') $O(A, B) = 1$ if there exists $i \in U$ such that $A(i) = B(i) = 1$

is called a normal overlap index.

Remark 2 In the original definition of overlap index [12], condition *(O2)* states that

$$O(A, B) = 1 \; if A(i) = 0 \; or B(i) = 1 \; or A(i)B(i) = 0$$

for all $i \in U$. For $A = \emptyset$ we obtain the following contradiction: *(O1)* implies that $O(A, A) = 0$ whereas *(O2)* implies $O(A, A) = 1$. Therefore we removed condition *(O2)* from the definition of an overlap index.

Example 2 1. The first example of overlap index in the literature is Zadeh's consistency index [22]:

$$O_Z(A, B) = \max_{i=1}^{n}(\min(A(i), B(i))).$$

Note that O_Z is normal.

2. Let $M : [0, 1]^2 \to [0, 1]$ be a symmetric aggregation function such that $M(x, y) = 0$ if and only if $xy = 0$. We have that

$$O_{M,Z}(A, B) = \max_{i=1}^{n}(M(A(i), B(i)))$$

is a normal overlap index that generalizes Zadeh's index.

Remark 3 For each overlap index $O : FS(U) \times FS(U) \to [0, 1]$, the function $M_O : [0, 1]^n \to [0, 1]$ given by

$$M_O(E) = \frac{O(E, U)}{O(U, U)}$$

with $E \in [0, 1]^n$, is an aggregation function without divisors of zero.

3.1 Modularity of Overlap Indexes

We are going to analyze several properties of overlap indexes that will be relevant for the construction of capacities from them. We first introduce the idea of symmetry for overlap indexes.

Definition 6 Let $O : FS(U) \times FS(U) \to [0, 1]$ be an overlap index and let $E \in FS(U)$. O is E-symmetric if for every $A, B \in FS(U)$ such that $card(supp(A)) = card(supp(B))$ it holds that:

$$O(A, E) = O(B, E).$$

Example 3 1. Every overlap index O is E-symmetric if $E = \emptyset \in FS(U)$.
2. Consider the strongest overlap index:

$$O_s(A, B) = \begin{cases} 0 & \text{if } A, B \text{ are disjoint fuzzy sets;} \\ 1 & \text{otherwise.} \end{cases}$$

We have that O_s is E-symmetric for every full set E.

Note that an overlap index can not be E-symmetric unless E is a full fuzzy set, as the next result shows.

Proposition 2 *If O is an overlap index which is E-symmetric with respect to some fuzzy set $E \in FS(U)$, $E \neq \emptyset$, then E is a full fuzzy set.*

Proof Assume that E is not a full fuzzy set and that

$$k = \min(card(supp(E)), n - card(supp(E))) > 0 .$$

Let $\widetilde{A} \subseteq supp(E)$ and $\widetilde{B} \subseteq U \backslash supp(E)$ with $card(\widetilde{A}) = card(\widetilde{B}) = k$. Consider the fuzzy sets

$$A = 1_{\widetilde{A}} \text{ and } B = 1_{\widetilde{B}} .$$

We have that $O(E, A) > 0$ (since E and A are not mutually disjoint) whereas $O(E, B) = 0$. Therefore, O can not be E-symmetric. □

Now we consider the notion of modularity.

Definition 7 Let $O : FS(U) \times FS(U) \to [0, 1]$ be an overlap index and let $E \in FS(U)$.

1. O is called E-supermodular if $O(E, A \cap B) + O(E, A \cup B) \geq O(E, A) + O(E, B)$ holds for all $A, B \in FS(U)$. Similarly, O is called E-submodular if $O(E, A \cap B) + O(E, A \cup B) \leq O(E, A) + O(E, B)$ for all $A, B \in FS(U)$.
2. If O is E-submodular and E-supermodular, then O is simply called E-modular.

Example 4 1. Every overlap index O is E-modular for $E = \emptyset \in FS(U)$.
2. The overlap index O_π is E-modular for every fuzzy set E.
3. O_Z is E-submodular but not E-modular.

The following construction method of overlap indexes by means of aggregation functions can be found in [13].

Theorem 1 *Let $M : [0, 1]^n \to [0, 1]$ be an aggregation function such that $M(x_1, \ldots, x_n) = 0$ if and only if $x_1 = \cdots = x_n = 0$ and let $G_O : [0, 1]^2 \to [0, 1]$ be an overlap function. The mapping $O : FS(U) \times FS(U) \to [0, 1]$ given by*

$$O(A, B) = M(G_O(A(1), B(1)), \ldots, G_O(A(n), B(n))) \tag{1}$$

is a normal overlap index in the sense of Definition 5.

Conversely, if G_O is an overlap function and $M : [0, 1]^n \to [0, 1]$ is an aggregation function such that O defined by Eq. 1 is an overlap index, then $M(x_1, \ldots, x_n) = 0$ if and only if $x_1 = \cdots = x_n = 0$.

4 Capacities from Overlap Indexes, and Overlap Functions

In this section, we present the core of the present chapter. Taking into account the usual construction of Bayesian conditional probabilities, we follow an analogous procedure to build a capacity from an overlap function.

We start introducing a bit of notation. Let $E \in FS(U)$ be a fixed non-empty fuzzy set (that is, with at least one membership different from zero). Given $\widetilde{A} \in 2^U$, let us define a fuzzy set $E_{\widetilde{A}}$ induced by E as follows:

$$E_{\widetilde{A}}(i) = \begin{cases} E(i) & \text{if } i \in \widetilde{A}; \\ 0 & \text{otherwise.} \end{cases}$$

Observe that $E_{\widetilde{A}}$ is the fuzzy intersection of the fuzzy set E and the crisp set $\widetilde{A}$, since

$$E_{\widetilde{A}}(i) = \min(1_{\widetilde{A}}(i), E(i)) .$$

Therefore, any aggregation function with no zero divisors could also be used instead of the minimum in this definition for the subsequent developments.

Now we are ready to introduce our main result.

Theorem 2 *If $E \in FS(U)$ is a fixed, non-empty fuzzy set, then the mapping $m_{O,E} : 2^U \to [0, 1]$ given by*

$$m_{O,E}(\widetilde{A}) = \frac{1}{O(E, E)} O(E, E_{\widetilde{A}})$$

is a capacity for every overlap index O.

Proof First of all observe that $m_{O,E}$ is well defined since $O(E, E) \neq 0$ and $O(E, E_{\widetilde{A}}) \leq O(E, E)$. If $\widetilde{A} = U$, then it follows that $E_{\widetilde{A}} = E$, so we have that $m_{O,E}(\widetilde{A}) = 1$. Moreover, if $\widetilde{A} = \emptyset$, then $E_{\widetilde{A}}(i) = 0$ for every $i \in U$. So, in particular, $O(E_{\widetilde{A}}, E) = 0$. Finally, if $\widetilde{A} \subset \widetilde{B}$, then it follows that $E_{\widetilde{A}} \subseteq E_{\widetilde{B}}$, so, in particular, $O(E, E_{\widetilde{A}}) \leq O(E, E_{\widetilde{B}})$ and hence $m_{O,E}(\widetilde{A}) \leq m_{O,E}(\widetilde{B})$. □

Recall that Benvenuti et al. [3] defined for each aggregation function $M : [0, 1]^n \to [0, 1]$ and $e \in]0, 1]$ such that $M(E) > 0$, where $E = (e, \ldots, e)$, a capacity $m_{M,e} : 2^U \to [0, 1]$ given by

$$m_{M,e}(\widetilde{A}) = \frac{M(e\widetilde{A})}{M(E)}$$

(for $e = 1$ see also Proposition 1). Obviously, in the terms of Theorem 2, $m_{O,E} = m_{M_O,e}$. Here M_O was defined in Remark 3.

Remark 4 Observe that, for a fixed full probability measure P on U, if we consider the overlap index O_P introduced in Remark 3, we recover the definition of Bayesian conditional probabilities, i.e., $m_{O_P,E} = P_{suppE}$.

Of course, the question of how two of these measures relate to each other arises.

Proposition 3 *Let O be an overlap index. For all full fuzzy sets $E_1, E_2 \in FS(U)$, the following statements are equivalent:*

1. $m_{O,E_1}(\widetilde{A}) \leq m_{O,E_2}(\widetilde{A})$ *for every* $\widetilde{A} \in 2^U$.
2. $\min_{\widetilde{A} \in 2^U} \frac{O(E_2, \widetilde{A}_{E_2})}{O(E_1, \widetilde{A}_{E_1})} = \frac{O(E_2, E_2)}{O(E_1, E_1)}$

Proof The inequality $m_{O,E_1}(\widetilde{A}) \leq m_{O,E_2}(\widetilde{A})$ implies that

$$\frac{O(E_2, E_2)}{O(E_1, E_1)} \leq \frac{O(E_2, \widetilde{A}_{E_2})}{O(E_1, \widetilde{A}_{E_1})} \tag{2}$$

for every $\widetilde{A} \in 2^U$, so (2) holds. Conversely, if (2) holds, then Eq. 2 is satisfied as well and we obtain (1). □

Corollary 1 *Let $E_1, E_2 \in FS(U)$ such that $O(E_1, E_1) = 1$. If $E_1 \subseteq E_2$, then $m_{O,E_1}(\widetilde{A}) \leq m_{O,E_2}(\widetilde{A})$ for every $\widetilde{A} \in 2^U$.*

Proof If $E_1 \subseteq E_2$, then we have that $1 \geq O(E_2, E_2) \geq O(E_1, E_1) = 1$. Consequently,

$$1 = \frac{O(E_2, E_2)}{O(E_1, E_1)} \leq \frac{O(E_2, \widetilde{A}_{E_2})}{O(E_1, \widetilde{A}_{E_1})}$$

for every $\widetilde{A} \in 2^U$, which implies that $m_{O,E_1}(\widetilde{A}) \leq m_{O,E_2}(\widetilde{A})$ for every $\widetilde{A} \in 2^U$. □

In order to reverse the construction method and get an overlap function from a measure, we need the concept of contraction.

Definition 8 Let $E \in FS(U)$. The contraction to E (or E-contraction) is the mapping $C_E : FS(U) \to FS(U)$ defined by:

$$C_E(A) = \{(i, E(i)A(i)) \mid i \in U\}.$$

Remark 5 The definition of contraction can be generalized by substituting the product with any other t-norm or even an overlap function. We postpone the analysis of the resulting operators to future works.

Let us continue by introducing some notations. For a fixed fuzzy set $E \in FS(U)$ and for $\widetilde{A} \in 2^U$, we define

$$Cl(E, \widetilde{A}) = \{A \in FS(U) \mid A \subseteq E_{\widetilde{A}} \text{ and } A \nsubseteq E_{\widetilde{B}} \text{ for every } \widetilde{B} \subset \widetilde{A}\}.$$

The proof of the following lemma is straightforward.

Lemma 1 *Let $E \in FS(U)$. Then $Cl(E, \widetilde{A}) = \emptyset$ for every $\widetilde{A} \in 2^U$ such that $\widetilde{A} \cap supp(E) = \emptyset$.*

Then we can state the following.

Proposition 4 *Let E be a full fuzzy set. The family $(Cl(E, \widetilde{A}))_{\widetilde{A} \in 2^U}$ is a partition of the set $\{A \in FS(U) \mid A \subseteq E\}$.*

Proof For any $A \subseteq E$ let's take $\widetilde{A} = supp(A)$. Then, and only then $A \subseteq E_{\widetilde{A}}$ and for any $\widetilde{B}$ which is a proper subset of $\widetilde{A}$, $A \nsubseteq E_{\widetilde{B}}$, i.e., $A \in Cl(E, \widetilde{A})$. The fact that $Cl(E, \widetilde{A}) \cap Cl(E, \widetilde{B}) = \emptyset$ for $\widetilde{A} \neq \widetilde{B}$ is trivial. Hence $\{Cl(E, \widetilde{A}) | \widetilde{A} \in 2^U\}$ is a partition of $\{A \in FS(U) \mid A \subseteq E\}$. □

Now we can show how to recover overlap indexes from capacities.

Theorem 3 *Let m be a capacity such that $m(\widetilde{A}) = 0$ if and only if $\widetilde{A} = \emptyset$. If E is a full fuzzy set, then the function $O_{E,m} : FS(U) \times FS(U) \to [0, 1]$ defined by:*

$$O_{E,m}(A, B) = \begin{cases} m(\widetilde{A}) & \text{if } A \cap B \in Cl(E, \widetilde{A}); \\ 1 & \text{otherwise} \end{cases}$$

is a normal overlap index such that the capacity induced by $O_{E,m}$ is equal to m.

Proof First of all, due to Proposition 4, $O_{E,m}$ is well defined. Let us prove that $O_{E,m}$ is an overlap index.

(O1) Assume that $O_{E,m}(A, B) = 0$. Since $m(\widetilde{A}) \neq 0$ for every $\widetilde{A} \neq \emptyset$, this happens if and only if $A \cap B \in Cl(E, \emptyset)$, i.e., if and only if A and B have disjoint supports.

(O3) Symmetry is obvious from the definition.

(O4) Let $A \in FS(U)$ be arbitrary, but fixed and let $B \subseteq C$. If $A \cap C \nsubseteq E$, then $O_{E,m}(A, C) = 1 \geq O_{E,m}(A, B)$. Now let us assume that $A \cap C \subseteq E$. From Proposition 4 and the fact that $A \cap B \subseteq A \cap C$, it follows that there exist $\widetilde{A}, \widetilde{B} \in 2^U$ with $\widetilde{B} \subseteq \widetilde{A}$ such that $A \cap C \in Cl(E, \widetilde{A})$ and $A \cap B \in Cl(E, \widetilde{B})$. Since m is a capacity, we have that $m(\widetilde{A}) \leq m(\widetilde{B})$ and therefore $O_{E,m}(A, B) \leq O_{E,m}(A, C)$.

(O2') Note that $U = U \cap U$. So if $E \neq U$ it follows that $O_{E,m}(U, U) = 1$, whereas if $E = U$, we have that $U \in Cl(U, U)$ and $O_{E,m}(U, U) = m(U) = 1$.

Finally, note that $E_{\widetilde{A}} = E_{\widetilde{A}} \cap E \in Cl(E, \widetilde{A})$ for every $\widetilde{A} \in 2^U$, which concludes the proof of the theorem since

$$m_{O_{E,m},E}(\widetilde{A}) = \frac{1}{O_{E,m}(E, E)} O_{E,m}(E, E_{\widetilde{A}}) = m(\widetilde{A}) \, .$$

This completes the proof. □

Theorem 3 can be extended to include non-strict measures as follows.

Corollary 2 *Consider a capacity m. Let $\widetilde{A}_0 = \{i \in U \mid m(\{i\}) = 0\}$. Suppose that E is a fuzzy set such that $E(i) \neq 0$ for $i \notin \widetilde{A}_0$. The function $O : FS(U) \times FS(U) \to [0, 1]$ given by*

$$O_{E,m,\widetilde{A}_0}(A,B) = \begin{cases} m(\widetilde{A}) & \text{if } A \cap B \in Cl(E, \widetilde{A} \backslash \widetilde{A}_0); \\ 1 & \text{otherwise} \end{cases}$$

is an overlap index.

Proof Symmetry and monotonicity are clear. We only need to check that (O1) holds. To see (O1), note that $O(A, B) = 0$ if and only if $A \cap B \in Cl(E, \widetilde{A} \backslash \widetilde{A}_0)$ for some $\widetilde{A} \in 2^U$ such that $m(\widetilde{A}) = 0$. But, $m(\widetilde{A}) = 0$ if and only if $\widetilde{A} \subseteq \widetilde{A}_0$, due to the monotonicity of capacities. So, $\min(A(i), B(i)) = 0$ if $i \in \widetilde{A}_0$. □

Example 5 For the bottom capacity m_*, we obtain $\widetilde{A}_0 = \{i \in U \mid m_*(\{i\}) = 0\} = U$. Thus, setting $E = \emptyset$ yields the following overlap function:

$$O_{E,m_*,U}(A,B) = \begin{cases} 0 & \text{if } A \cap B = \emptyset; \\ 1 & \text{otherwise,} \end{cases}$$

which is the strongest overlap index.

Corollary 3 *For every capacity m there exists a fuzzy set E and a continuous overlap index $O_{E,m}$ such that the measure induced by $O_{E,m}$ is equal to m.*

Proof It is just a matter of using the overlap index defined by means of Theorem 3 and Corollary 2. □

4.1 Construction of Capacities from Overlap Indexes Based on Overlap Functions

We can make use of Theorem 1 to get capacities from overlap functions and aggregation functions.

Proposition 5 *Let $M : [0, 1]^n \to [0, 1]$ be an aggregation function such that $M(x_1, \ldots, x_n) = 0$ if and only if $x_1 = \cdots = x_n = 0$, let $G_O : [0, 1]^2 \to [0, 1]$ be an overlap function and let $E \in FS(U)$ be a non-empty fuzzy set. The mapping $m_{E,M,G_O} : 2^U \to [0, 1]$ is given by*

$$m_{E,M,G_O}(\widetilde{A}) = \frac{1}{M(G_O(E))} M(G_O(E(1), E_{\widetilde{A}}(1)), \ldots, G_O(E(n), E_{\widetilde{A}}(n))),$$

where $M(G_O(E)) = M(G_O(E(1), E(1)), \ldots, G_O(E(n), E(n))$ is a capacity.

Note that if we take $E = U$, then we have $m_{U,M,G_O}(\widetilde{A}) = M(1_{\widetilde{A}})$.

Proposition 6 *Let M be an aggregation function as in Proposition 5. For any non-empty fuzzy set E, we have:*

1. $m_{E,M,G_O}(\widetilde{A}) = 0$ *if and only if* $E(i) = 0$ *for every* $i \in \widetilde{A}$;
2. $m_{E,M,G_O}(\widetilde{A}) = 1$ *whenever* $E(i) \neq 0$ *for every* $i \in \widetilde{A}$.

Proof 1. If $m_{E,M,G_O}(\widetilde{A}) = 0$, then

$$\frac{1}{M(G_O(E))} M(G_O(E(1), E_{\widetilde{A}}(1)), \ldots, G_O(E(n), E_{\widetilde{A}}(n))) = 0 .$$

Since M is an aggregation function, this implies that $G_O(E(i), E_{\widetilde{A}}(i)) = 0$ for every $i = 1, \ldots, n$. From the definition of an overlap function, this happens only if $E(i)E_{\widetilde{A}}(i) = 0$ for every $i = 1, \ldots, n$. If $E_{\widetilde{A}}(i) \neq 0$ it follows that $E(i) = 0$, which is impossible due to the definition of $E_{\widetilde{A}}$. Therefore, we infer that $E_{\widetilde{A}}(i) = 0$ for every $i \in \widetilde{A}$, that is, if $i \in \widetilde{A}$ then $E(i) = 0$. The other direction follows from the fact that $E(i) = 0$ for every $i \in \widetilde{A}$ implies that $E_{\widetilde{A}}(i) = 0$ for every $i \in U$.

2. If $\widetilde{A}$ is as in the statement of this property, then we obtain $E_{\widetilde{A}} = E$, and the result follows from the monotonicity of aggregation and overlap functions. □

The following corollary is a straightforward consequence of the previous result.

Corollary 4 *Let M be an aggregation function as in Proposition 5. For any non-empty fuzzy set E, we have:*

1. m_{E,M,G_O} *satisfies the property*

$$m_{E,M,G_O}(\widetilde{A}) = 0 \text{ if and only if } \widetilde{A} = \emptyset$$

if and only if E is a full fuzzy set.

2. $m_{E,M,G_O}(supp(E)) = 1$.

Theorem 4 *For a fixed overlap function G_O and an n-ary aggregation function M as in Proposition 5, the following claims are equivalent:*

1. *For each non-empty fuzzy subset $E \in FS(U)$, the measure m_{E,M,G_O} is additive.*
2. *M is modular, i.e., $M(\max(x, y)) + M(\min(x, y)) = M(x) + M(y)$ for all $x, y \in [0, 1]^n$.*

Proof Observe first that any modular aggregation function M such that $M(x_1, .., x_n) = 0$ only if $x_1 = \cdots = x_n = 0$ has the form $M(x_1, \ldots, x_n) = \sum_{i=1}^{n} f_i(x_i)$, where each $f_i : [0, 1] \to [0, 1]$ is increasing, $f_i(x) = 0$ only if $x = 0$, and $\sum f_i(1) = 1$, see, e.g., [19]. To see the necessity, observe that the additivity of m_{E,M,G_O} implies that

$$M(G_O(E(1), E_{\widetilde{A}}(1)), \ldots, G_O(E(n), E_{\widetilde{A}}(n))) =$$
$$M(G_O((E(1), E_{\widetilde{A}}(1)), 0, \ldots, 0)) + M(0, \ldots, 0, G_O(E(n), E_{\widetilde{A}}(n))).$$

In view of the mean value theorem for overlap functions, this equality is equivalent to

$$M(x_1,\ldots,x_n) = M(x_1,0,\ldots,0) + M(0,x_2,0,\ldots,0) + \cdots + M(0,\ldots,0,x_n)$$

for every $x_1,\ldots,x_n \in [0,1]$. Defining $f_i(x) = M(0,0,\ldots,0,x,0,\ldots,0)$, where the x is in the i-th position, the result follows. To see the converse, observe that $M(x_1,\ldots,x_n) = \sum_{i=1}^{n} f_i(x_i)$ is an aggregation function such that $M(x_1,\ldots,x_n) = 0$ if and only if $x_1 = \cdots = x_n = 0$. □

Observe that if M satisfies the requirements of the previous theorem, then m_{E,M,G_O} is additive for all G_O and all $E \neq \emptyset$.

5 Conclusions

In this chapter we have made a summary of the developments in [8] and discussed a method to build capacities from overlap indexes and overlap measures. In this way, relationships among data in the problem can be taken into account in order to build an appropriate measure which will later be used to determine the aggregation procedure.

In future research, we intend to analyse aggregations that make use of our measures, such as Sugeno, Shilkret, Choquet or copula-based integrals, and compare the obtained results using other aggregations in different problems, such as, for instance, digital fingerprint recognition or decision making.

Acknowledgements The work has been supported by projects TIN2013-40765-P and TIN2015-66471-P of the Spanish Ministry of Science, by grant VEGA 1/0420/15 and by grant VEGA 1/0419/13.

References

1. A. Amo, J. Montero, E. Molina, Representation of consistent recursive rules. *European Journal of Operational Research*, **130**, 29–53, 2001.
2. P. Benvenuti, R. Mesiar, Pseudo-additive measures and triangular-norm-based conditioning. *Annals of Mathematics and Artificial Intelligence*, **35**, 63–69, 2002.
3. P. Benvenuti, D. Vivona, M. Divari, Aggregation operators and associated fuzzy measures. *International Journal of Uncertainty, Fuzziness and Knowledge-Based Systems*, **9**, 197–204, 2001.
4. H. Bustince, M. Pagola, R. Mesiar, E. Hüllermeier, F. Herrera, Grouping, Overlap, and Generalized Bientropic Functions for Fuzzy Modeling of Pairwise Comparisons. *IEEE Transactions on Fuzzy Systems*, **20**(3), 405–415, 2012.
5. H. Bustince, E. Barrenechea, M. Pagola, F. Soria, Weak fuzzy S-subsethood measures: Overlap index. *International Journal of Uncertainty Fuzziness and Knowledge-Based Systems*, **14**(5), 537–560, 2006.

6. H. Bustince, J. Montero, E. Barrenechea, M. Pagola, Semiautoduality in a restricted family of aggregation operators. *Fuzzy Sets and Systems*, **158**, 1360–1377, 2007.
7. H. Bustince, J. Fernandez, R. Mesiar, J. Montero, R. Orduna, Overlap functions. *Nonlinear Analysis: Theory, Methods & Applications*, **72**(3-4), 1488–1499, 2010.
8. H. Bustince, D. Paternain, M. Pagola, P. Sussner, A. Kolesárová, R. Mesiar, Capacities and Overlap Indexes with an Application in Fuzzy Rule- Based Classification Systems, *Fuzzy Sets and Systems*, accepted for publication, in press.
9. T. Calvo, A. Kolesárová, M. Komorníková and R. Mesiar: Aggregation Operators: Properties, Classes and Construction Methods. In T. Calvo, G. Mayor and R. Mesiar (Eds.): *Aggregation Operators. New Trends and Applications*. Physica-Verlag, Heidelberg, pp. 3–104, 2002.
10. G. Choquet, Theory of capacities. *Ann. Inst. Fourier* **5**, 131–295, 1953-54.
11. V. Cutello, J. Montero, Recursive connective rules. *International Journal of Intelligent Systems*, **14**, 3–20, 1999.
12. D. Dubois, W. Ostasiewicz, H. Prade, Fuzzy Sets: History and Basic Notions. In D. Dubois, H. Prade (Eds.), *Fundamentals of Fuzzy Sets*. Kluwer, Boston, MA, pp. 21-124, 2000.
13. S. Garcia, H. Bustince, E. Hüllermeier, R. Mesiar, N. Pal, A. Pradera, Overlap Indices: Construction of and Application to Interpolative Fuzzy Systems, *IEEE Transactions on Fuzzy Systems* **23**, 1259–1273, 2015.
14. D. Gomez, J.T. Rodriguez, J. Montero, H. Bustince, E. Barrenechea, N-dimensional overlap functions, *Fuzzy Sets and Systems* **287**, 57–75, 2016.
15. M. Grabisch, J.-L. Marichal, R. Mesiar, E. Pap, *Aggregation Functions*. Cambridge University Press, Cambridge, 2009.
16. A. Jurio, H. Bustince, M. Pagola, A. Pradera, R.R. Yager, Some properties of overlap and grouping functions and their application to image thresholding. *Fuzzy Sets and Systems*, **229**, 69–90, 2013.
17. D. Gómez, J. Montero, A discussion on aggregation operators. *Kybernetika* **40**, 107–120, 2004.
18. G.J. Klir, T.A. Folger, *Fuzzy Sets, Uncertainty and Information*. Prentice Hall, Englewood Cliffs, NJ, 1988.
19. R. Mesiar, A. Mesiarová-Zemánková, The Ordered Modular Averages. *IEEE Transactions on Fuzzy Systems*, **19**, 42–50, 2011.
20. E. Pap, *Handbook of Measure Theory*, Part I, PartII. Elsevier, Amsterdam, 2002.
21. V. Torra, Y. Narukawa, M. Sugeno, Eds., Non-Additive Measures, Theory and Applications. *Studies in Fuzziness and Soft Computing*, **310**, 2014.
22. L.A. Zadeh, Fuzzy sets. *Information Control*, **8**, 338–353, 1965.

Clustering Alternatives and Learning Preferences Based on Decision Attitudes and Weighted Overlap Dominance

Camilo Franco, Jens Leth Hougaard and Kurt Nielsen

Abstract An initial assessment on a given set of alternatives is necessary for understanding complex decision problems and their possible solutions. Attitudes and preferences articulate and come together under a decision process that should be explicitly modeled for understanding and solving the inherent conflict of decision making. This paper revises multi-criteria modeling of imprecise data, inferring outranking and indifference binary relations and classifying alternatives according to their similarity or *dependency*. After the initial assessment on the set of alternatives, preference orders are built according to the attitudes of decision makers, aiding the decision process by identifying solutions with minimal dissention.

Keywords Decision attitudes · Dependency-based clustering · Preference learning · Consensus and dissention

1 Introduction

Outranking methodologies are usually implemented for ordering a given set of objects (alternatives), based on its associated multidimensional data (see e.g. [4, 10, 12]). Here we explore an outranking approach for imprecise data, measuring the volume of imprecision on all dimensions (criteria/attributes), and computing the multi-criteria likelihood of outranking among pairs of alternatives. Based on the weighted overlap dominance (WOD) model [10], a pairwise comparison process is developed by means of criteria weights and decision attitudes, understanding the different relations among alternatives and building solutions for intelligent decision support (see e.g. [7]).

The WOD model proposes a volume-based pairwise methodology for dealing with interval information. Due to the interval nature of data, decision attitudes are used to handle the uncertainty in estimating preference relations. In this way, attitudes have an impact on preferences according to the amount of evidence that is required for affirming an outranking dominance relation, evaluating the amount of

C. Franco (✉) · J.L. Hougaard · K. Nielsen
IFRO, Faculty of Science, Copenhagen University, Frederiksberg, Denmark
e-mail: cf@ifro.ku.dk

V. Torra et al. (eds.), *Fuzzy Sets, Rough Sets, Multisets and Clustering*,
Studies in Computational Intelligence 671, DOI 10.1007/978-3-319-47557-8_20

overlap existing among intervals. Besides, attitudes are also related to the purpose of aggregation in obtaining an overall value from the different pieces of evidence. Hence, both the verification of sufficient evidence and its means of aggregation are related to decision attitudes, having a determinant role for classifying alternatives according to their similarity or *dependency*. After the initial screening of the decision problem, decision support is offered by learning preferences and minimizing dissention among attitudes, aiding the decision process for reaching a satisfactory solution.

The focus of this paper consists in offering a multi-criteria framework for classifying alternatives according to their similarity, which is modeled by means of a *dependency* relation arising from the outranking-preference order. Then, dependent alternatives are grouped together, conforming a clustered set of alternatives which can be totally ordered according to the WOD procedure. Nonetheless, the methodology should also allow learning a preference order on the initial set of alternatives. For this purpose, relevance measures are examined, and decision support is offered on attitude-based solutions with minimum dissention.

This paper is organized as follows. Section 2 revises the WOD outranking procedure. Section 3 extends the WOD model for imprecision, interval weights and decision attitudes. Section 4 presents the whole clustering and preference learning methodology for decision support, concluding with some final comments for future research.

2 Articulation of Outranking-Preferences from Interval Data

The WOD procedure [10] is a multi-criteria outranking model that has been initially proposed to handle interval-valued scores for a given set of alternatives. This procedure makes use of criteria weights and risk attitude parameters to identify the outranking-preference relations holding among alternatives.

Consider a set of decision makers (DMs) D, which can be composed of individual DMs or coalitions of DMs, a set of alternatives N, and a set of criteria M. For every alternative $a \in N$ and criterion $i \in M$, there is a minimum and maximum score, respectively given by $x_{ai}^L, x_{ai}^U \in [0, 1]$, such that $x_{ai}^L \leq x_{ai}^U$, expressing the suitability of a with respect to i.

The nature of interval scores can be grounded on interval fuzzy sets [2, 9], as initially proposed in [5, 6], where an interval-valued fuzzy set A is defined by its (specifically designed) membership function $\mu_A : N \to [0, 1]^2$. Therefore, under the framework for multi-criteria decision support and the WOD procedure, the fuzzy interval degree

$$\mu_{ai} = [x_{ai}^L, x_{ai}^U] \in [0, 1]^2,$$

represents the extent up to which a verifies (fuzzy) criterion i.

In this way, for every alternative $a \in N$, the multi-dimensional suitability degree of a regarding the set of criteria M, $|M| = m$, is given by the hypercube,

$$c_a = \left[x_{a1}^L, x_{a1}^U\right] \times \cdots \times \left[x_{am}^L, x_{am}^U\right].$$

Each DM ($e \in D$) has an associated vector of weights $w^e \in [0, 1]^m$, expressing the subjective relative importance of criteria, and a couple of subjective decision attitudes given by $\beta_e \in [0, 1]$ and $\gamma_e \in \mathbb{R}^+$. In this way, hypercubes, weights and attitude parameters are the basic input for the WOD procedure.

Based on this information, a pairwise comparison process is developed among alternatives $a, b \in N$, being previously labeled such that,

$$w^e \cdot c_a^U = \sum_{i=1}^{m} w_i^e x_{ai}^U \geq w^e \cdot c_b^U = \sum_{i=1}^{m} w_i^e x_{bi}^U. \tag{1}$$

Examining the alternatives' hypercubes, together with the vector of weights w^e, the sets $\hat{Z}$, $\check{Z}$ and $\tilde{Z}$ can be respectively defined for all pairs (a, b), such that

$$\hat{Z}(a, b) = \{c_j \subseteq c_a | w^e \cdot c_j^L \geq w^e \cdot c_b^U\}, \tag{2}$$

$$\check{Z}(a, b) = \{c_j \subseteq c_a | w^e \cdot c_j^U \leq w^e \cdot c_b^L\}, \tag{3}$$

$$\tilde{Z}(a, b) = \{c_j \subseteq c_b | w^e \cdot c_j^L \geq w^e \cdot c_a^L\}. \tag{4}$$

Hence, $\hat{Z}(a, b)$ contains the subset of c_a which is above (the upper bound of) c_b, $\check{Z}(a, b)$ contains the subset of c_a which is below (the lower bound of) c_b, and $\tilde{Z}(a, b)$ contains the subset of c_b which is covered by (above the lower bound of) c_a.

Examining the amount of overlap existing between c_a and c_b, the WOD procedure infers the preference relation holding among a and b. There are three kinds of overlap, namely *no overlap*, *partial overlap* and *complete overlap*, which can be examined through sets $\hat{Z}$, $\check{Z}$ and $\tilde{Z}$ (2)–(4).

In the first place, if $\tilde{Z}(a, b)$ is empty, then there is no overlap between a and b, being there a *sure* outranking situation where no uncertainty exists that a outranks b, denoted by $a \succ b$. However, if it holds that $\tilde{Z}(a, b) \neq \emptyset$, then there is some overlap between c_a and c_b, being there some uncertainty on the outranking situation that holds among (a, b).

In this way, if $\check{Z} = \emptyset$, then there is partial overlap in the sense that $w^e \cdot c_b^U \geq w^e \cdot c_a^L \geq w^e \cdot c_b^L$, and an outranking situation holds such that,

$$a \succ b \Leftrightarrow P(a, b) > \beta_e, \tag{5}$$

where P is a proxy for the probability that a value randomly drawn from c_a is higher than a value (randomly) drawn from c_b. Here P is defined such that,

$$P\,(a,b) = \frac{V(\hat{Z})}{V(c_a)} + \frac{V(c_a \setminus \hat{Z})}{V(c_a)} \frac{V(c_b \setminus \tilde{Z})}{V(c_b)},$$

with $V(\cdot)$ being a volume operator such that $V(\emptyset)=0$ and $\forall x \in [0, 1]^{2m}$,

$$V(x) = \prod_{i=1}^{m} (x_i^U - x_i^L).$$

The idea is that P has to be higher than β_e in order for a to outrank b. In this sense, β_e is a risk attitude parameter for establishing the outranking of a over b, where the higher it is, the lower the risk of affirming outranking without sufficient evidence.

Now, in the case that $P\,(a,b) \leq \beta_e$, then both alternatives are considered to be indifferent, i.e., $a \sim b$ holds.

On the other hand, if $\check{Z} \neq \emptyset$, then there is complete overlap in the sense that $w^e \cdot c_b^L \geq w^e \cdot c_a^L$. In this case, it holds that

$$a \succ b \;\Leftrightarrow\; C\,(a,b) > \gamma_e, \tag{6}$$

where C is a proxy for the probability that a value randomly drawn from c_a is higher than a value (randomly) drawn from c_b. Here C is defined such that,

$$C\,(a,b) = \frac{V(\hat{Z})}{V(\check{Z})}.$$

Then, if $C\,(a,b) = \gamma_e$ or $C\,(a,b) < \gamma_e$ holds, then it respectively holds that $a \sim b$ or that $b \succ a$. Here the outranking situation is determined by the ratio between the volume of c_a that is over c_b ($\hat{Z}$), and the volume of c_a that is below c_b ($\check{Z}$). Thus, γ_e expresses the risk attitude towards affirming an outranking relation of a over b, such that the higher it is, the lower the risk of affirming outranking without sufficient evidence. Nonetheless, as it is defined here, the inverse outranking relation holds if no initial outranking is verified, which somehow passes the risk to the other (inverse) relation. That is, the reciprocal outranking relation for this overlapping case entails a reciprocal risk attitude which forces *risk proneness* no matter the value of γ_e. This is a problematic issue which is addressed in the following Sect. 3, together with the *crisp* estimation of criteria weights w^e.

Concluding the WOD process, and examining the outcome of this procedure, alternatives are ordered according to their interval scores. As a result, two types of binary relations hold, i.e., outranking, ($\succ$) or indifference ($\sim$).

As it has been shown in [10], the outranking relation $\succ$ is a *semi-transitive* relation, such that for every $a, b, c \in N$, if $a \succ b$ and $b \succ c$ hold, then it cannot hold that $c \succ a$. Hence, $\succ$ is *irreflexive*, i.e., it does not hold that $a \succ a$, and *asymmetric*, i.e., if it holds that $a \succ b$, then it does not hold that $b \succ a$.

Even more, the indifference relation is a non-transitive *equivalence* relation, i.e., being reflexive and symmetric.

For example, consider three alternatives $a, b, c \in N$, such that $a \succ b$ and $b \succ c$ hold. Then, due to the semi-transitivity of $\succ$, it holds either that $a \succ c$ or that $a \sim c$. But if $a \sim c$ holds, then it cannot be that $\sim$ is transitive while both $a \succ b$ and $b \succ c$ are true.

Therefore, the WOD procedure assigns binary pairwise relations that conform an (semi-transitive) outranking order on N, which can be used to cluster alternatives into different classes of *dependency*, as it will be examined in Sect. 4.

3 Extending WOD for Imprecision, Interval Weights and Decision Attitudes

3.1 Imprecision

The WOD framework can be extended for representing the volume of imprecision for both crisp data (i.e., data in the form of $\mu = [x^L, x^U]$, such that $x^L = x^U$) and interval data (where $x^L < x^U$ holds). This can be done by imprecision functions $\delta : [0, 1]^2 \rightarrow [\varepsilon, 1]$, such as [6],

$$\delta(\mu) = x^U - x^L + \varepsilon.$$

Imprecision functions [6] characterize imprecision by a minimum amount of imprecision equal to ε, if and only if the lower and upper bounds are the same, and a maximum amount of imprecision equal to 1, if and only if the lower bound is minimum, i.e. $x^L = 0$, and the upper bound is maximum, i.e. $x^U = 1$. Then, the measure for imprecision will be positive and greater than ε every time the upper bound of the interval is greater than its lower bound, and the greater the difference between the lower and upper bounds, the greater it will be the uncertainty due to imprecision.

As a result, $\forall a \in N$, the volume of imprecision associated to alternative a is given by,

$$V^{\delta}(c_a) = \prod_{i=1}^{m} \delta(\mu_{ai}).$$

3.2 Interval Weights

Now, it is necessary to address the estimation of the vector of weights w^e. Criteria weights allow establishing the trade-off between criteria for their complete comparability. Up to now we have been assuming the existence of a crisp vector of weights $w^e \in [0, 1]^m$.

These weights are essential for the overall aggregation of criteria, establishing a hierarchy among criteria for arriving at an informed judgment over the decision problem. Considering that weights are based on the DMs' beliefs or *a-priori* knowledge regarding the relative importance of the criteria in M, their estimation can be aided by allowing their interval valuation. Thus, the computation of an outranking situation could be developed on a set of lower and upper bounds for the possible values of weights.

In this way, $\forall e \in D$, there is an initial set of interval weights $\hat{w}^e \in [0, 1]^{2m}$, with lower and upper bounds respectively given by $\hat{w}^L, \hat{w}^U \in [0, 1]^m$, such that (2)–(4) can be reformulated as follows,

$$\hat{Z}^e(a, b) = \{c_j \subseteq c_a | \hat{w}^L \cdot c_j^L \geq \hat{w}^U \cdot c_b^U\}, \tag{7}$$

$$\check{Z}^e(a, b) = \{c_j \subseteq c_a | \hat{w}^U \cdot c_j^U \leq \hat{w}^L \cdot c_b^L\}, \tag{8}$$

$$\tilde{Z}^e(a, b) = \{c_j \subseteq c_b | \hat{w}^L \cdot c_j^L \geq \hat{w}^L \cdot c_a^L\}. \tag{9}$$

Hence, interval weights extend the WOD method, computing the respective upper and lower bounds of weights with the ones of overlapping hypercubes, properly identifying the amount of (imprecise) weighted overlap for every pair (a, b).

3.3 Decision Attitudes on (inverse) Outranking and Indifference

As mentioned above, there is a certain difficulty when determining a unique value for γ_e, because the higher its value, the less risk there is for affirming outranking without sufficient evidence, but at the same time, the higher the risk in affirming the inverse outranking situation. Hence, the attitude parameter γ_e could refer to an interval-valued parameter expressing in a more general and faithful manner the attitudes of DMs.

In this way, defining $\gamma_e' = [\gamma^L, \gamma^U] \in [0, 1]^2$, it is possible to extend the WOD procedure for the parameter γ_e' to faithfully represent an attitude parameter. Hence, re-visiting the complete overlap case, the definition for an outranking relation can be reformulated by,

$$a \succ b \Leftrightarrow C\,(a, b) > \gamma_e^U, \tag{10}$$

and indifference can be explicitly defined such that,

$$a \sim b \Leftrightarrow \gamma_e^U \geq C\,(a, b) \geq \gamma_e^L. \tag{11}$$

Thus, if neither (10) nor (11) holds, then the inverse outranking situation holds, such that $b \succ a$. As a result, $\forall a, b \in N$, indifference exists every time that $C(a, b)$

lies inside the region given by $\gamma'_e = [\gamma^L, \gamma^U]$, and a outranks b only if $C(a, b)$ is greater than γ^U, while b outranks a only if $C(a, b)$ is less than γ^L. In this way, the reciprocity of the attitude parameter is avoided, allowing DMs to be truly *risk averse* by setting γ^L sufficiently low and γ^U sufficiently high. Otherwise, *risk proneness* is expressed by setting γ'_e such that $\gamma^L = \gamma^U$.

3.4 Aggregation Attitudes

Now, in order to model decision attitudes towards the purpose of multi-criteria aggregation, consider the aggregation function $F : [0, 1]^m \to [0, 1]$, such that $F(0, \dots, 0) = 0$, $F(1, \dots, 1) = 1$, and $\forall x_i, y_i \in [0, 1]$, $F(x_1, \dots, x_m) > F(y_1, \dots, y_m)$ whenever $\forall i, x_i > y_i$.

Then, three attitudes for aggregation can be considered, which can be somehow related to different degrees of *orness* and *andness* of aggregation operators (see e.g. [3]). On the one hand, a conjunctive aggregation is related to a *demanding* attitude, which requires that all criteria are completely satisfied. This attitude is represented through an operator $F = F^t$, such that $F^t(0, 1) = F^t(1, 0) = 0$ holds. Hence, the aggregation operator F^t can be associated with an *overlapping* operator [1], e.g., $F^t = \min$.

On the other hand, a disjunctive aggregation is related to a *soft* attitude, which requires that at least one criterion is satisfied. This attitude is represented by an operator $F = F^s$, such that $F^s(0, 1) = F^s(1, 0) = 1$ holds. Therefore, the aggregation operator F^s can be associated with a *grouping* operator [1], e.g., $F^s = \max$.

Otherwise, taking an *averaging* attitude, it is required that in average all criteria are satisfied, allowing some trade-off or compensation among them. Its associated operator is given by $F = F^b$, such that $0 < F^b(0, 1) = F^b(1, 0) < 1$ holds. This aggregation operator F^b can be computed by the Choquet integral, generalizing the (weighted) arithmetic mean (see e.g., [8, 11]).

Therefore, taking different attitudes on the purpose of aggregation, the WOD procedure can be formulated through the set of aggregation operators

$$F = \{F^t, F^s, F^b\},$$

inferring an outranking order for each attitude. For this purpose, $\forall a \in N$, the product of $\hat{w}^e$ and c_a is computed by

$$\hat{w}^e \otimes c_a = \left[\hat{w}^L \otimes c_a^L, \hat{w}^U \otimes c_a^U\right] = \left[\underset{\forall i \in M}{F}(\hat{w}^L x_{ai}^L), \underset{\forall i \in M}{F}(\hat{w}^U x_{ai}^U)\right]. \tag{12}$$

After labeling alternatives such that $\hat{w}^U \otimes c_a^U \geq \hat{w}^U \otimes c_b^U$, the different overlapping sets (7)–(9) can be reformulated by,

$$\hat{Z}^e(a,b) = \{c_j \subseteq c_a | \hat{w}^L \otimes c_j^L \geq \hat{w}^U \otimes c_b^U\}, \tag{13}$$

$$\check{Z}^e(a,b) = \{c_j \subseteq c_a | \hat{w}^U \otimes c_j^U \leq \hat{w}^L \otimes c_b^L\}, \tag{14}$$

$$\tilde{Z}^e(a,b) = \{c_j \subseteq c_b | \hat{w}^L \otimes c_j^L \geq \hat{w}^L \otimes c_a^L\}. \tag{15}$$

Then, no overlap exists if it holds that

$$\hat{w}^L \otimes c_a^L > \hat{w}^U \otimes c_b^U,$$

partial overlap exists if it holds that

$$\hat{w}^L \otimes c_b^L < \hat{w}^L \otimes c_a^L \leq \hat{w}^U \otimes c_b^U,$$

and there is complete overlap if it holds that

$$\hat{w}^L \otimes c_a^L < \hat{w}^L \otimes c_b^L.$$

Then, following (13)–(15), $P(a,b)$ and $G(a,b)$ express the attitude-based likelihood that any point belonging to c_a is greater than any other point in c_b, inferring outranking and indifference binary relations as in (5) and (10)–(11).

4 Preference-Based Clustering and Decision Support

4.1 Building Clusters and Preference Orders

Following the multi-criteria WOD procedure, a certain dependency may arise among alternatives due to the semi-transitivity of the outranking order. Hence, alternatives can be clustered together into *semi-equivalent* classes, addressing the formation of new (clustered) alternatives along the decision process. These clusters can then be used to simplify the decision problem, or can be refined until a preference order (weak or total order) is identified.

First of all, under the specification of attitude parameters $\beta_e = 1$ and $\gamma_e' = [0, K]$, where $K \in \mathbb{R}^+$ is sufficiently high, all overlapping alternatives are clustered together under a semi-equivalence class, due to the verification of indifference on all pairs (a, b). Under such scenario, there is an absolutely *risk-averse* attitude towards outranking, such that no outranking relation can hold (among overlapping pairs (a, b)).

Such a risk averse scenario can be relaxed by more *risk-neutral* or *risk-prone* attitudes, identifying a higher number of clusters. These clusters represent semi-equivalent classes, which are defined by (at least) any three alternatives $a, b, c \in N$ such that $a \succ b$, $a \sim c$ and $b \succsim c$ hold.

In this way, the semi-transitivity of the outranking order allows semi-equivalent classes to emerge. On formal terms, given a subset of alternatives $E \subseteq N$, E consists in a semi-equivalence class if its elements are either related through the indifference relation $\sim$, or in case $a \succ b$ for some pair $a, b \in E$, then a is dependent with b by a third element $c \in E$, such that either $a \sim c$ and $c \nsucc b$ or $c \succ a$ and $b \sim c$.

As a result, considering the hierarchy of semi-equivalence classes $E_t \subseteq N$, such that $E_1 \succ E_2 \succ \cdots \succ E_T$, and the individual alternatives $a \in N$ such that $a \notin E_t$, it follows that a total order holds on $N' = \{E_1 \cup \cdots \cup E_T \cup N \setminus \mathbb{E}\}$, where $\mathbb{E}$ is the set of all semi-equivalent classes. That is, taking the clusters of dependent alternatives as new alternatives, a total order arises which can be used to assess the decision problem and identify initial solutions together with their associated attitudes.

Therefore, the existence of semi-equivalence classes highlights the complexity of assigning a weak or total order on the initial set of alternatives N. In fact, alternatives may hold a conflicting dependency between them, making it impossible to establish a ranking unless a specific rank-construction procedure is applied.

Following the initial proposal of [6], alternatives can be ranked according to their *relevance*. In this way, alternatives are scored (and accordingly ranked) with respect to the number and the importance of the alternatives that they outrank. That is, $\forall a \in N$, the relevance of a is given by,

$$\sigma(a) = s_a + \alpha \sum_{\forall b \in S_a} s_b, \tag{16}$$

where S_a is the set of alternatives that are outranked by a, $s_a = |S_a|$, and $\alpha \in [0, 1]$ is a parameter that can be calibrated for minimizing the number of (tied) rank-equivalent alternatives. Thus, the optimal value for α obtains a weak order that is as close as possible to a total order on N.

In summary, the whole procedure consists in taking the attitudes and criteria weights for every DM $e \in D$, and for every attitude and set of weights, compute an (outranking semi-transitive) order on N, resulting from the WOD evaluation of data. Then, a total order can be identified on the new set of alternatives N', or a preference order can be learnt by means of (16), under an optimal value of α. In consequence, for all the given combinations of decision attitudes and criteria weights, the system returns a total order on the new set of alternatives N', together with a (weak or total) preference order on the initial set of alternatives N.

4.2 Attitude-Based Decision Support

Based on the outcome of the extended-WOD procedure, decision support can be offered on the different attitudes and the amount of consensus existing on the decision problem. In this way, for all the combinations of attitudes and weights, a total preference order ρ' is identified for N', and a preference order ρ is identified for N. Once these solutions are known, the system can offer support for resolving conflict among them, aiming at reducing dissention. The goal of the system focuses on interacting with DMs for maximizing consensus and arriving at a satisfactory solution.

The decision process is guided towards reducing discrepancies among attitudes and different DMs, suggesting decisions with maximal consensus (see e.g. [7]). Given multiple (weak or total) preference orders, *dissention degrees* are introduced here to measure the amount of disagreement for all pairs of orders ρ_1, ρ_2, computing the position of alternative $a \in \rho_1, \rho_2$, respectively given by θ_1^a and θ_2^a, and obtaining the pairwise dissention degree $\mathbb{D}(\rho_1, \rho_2)$, such that,

$$\mathbb{D}(\rho_1, \rho_2) = \sum_{\forall a \in N} |\theta_1^a - \theta_2^a|. \quad (17)$$

The decision process aims at maximizing consensus, based on the previous calculation of pairwise dissention degrees among all orders. Then, the system identifies all pairs ρ_1, ρ_2 with minimal dissention, suggesting a common set of attitudes and weights that increases the general consensus on a satisfactory solution. This process is iterated by eliciting new attitudes and/or weights, where DMs can rectify their input until no further consensus can be reached (i.e., until the amount of minimal dissention remains unchanged). In consequence, decision attitudes guide the articulation of preferences through the interaction between the system and DMs, also supporting a hypothetical negotiation process among different DMs.

5 Conclusion

The WOD outranking procedure has been examined and extended for considering imprecise data together with decision attitudes and interval weights, understanding and reconfiguring alternatives for building solutions and intelligent decision support. In this way, a clustering methodology has been provided with the purpose of aiding the decision process, grouping dependent alternatives and learning preference orders, aiming at supporting the consensus between different attitudes and DMs.

For further research, experimental simulations and/or case studies should be carried out, testing the extended WOD procedure for attitude-based clustering, preference learning and decision support. From a theoretical perspective, it could be studied how reciprocal preference structures entail equally reciprocal (risk) attitudes on the affirmation of preference relations, exploring in detail the relation between preferences and attitudes. Even more, the linguistic representation of non-reciprocal (risk) attitudes should be examined, modeling the negotiation process of DMs, focusing on the strategical behavior of DMs and their possible characterization for optimal decision making.

Acknowledgments This research has been partially supported by the Danish Industry Foundation and the Center for research in the Foundations of Electronic Markets (CFEM), funded by the Danish Council for Strategic Research.

References

1. Bustince H, Barrenechea E, Pagola M, Fernandez J.: The notions of overlap and grouping functions. Studies in Fuzziness and Soft Computing 336,137-156 (2016)
2. Bustince H, Montero J, Pagola M, Barrenechea E, Gómez D.: A survey of interval-valued fuzzy sets. In: Pedrycz W, Skowron A, Kreinovich V (eds.), Handbook of Granular Computing, Chichester: John Wiley & Sons; 2008, p. 491-515.
3. Fernández JM, Murakami S.: Extending Yager's orness concept for the OWA aggregators to other mean operators. Fuzzy Sets and Systems 139, 515-542 (2003)
4. Fodor J, Roubens M.: Fuzzy Preference Modelling and Multicriteria Decision Support. Kluwer Academic Publishers, Dordrecht (1994)
5. Franco C, Hougaard JL, Nielsen K.: A fuzzy approach to the Weighted Overlap Dominance Model. Proceedings of the 15th Conference of the Spanish Association for Artificial Intelligence, Madrid, Spain, September 17-20, 2013, p. 1240-1249.
6. Franco C, Hougaard JL, Nielsen K.: Ranking alternatives based on imprecise multi-criteria data and pairwise overlap dominance relations. MSAP Working Papers Series (2014)
7. Franco C, Hougaard JL, Nielsen K.: Handling risk attitudes for preference learning and intelligent decision support. Lecture Notes in Computer Science 9321, 78-89 (2015)
8. Grabisch M, Labreuche Ch.: A decade of application of the Choquet and Sugeno integrals in multi-criteria decision aid. Annals of Operations Research 175, 247-290 (2010)
9. Grattan-Guinness I.: Fuzzy membership mapped onto interval and many-valued quantities. Mathematical Logic Quarterly 22, 149-60 (1976)
10. Hougaard JL, Nielsen K.: Weighted Overlap Dominance - A procedure for interactive selection on multidimensional interval data. Applied Mathematical Modelling 35, 3958-3969 (2011)
11. Y. Narukawa, V. Torra. Fuzzy measures and Choquet integral on discrete spaces. In: B. Reusch (ed.), Computational Intelligence, Theory and Applications, Berlin: Springer; 2004, p. 573-581.
12. Roy B.: The outranking approach and the foundation of ELECTRE methods. Theory and Decision 31, 49-73 (1991)

Printed in the United States
By Bookmasters